To get extra value from this book for no additional cost, go to:

http://www.thomson.com/duxbury.html

thomson.com is the World Wide Web site for Duxbury/ITP and is your direct source to dozens of on-line resources. *thomson.com* helps you find out about supplements, experiment with demonstration software, search for a job, and send e-mail to many of our authors. You can even preview new publications and exciting new technologies.

thomson.com: *It's where you'll find us in the future.*

Datasets to accompany this book are available from the publisher. Contact your bookstore for more information or go to http://www.thomson.com/duxbury.html and access the dataset archive.

Statistics with Stata® 5

Lawrence C. Hamilton
University of New Hampshire

An Imprint of Brooks/Cole Publishing Company

I(T)P® **An International Thomson Publishing Company**

Pacific Grove ▪ Albany, NY ▪ Belmont, CA ▪ Bonn ▪ Boston ▪ Cincinnati ▪ Detroit ▪ Johannesburg ▪ London
Madrid ▪ Melbourne ▪ Mexico City ▪ New York ▪ Paris ▪ Singapore ▪ Tokyo ▪ Toronto ▪ Washington

Sponsoring Editor: Curt Hinrichs
Marketing Communications: Joy Westberg
Editorial Assistant: Rita Jaramillo
Production: Forbes Mill Press
Production Coordinator: Kirk Bomont
Manuscript Editor: Robin Gold
Interior Design: Robin Gold
Cover Design: Image House
Typesetting: Patricia K. Bonn
Printing and Binding: Malloy Lithographing, Inc.

For more information, contact Duxbury Press at Brooks/Cole Publishing Company:

BROOKS/COLE PUBLISHING COMPANY
511 Forest Lodge Road
Pacific Grove, CA 93950
USA

International Thomson Publishing Europe
Berkshire House 168-173
High Holborn
London WC1V 7AA
England

Thomas Nelson Australia
102 Dodds Street
South Melbourne, 3205
Victoria, Australia

Nelson Canada
1120 Birchmount Road
Scarborough, Ontario
Canada M1K 5G4

International Thomson Editores
Seneca 53
Col. Polanco
11560 México, D. F., México

International Thomson Publishing GmbH
Königswinterer Strasse 418
53227 Bonn
Germany

International Thomson Publishing Asia
221 Henderson Road
#05-10 Henderson Building
Singapore 0315

International Thomson Publishing Japan
Hirakawacho Kyowa Building, 3F
2-2-1 Hirakawacho
Chiyoda-ku, Tokyo 102
Japan

Printed in the United States of America

10 9 8 7 6 5 4 3 2 1

Library of Congress Cataloging-in-Publication Data
Hamilton, Lawrence C.
Statistics with Stata 5 / Lawrence C. Hamilton.
p. cm.
Includes bibliographical references (p. -) and index.
ISBN 0-534-26559-6
1. Statistics—Data processing. 2. Stata. I. Title.
QA276.4.H355 1998
519.5'0285'53042—dc21 97-17794

Contents

Preface

Statistics with Stata 5 is intended for students and practicing researchers, to bridge the gap between statistical textbooks and Stata's own documentation. In this intermediate role, it does not provide the detailed expositions of a proper textbook, nor does it attempt to describe all of Stata's features. Instead, it demonstrates how to use Stata to accomplish some of the most common tasks of data management and statistical analysis. Chapter topics follow conceptual themes rather than focusing on particular Stata procedures or commands, which gives *Statistics with Stata 5* a quite different structure from the Stata Reference Manuals. The chapter on "Fitting Curves," for example, encompasses several methods that are mathematically independent, and involve unrelated Stata commands. For analysts considering what method to use, however, it may help to find these alternative ways to fit curves described together in one place.

The general topics of the first six chapters (through ordinary least squares regression) roughly parallel an introductory course in applied statistics, but with digressions covering tasks often needed in actual research—how to aggregate data, create dummy variables, or translate ANOVA into regression, for instance. In Chapter 7 (Regression Diagnostics) and beyond, we move into the territory of advanced courses or more technical research. Here readers can find how to obtain diagnostic statistics and graphs; perform robust, quantile, nonlinear, Poisson, ordered logit, or multinomial logit regression; fit survival-time and event-count models; or construct composite variables through factor analysis and principal components. Stata has worked hard in recent years to maintain its state-of-the-art standing, and this effort is particularly apparent in the wide range of regression and model-fitting commands it now offers.

Finally, we conclude with a look at programming in Stata. Many readers will find that Stata does everything they need already, so they have no need to write original programs. For an active minority, however, programmability is one of Stata's principal attractions, and it is certainly one reason for Stata's currency and rapid advance. This chapter opens the door for new users to explore Stata programming, whether for specialized data management purposes, to establish a new statistical capability, or for teaching.

Generally similar versions of Stata run on Windows, Macintosh, and Unix computers. Across all platforms, Stata uses the same commands, data files, and output. The versions do differ in some details of screen appearance, menus, and file handling, where Stata follows the conventions native to that platform, such as \directory\filename file specifications under Windows in contrast with the /directory/filename specifications under Unix. Rather than try to display all three, I employ Windows conventions, but users with other systems should find that only minor translations are needed.

Anyone familiar with this book's predecessor, *Statistics with Stata 3*, will discover substantial new material here. Internet resources, notably Stata's Web site, technical support via e-mail, and the independent discussion forum STATALIST, are described in Chapter 1. Chapter 2 contains greatly expanded coverage of data management, reflecting Stata's improvements in this area and its importance to data analysis. Chapter 11, on survival analysis, and Chapter 13, on programming, are entirely new. Other, less obvious changes occur throughout each chapter, where new or modified commands have enhanced Stata's capabilities.

Acknowledgments

Stata's architect, William Gould, deserves credit for originating the elegant program that this book describes. Over the years he has patiently helped me keep up to date. Joseph Hilbe reviewed chapter drafts and contributed the trimmed-mean and linear regression programs in Chapter 13. Steve Selvin shared several of his examples from *Practical Biostatistical Methods* for Chapter 11, and read the resulting draft. Jacqueline Kiffe at Stata Corporation also reviewed drafts, and offered key advice when the three-fourths-completed manuscript had to change from *Statistics with Stata 4* to *Statistics with Stata 5*. Two editors, first Stan Loll and subsequently Curt Hinrichs, helped over the duration of this unexpectedly long project. Final layout and production editing were done admirably by the same person, Robin Gold of Forbes Mill Press.

A book like this could not exist without data. To avoid endlessly recycling my own old examples, I relied on fresh research and contributions from others. The list of debts has grown long, but I would especially like to mention my co-authors in Arctic research—Cynthia M. Duncan, Nicholas Flanders, Per Lyster, Oddmund Otterstad, Rasmus Rasmussen, and Carole Seyfrit. Others who lent their efforts or support to these projects include Noel Broadbent, Sydney Callahan, Kimberly Hill, Marlies Kruse, Michael Ledbetter, Dennis Meadows, Brigid Murray and Birger Poppel. Still others contributed their own data or expert advice, including Paul Mayewski, Loren D. Meeker, Heather Turner, and Sally Ward.

Dedication

To Leslie, who stood with me, and to Sarah and Dave, who grew up while I was writing.

1

Stata and Stata Resources

Stata is a full-featured statistical program for Windows, Macintosh or UNIX computers. It combines ease of use with speed, a library of pre-programmed analytical and data-management capabilities, and programmability that allows users to invent and add further capabilities as needed. Optional menus provide a painless entry for beginners. Beyond the menus, Stata's consistent, intuitive command syntax simplifies both learning and use.

After introductory information, we'll begin with an example Stata session to give you a sense of the "flow" of data analysis. Later chapters explain in more detail. Even without explanations, however, you can see how straightforward the commands are—**use *filename*** to retrieve dataset *filename*, **summarize** when you want summary statistics, **correlate** to get a correlation matrix, and so forth.

Stata users have a variety of resources to help them learn about Stata and solve problems at any level of difficulty. These resources come not just from Stata Corporation but also from an actively communicating community of other users. Sections of this chapter introduce some key resources—Stata's documentation; where to phone, fax, write, or e-mail for technical help; Stata's World Wide Web page; the Statalist Internet forum; and the bimonthly updates in the *Stata Technical Bulletin.*

A Typographical Note

This book employs several typographical conventions as a visual clue to how words are used:

```
. command varname, options
```

Commands typed by the user appear in a **bold Courier** font. When the whole command line is given, it starts with a period—as seen in a Stata Results window or log (output) file. Variable or file ***names*** within these commands appear in italics, to emphasize the fact that they are arbitrary and not a fixed part of the command.

Names of variables or files also appear in italics within the main text, to distinguish them from ordinary words.

Stata output as seen in the Results window, printed out or saved in a log file is shown in a small Courier font. The small font is necessary to fit Stata's 80-column output within the margins of this book.

Thus, we show the calculation of summary statistics for a variable named *penalty* as follows:

```
. summarize penalty
Variable |      Obs        Mean   Std. Dev.         Min          Max
---------+-----------------------------------------------------------
 penalty |       10          63   59.59493           11          183
```

These typographic conventions exist only in this book, and not within the Stata program itself. Stata can display a variety of on-screen fonts, but it does not use italics in commands or elsewhere. When importing Stata log files into a word processor, you might want to specify a Courier 10 point font.

In its commands and variable names, Stata is case sensitive. Thus **summarize** is a command, but Summarize and SUMMARIZE are not. *Penalty* and *penalty* would be two different variables.

An Example Stata Session

As a preview of Stata at work, this section retrieves and analyzes a previously created dataset named *lofoten.dta.* Jentoft and Kristoffersen (1989) originally published these data in an article about self-management by fishermen on Norway's arctic Lofoten Islands. There are 10 observations (years) and 5 variables, including *penalty,* a count of how many fishermen were cited each year for violating Lofoten fisheries regulations.

In this session, the **bold statements preceded by a period** are commands typed directly by the user in Stata's Command window. Because we eventually want a printed copy of our results from this session, we begin by opening a "log file," here arbitrarily named *friday1.log,* that will record all of these commands and the resulting output:

```
. log using friday1
```

Next we **use** or retrieve our dataset, if necessary telling Stata where to find it. In this example, file *lofoten.dta* resides in the *data* directory of drive C. (Here as elsewhere in the book, I follow Windows file and directory conventions. Users of other systems will need slightly different commands. Alternatively, on any system you can select files through the menus.)

```
. use c:\data\lofoten
(Jentoft & Kristoffersen '89)
```

To see a brief description of the dataset now in memory, type the following:

```
. describe
Contains data from c:\data\lofoten.dta
  obs:            10                           Jentoft & Kristoffersen '89
 vars:             5                           27 Apr 1996 12:01
 size:           130 (96.5% of memory free)
-------------------------------------------------------------------------------
   1. year        int     %9.0g                Year
   2. boats       int     %9.0g                Number of fishing boats
   3. men         int     %9.0g                Number of fishermen
   4. penalty     int     %9.0g                Number of penalties issued
   5. decade      byte    %9.0g      decade    Early 1970s or early 1980s
-------------------------------------------------------------------------------
Sorted by:  decade  year
```

Many Stata commands can be abbreviated to their first few letters. For example, we could shorten **describe** to just the letter **d**.

This dataset has only 10 observations and 5 variables, so we can easily list it by typing the command **list** (or just the letter **l**):

```
. list

         year      boats        men    penalty     decade
  1.     1971       1809       5281         71      1970s
  2.     1972       2017       6304        152      1970s
  3.     1973       2068       6794        183      1970s
  4.     1974       1693       5227         39      1970s
  5.     1975       1441       4077         36      1970s
  6.     1981       1540       4033         11      1980s
  7.     1982       1689       4267         15      1980s
  8.     1983       1842       4430         34      1980s
  9.     1984       1847       4622         74      1980s
 10.     1985       1365       3514         15      1980s
```

Analysis could begin with a table of means, standard deviations, minimum and maximum values (**summarize** or **su**):

```
. summarize

Variable |     Obs        Mean   Std. Dev.        Min        Max
---------+-----------------------------------------------------
    year |      10        1978   5.477226       1971       1985
   boats |      10      1731.1   232.1328       1365       2068
     men |      10      4854.9   1045.577       3514       6794
 penalty |      10          63   59.59493         11        183
  decade |      10          .5   .5270463          0          1
```

Did the number of penalties for fishing violations change over the two decades here? A table containing summary statistics for *penalty* at each value of *decade* shows there were more penalties in the 1970s:

```
. tabulate decade, summ(penalty)

Early 1970s | Summary of Number of penalties issued
or early    |
1980s       |        Mean   Std. Dev.        Freq.
------------+-----------------------------------
      1970s |        96.2    67.41439            5
      1980s |        29.8   26.281172            5
------------+-----------------------------------
      Total |          63   59.594929           10
```

Perhaps the number of penalties declined because fewer people were fishing in the 1980s. The number of penalties correlates strongly ($r > .8$) with the number of boats and fishermen:

```
. correlate boats men penalty
(obs=10)

        |   boats      men  penalty
--------+---------------------------
   boats|  1.0000
     men|  0.8748   1.0000
 penalty|  0.8259   0.9312   1.0000
```

A graph might help clarify these interrelationships. Figure 1.1 shows a plot of *men* and *penalty* against *year*, produced through the **graph** command. Options after the variable list in this command specify that both *y*-variable series should be connected

Figure 1.1

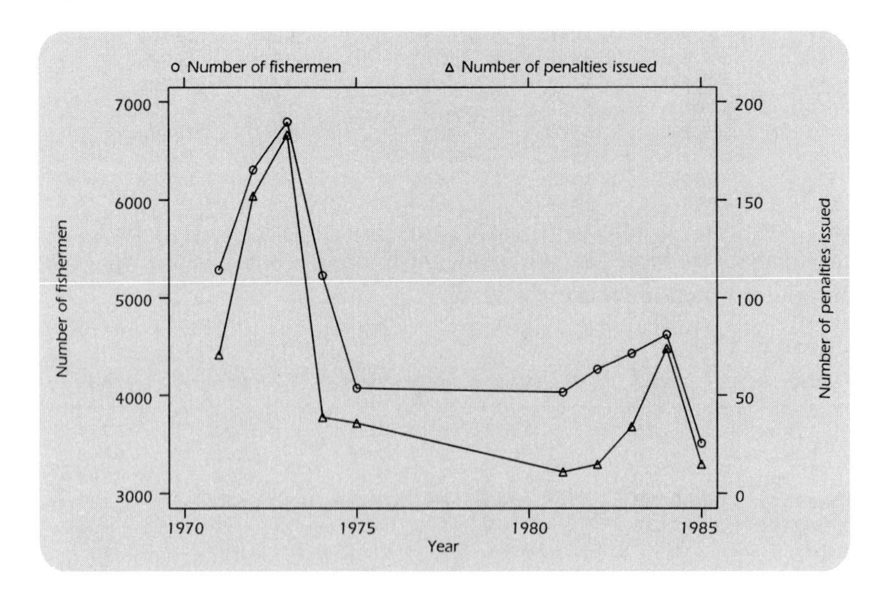

by line segments, with separately scaled vertical axes for each. **`ylabel xlabel rlabel`** tells Stata to choose round numbers when labeling the left vertical, horizontal, and right vertical axes.

```
. graph men penalty year, connect(ll) rescale ylabel
      xlabel rlabel
```

Because the years 1976 to 1980 are missing in these data, Figure 1.1 shows 1975 connected to 1981. For some purposes we might hesitate to do this. Instead, we could either find the missing values or else leave the gap unconnected by issuing a slightly more complicated set of commands.

To print this graph or save it as a disk file, pull down the File→Save graph or the File→Print graph menus once the graph is displayed on screen. As with any of Stata's windows, we can move or resize a Graph window as desired and save our preferred windowing configuration.

Finally, having concluded our analysis, we need to close the log file that contains our results.

```
. log close
```

This important step completes *friday1.log*, an ASCII (text) format file that we can later import into our word processor for editing or printing.

Stata's Documentation and Help Files

Professional versions of Stata come with a slim *Getting Started* manual (for example, *Getting Started with Stata for Windows*), the more extensive *User's Guide*, and the encyclopedic *Stata Reference Manual*. *Getting Started* helps you do just that, with the basics of installation, window management, data entry, printing, and so on. The *User's Guide* contains an extended discussion of general topics including resources and troubleshooting. The multivolume *Stata Reference Manual* lists all Stata commands alphabetically. Entries for each command include the full command syntax, descriptions of all available options, examples, technical notes regarding formulas and rationale, and references for further reading.

But when you are in the midst of a Stata session, it is often simpler to ask for on-screen help instead of consulting the manuals. Selecting Help from Stata's menus opens a second menu that asks whether you want to search for information on a general topic (for example, "date functions" or "regression diagnostics") or on a specific command such as **`tabulate`** or **`graph`**. Like the *Stata Reference Manual*, on-screen help provides command syntax diagrams and complete lists of options. It also includes some examples, although they are generally less detailed and without the technical discussions found in the *SRM*. Besides immediacy, on-screen help has two other advantages over the manuals. Hypertext links take you directly to related entries, and the on-screen help includes material on "unofficial" Stata programs that users might have installed from the *Stata Technical Bulletin.*

Stata Corporation

For orders, licensing, and upgrade information, you can contact Stata Corporation directly or visit their World Wide Web site. Their physical address is

Stata Corporation
702 University Drive East
College Station, TX 77840 USA

Stata Corporation can also be reached by fax or telephone, including an easily remembered 800 number:

Telephone: 409-696-4600 Fax: 409-696-4601
800-782-8272
(800-STATAPC)

Technical support can be obtained by sending e-mail messages *with your Stata serial number on the subject line* to

```
tech_support@stata.com
```

After the first 14 days, only people with registered copies are eligible for technical support.

Before calling or writing for technical help, though, you might want to look at the Web site to see whether your question is an "FAQ." Stata's site on the World Wide Web is

```
http://www.stata.com
```

The site provides product, ordering, and help information; international notes; and assorted news and announcements. More than half of the Web site is devoted to user support, including the following:

Stata FAQs Frequently Asked Questions, and their answers. If you are puzzled by something and can't find the answer in the manuals, check here next—it might be a FAQ. Example questions range from basic—for example, "How do I read data from my Excel file into Stata?"—to more technical queries such as "How can I simulate random multivariate normal observations from a given correlation matrix?"

Cool ado-files Selected new Stata programs ("ado-files") from Statalist. These programs are too new or too minor to have been published in *STB,* but

	they look useful and have been tried out and recommended by at least one user besides their author.
STB	Contents and a selected article from the most recent issue of the *Stata Technical Bulletin*, a bimonthly journal of new programs for Stata, can be read on line. Stata Corporation hopes these will entice you to subscribe.
STB **FTP**	Easy-to-follow instructions tell you how to download free *STB* programs from the Harvard FTP (File Transfer Protocol) site.
Netcourses	On-line courses are periodically scheduled on topics such as how to write Stata programs.
Statalist	Buttons allow you to subscribe or unsubscribe to Statalist, an Internet forum popular with serious Stata users.
Links to other sites	You can connect from here to the Web sites of other statistical software providers, including Stata's main competitors.

Statalist

Statalist provides a valuable on-line forum for communication among active Stata users. It is independent of Stata Corporation, although Stata programmers monitor it and often contribute to the discussion. To subscribe to Statalist, send an e-mail message to

```
majordomo@hsphsun2.harvard.edu
```

The body of this message should contain only the following words:

```
subscribe statalist
```

The list processor will acknowledge your message and send instructions for using the list, including how to post messages of your own. Any message sent to the following address goes out to all the current subscribers:

```
statalist@hsphsun2.harvard.edu
```

Do *not* try to subscribe or unsubscribe by sending messages directly to the Statalist address. This does not work, and your mistake goes to hundreds of subscribers. To unsubscribe from the list, use the same majordomo address used to subscribe:

```
majordomo@hsphsun2.harvard.edu
```

But send only the message:

```
unsubscribe statalist
```

Or send the equivalent message:

```
signoff statalist
```

If you plan to be traveling or off line for a while, unsubscribing will keep your mailbox from filling up with Statalist messages. You can always re-subscribe.

Material on Statalist includes requests for programs, solutions, or advice, as well as answers and general discussion. Topics range from broad ("What are the advantages and disadvantages of Stata compared with SPSS or SAS?") to specialized ("Has anyone written programs that can be used with time-varying covariates to test the reasonableness of using Cox hazard models?"). Along with the *Stata Technical Bulletin* (discussed later), Statalist plays a major role in extending the capabilities both of Stata and of serious Stata users.

Stata Technical Bulletin

The *Stata Technical Bulletin*, a printed journal that includes a disk, supplies its subscribers with bimonthly updates, including both official and user-written new programs. These programs range from bug fixes or enhancements of existing Stata commands to contributions that add important new capabilities. Generalized linear modeling, nonlinear regression, and a library of time series routines were among the many Stata features originally circulated (and subsequently refined) through the *STB*. Other articles have discussed issues such as types of weighting or the properties of different indexes of inequality. Often the authors explicitly compare their programs with similar features of other major statistical software packages. The *STB* also disseminates a variety of specialized programs, as well as readers' solutions to particular analytic or data management problems.

Accumulated *STB* articles are published in book form each year as *Stata Technical Bulletin Reprints*, which makes a convenient reference. *STB* subscriptions or back issues of the *Stata Technical Bulletin Reprints* can be ordered directly from Stata Corporation.

Contact Stata Corporation to subscribe. *STB* programs can also be downloaded freely via anonymous FTP; see Stata's Web page for instructions.

StataQuest

StataQuest is a student-oriented subset of Stata that retains many of Stata's core data-analysis capabilities, from basic tables and graphs through ANOVA, multiple regression, and nonparametric tests. Other capabilities include a statistical calculator (with probability tables), random data generation, robust regression, and simple logistic regression. All these capabilities are accessible through an easy-to-learn set of menus, so students can immediately begin working through assignments for an introductory statistics course without ever typing Stata commands. When menu choices are made, however, the corresponding Stata commands unobtrusively appear above the output in the Results window. Thus StataQuest can also serve as "training wheels" for learning Stata commands gradually. Users can type commands directly in StataQuest's Command window any time they no longer need a menu. StataQuest works with Stata-format datasets containing as many as 600 observations, or 25 variables, or 6,000 data cells (observations times variables)—whichever is smaller.

Several books introduce StataQuest or build it into an introductory statistics course:

Anagnoson, J. Theodore and Richard E. DeLeon, 1996. *StataQuest 4.* Belmont, CA: Duxbury. The most complete StataQuest book, covering the program's main features. Examples show how to perform and interpret statistical analyses. An abbreviated version, *StataQuest 4 Text Companion*, is also available.

Hamilton, Lawrence C. 1996. *Data Analysis for Social Scientists.* Belmont, CA: Duxbury. An introductory statistics text covering topics through multiple regression, with more than 400 real-data problems for student computer work.

Hamilton, Lawrence C. 1996. *Data Analysis for Social Scientists with StataQuest.* Belmont, CA: Duxbury. A compact introduction to StataQuest, for use with an introductory statistics text such as *Data Analysis for Social Scientists.*

Textbooks Using Stata

A growing number of statistics textbooks incorporate Stata output or commands in their presentations. These include

Cryer, Jonathan B. and Robert B. Miller (1994). *Statistics for Business: Data Analysis and Modeling*, 2nd edition. Belmont, CA: Duxbury.

Davis, Duane. (1996). *Business Research for Decision Making*, 4th edition. Belmont, CA: Duxbury.

Hamilton, Lawrence C. (1992). *Regression with Graphics: A Second Course in Applied Statistics.* Belmont, CA: Duxbury.

Howell, David C. (1995). *Fundamental Statistics for the Behavioral Sciences,* 3rd edition. Belmont, CA: Duxbury.

Howell, David C. (1992). *Statistical Methods for Psychology,* 3rd edition. Belmont, CA: Duxbury.

Keller, Gerald, Brian Warrack, and Henry Bartel. (1994). *Statistics for Management and Economics,* 4th edition. Belmont, CA: Duxbury.

Pagano, Marcello and Kim Gauvreau. (1993). *Principles of Biostatistics.* Belmont, CA: Duxbury.

Selvin, Steve. (1995). *Practical Biostatistical Methods.* Belmont, CA: Duxbury.

2

Data Management

The first step in data analysis involves organizing the raw data into a format usable by Stata. We can bring new data into Stata in several ways: type them from the keyboard (**`input`** or **`edit`**); read a text or ASCII file containing the raw data (**`infile`**, **`insheet`**, or **`infix`**); or, using a third-party data transfer program, translate data directly from a system file created by another spreadsheet, database, or statistical program. Once Stata has the data in memory, we can save the data in Stata format for easy retrieval and updating in the future.

Data management encompasses the initial tasks of creating a dataset, editing to correct errors, and adding internal documentation such as variable and value labels. It also encompasses many other jobs required by ongoing projects, such as reorganizing, simplifying, or sampling from the data; adding further observations or variables; separating, combining, or collapsing datasets; converting variable types; and creating new variables through algebraic or logical expressions. Although Stata is best known for its analytical capabilities, it possesses a broad range of data-management features as well. This chapter introduces some of the basics.

Example Commands

```
. edit
```

Opens a spreadsheet-like editor where data can be entered or edited.

```
. use oldfile
```

Retrieves previously-saved, Stata-format dataset *oldfile.dta* from disk and places it in memory. If other data are currently in memory, and you want to discard those data without saving them, type **use *oldfile*, clear**. Alternatively, these tasks can be accomplished through menus: File→Open→(file specification).

. save *newfile*

Saves the data currently in memory, as a file named *newfile.dta*. If *newfile.dta* already exists, and you want to write over the previous version, type **save *newfile*, replace**. To save *newfile.dta* in a format readable by older (pre-4.x) versions of Stata, type **save *newfile*, old**. Alternatively, use the menus: File→Save or File→Save As.

. infile *x y z* using *data.raw*

Reads an ASCII file named *data.raw*, containing data on three variables we want to name *x*, *y*, and *z*. Values of these variables are separated by at least one space. Other commands exist for reading tab-delimited, comma-delimited or fixed-column raw data.

. compress

Automatically changes all variables to their most efficient storage types to conserve memory and disk space. Subsequently typing the command **save, replace** will make these changes permanent.

. sort *x*

Sorts the data from lowest to highest values of *x*. Observations with missing *x* values appear last after sorting. Type **help gsort** for a more general sorting command that can arrange values in either ascending or descending order and can optionally place the missing values first.

. list *x y z* in 5/20

Lists the *x*, *y*, and *z* values of the 5th through 20th observations, as the data are presently sorted. The **in** qualifier works in similar fashion with most other Stata commands as well.

. tabulate *x* if *y* > 65

Produces a frequency table for *x*, using only those observations that have *y* values above 65. The **if** qualifier works similarly with most other Stata commands, as does **in**.

. generate *newvar* = (*x* + *y*)/100

Creates a new variable named *newvar*, equal to *x* plus *y* divided by 100.

. generate *newvar* = uniform()

Creates a new variable with values sampled from a uniform random distribution over the interval [0,1).

```
. replace oldvar = 100 * oldvar
```

Replaces the values of *oldvar* with 100 times their previous values.

```
. encode stringvar, gen(numvar)
```

Creates a new variable named *numvar*, with labeled numerical values based on the first eight characters of string (non-numeric) variable *stringvar*.

```
. append using olddata
```

Reads previously saved dataset *olddata.dta* and adds all its observations to the data currently in memory. Subsequently typing

```
. save newdata, replace
```

would save the combined dataset as *newdata.dta*.

```
. merge id using olddata
```

Reads the previously-saved dataset *olddata.dta* and matches observations from *olddata* with observations in memory that have identical *id* values. Both *olddata* (the "using data") and the data currently in memory (the "master data") must already be sorted in *id* order.

```
. sample 10
```

Drops all the observations in memory except for a 10% random sample.

Creating a New Dataset

Data that were previously saved to disk in Stata format can be retrieved into memory either by typing a command with the form **use *filename***, or by using the File→Open menus. This section describes basic methods for creating a Stata-format dataset in the first place, using as our example the Canadian data (from the Federal, Provincial and Territorial Advisory Committee on Population Health, 1996) listed in Table 2.1.

The simplest way to create a dataset from Table 2.1 is through Stata's spreadsheet-like editor, invoked either by clicking on the Editor button or by typing the command **edit**. Then begin typing values for each variable, in columns that Stata automatically calls *var1*, *var2*, and so forth. Thus *var1* contains place names (Canada, Newfoundland, and so on); *var2* contains populations, and so forth. Cells left empty, such as employment rates for the Yukon and Northwest Territories, will automatically be assigned Stata's missing value code, a period. At any time we can close the Editor, then save the dataset to disk. Clicking on the Editor again allows us to resume editing, to type further values, or to change any of the values entered previously.

Table 2.1: Data on Canada and Its Provinces

Place	1995 Pop. (1000s)	Unemployment Rate (percent)	Male Life Expectancy	Female Life Expectancy
Canada	29606.1	10.6	75.1	81.1
Newfoundland	575.4	19.6	73.9	79.8
Prince Edward Island	136.1	19.1	74.8	81.3
Nova Scotia	937.8	13.9	74.2	80.4
New Brunswick	760.1	13.8	74.8	80.6
Quebec	7334.2	13.2	74.5	81.2
Ontario	11100.3	9.3	75.5	81.1
Manitoba	1137.5	8.5	75.0	80.8
Saskatchewan	1015.6	7.0	75.2	81.8
Alberta	2747.0	8.4	75.5	81.4
British Columbia	3766.0	9.8	75.8	81.4
Yukon	30.1		71.3	80.4
Northwest Territories	65.8		70.2	78.0

If the first value entered for a variable is a number, as with population, unemployment, and life expectancy, then Stata assumes that this column is a "numerical variable," and it will thereafter permit only numerical values. Numerical values can also begin with a plus or minus sign, include decimal points, or be expressed in scientific notation. For example, we could represent Canada's population as 2.96061e+7, which means the same thing as 2.96061×10^7 or about 29 million. Numerical values *should not include any commas,* such as 29,606,100. If we did happen to put commas within the first value typed in a column, Stata would interpret this as a "string variable" (next paragraph) rather than a number.

If the first value entered for a variable includes nonnumber characters, as did the place names earlier, then Stata considers this column to be a string variable. String variable values can be almost any combination of letters, numbers, symbols, or spaces up to 80 characters long. We can thus store names, quotations, or other descriptive information. Because string variable values can be counted but not otherwise analyzed statistically, we might eventually also want to encode their information numerically.

After typing the information from Table 2.1 in this fashion, we close the Editor and save our data, perhaps with the name *canada0.dta:*

```
. save canada0
```

Stata automatically adds the extension `.dta` to any dataset name, if we do not tell it otherwise. (Similarly, it adds the extension `.gph` to a saved graph's name, and `.log` to a log file.) If we already had saved and named an earlier version of this file, it is possible to write over that with the newest version by typing

```
. save, replace
```

At this point, our new dataset looks like this:

```
. describe
Contains data from c:\data\canada0.dta
 obs:              13
 vars:              5                             7 Nov 1996 11:59
 size:            533 (95.4% of memory free)
---------------------------------------------------------------------------
   1. var1         str21  %21s
   2. var2         float  %9.0g
   3. var3         float  %9.0g
   4. var4         float  %9.0g
   5. var5         float  %9.0g
---------------------------------------------------------------------------
Sorted by:
```

```
. list
                        var1        var2        var3        var4        var5
  1.                  Canada     29606.1        10.6        75.1        81.1
  2.            Newfoundland       575.4        19.6        73.9        79.8
  3.    Prince Edward Island       136.1        19.1        74.8        81.3
  4.             Nova Scotia       937.8        13.9        74.2        80.4
  5.           New Brunswick       760.1        13.8        74.8        80.6
  6.                  Quebec      7334.2        13.2        74.5        81.2
  7.                 Ontario     11100.3         9.3        75.5        81.1
  8.                Manitoba      1137.5         8.5          75        80.8
  9.            Saskatchewan      1015.6           7        75.2        81.8
 10.                 Alberta        2747         8.4        75.5        81.4
 11.        British Columbia        3766         9.8        75.8        81.4
 12.                   Yukon        30.1           .        71.3        80.4
 13. Northwest Territories        65.8           .        70.2          78
```

```
. summarize
Variable |     Obs        Mean    Std. Dev.       Min        Max
---------+-----------------------------------------------------
    var1 |       0
    var2 |      13    4554.769    8214.304       30.1    29606.1
    var3 |      11    12.10909    4.250048          7       19.6
    var4 |      13    74.29231    1.673052       70.2       75.8
    var5 |      13    80.71539    .9754027         78       81.8
```

Examining such output tables gives us a chance to look for errors that should be corrected. The **summarize** table, for instance, provides several numbers useful in proofreading, including the count of nonmissing observations (always 0 for string variables) and the minimum and maximum for each variable. Substantive interpretation of the summary statistics would be premature at this point, because our dataset contains one observation (Canada) that represents a combination of the other 12 provinces and territories.

The next step is to make our dataset more self-documenting. The variables could be given more descriptive names with as many as 8 characters, such as the following:

```
. rename var1 place

. rename var2 pop

. rename var3 unemp

. rename var4 mlife

. rename var5 flife
```

Stata also permits us to add three kinds of labels to the data. **label data** describes the dataset as a whole (as many as 31 characters). For example:

```
. label data "Canadian dataset 0"
```

label variable describes an individual variable (as many as 31 characters). For example:

```
. label variable place "Place name"

. label variable pop "Population in 1000s 95"

. label variable unemp "% all 15+ unemployed 95"

. label variable mlife "Male life expectancy years"

. label variable flife "Female life expectancy years"
```

label values permits us to assign labels to specific values of a numeric variable. This capability was not needed with the previous example, but will be illustrated later in this chapter.

By labeling data, variables, and values (if appropriate), we obtain a dataset that is more self-explanatory:

```
. describe
Contains data from c:\data\canada0.dta
  obs:          13                          Canadian dataset 0
 vars:           5                          7 Nov 1996 12:58
 size:         533 (95.4% of memory free)
-------------------------------------------------------------------------------+
   1. place     str21  %21s                 Place name
   2. pop       float  %9.0g                Population in 1000s 95
   3. unemp     float  %9.0g                % all 15+ unemployed 95
   4. mlife     float  %9.0g                Male life expectancy years
   5. flife     float  %9.0g                Female life expectancy years
-------------------------------------------------------------------------------+
Sorted by:
```

Once labeling is completed, we should type the following again to save the data on disk:

```
. save, replace
```

We can later retrieve these data any time by typing the following:

```
. use c:\data\canada0
(Canadian dataset 0)
```

We can then proceed with analysis. We might notice, for instance, that male and female life expectancies correlate positively with each other and also negatively with the unemployment rate. The life expectancy–unemployment rate correlation is slightly stronger for males.

```
. correlate unemp mlife flife
(obs=11)

       |    unemp    mlife    flife
-------+---------------------------
  unemp|   1.0000
  mlife|  -0.7440   1.0000
  flife|  -0.6173   0.7631   1.0000
```

The order of cases within a dataset can be changed through the **sort** command. For example, to rearrange observations from the lowest to highest in population,

```
. sort pop
```

String variables are sorted alphabetically instead of numerically. Typing the following will rearrange the observations with Alberta first, then British Columbia, and so on.

```
. sort place
```

We can control the order of variables listed in the data, using the **order** command. For example, we could make unemployment rate the second variable and population last:

```
. order place unemp mlife flife pop
```

The Editor also has buttons that perform these functions. The Sort button applies to the column currently highlighted by the cursor. The << and >> buttons move the current variable to the beginning or end of the variable list, respectively. As with any other editing, these changes only become permanent if we subsequently save our data.

The Editor's Hide button does not rearrange the data but, rather, makes a column temporarily invisible on the spreadsheet. This feature is convenient if, for example, we need to type more variables and want to keep the province names or some other case identification column in view, adjacent to the "active" column where we are entering data.

Specifying Subsets of the Data: `in` and `if` Qualifiers

Many Stata commands can be restricted to a subset of the data, by adding an **`in`** or **`if`** qualifier. **`in`** specifies the observation numbers to which the command applies. For example, **`list in 5`** tells Stata to list only the 5th observation. To list the 5th through 10th observations:

```
. list in 5/10
```

The letter **`l`** denotes the last case, and **`-3`**, for example, the third-from-last. Thus, we could list the four most populous Canadian places (which will include Canada itself) as follows:

```
. sort pop
```

```
. list place pop in -4/l
```

The **`in`** qualifier works in a similar way with most other analytical or data-editing commands. It always refers to the data *as presently sorted.*

The **`if`** qualifier also has broad applications, but it selects observations based on specific variable values. For example, the observations in *canada0.dta* include not only 12 Canadian provinces or territories but also Canada as a whole. For many purposes, we might want to exclude Canada from analyses involving the twelve territories and provinces. One way to do this is to restrict the analysis to only those places with populations below 20 million (20000 thousand), that is, every place except Canada:

```
. summarize if pop < 20000

Variable |     Obs        Mean   Std. Dev.       Min        Max
---------+-----------------------------------------------------
   place |       0
     pop |      12    2467.158   3435.521       30.1    11100.3
   unemp |      10       12.26    4.44877          7       19.6
   mlife |      12      74.225   1.728965       70.2       75.8
   flife |      12    80.68333     1.0116         78       81.8
```

Compare this with the earlier **`summarize`** output to see how much has changed. The previous mean of population, for example, was grossly misleading because it counted every person twice.

The " < " (is less than) sign is one of Stata's six *comparison operators:*

`==`	is equal to
`!=`	is not equal to (`~=` also works)
`>`	is greater than
`<`	is less than

`>=`	is greater than or equal to
`<=`	is less than or equal to

Note the use of a double equals sign " == " to denote the logical test, "Is the value on the left side the same as the value on the right?"

Any of these comparison operators can be used to select observations based on their values for numerical variables. Only two comparison operators, `==` and `!=`, make sense with string variables. To use string variables in an **`if`** qualifier, enclose the target value in quotes. For example, we could get a summary table for Canada only (leaving out the twelve provinces and territories):

```
. summarize if place == "Canada"
Variable |     Obs        Mean   Std. Dev.       Min        Max
---------+-----------------------------------------------------
   place |       0
     pop |       1     29606.1          .    29606.1    29606.1
   unemp |       1        10.6          .       10.6       10.6
   mlife |       1        75.1          .       75.1       75.1
   flife |       1        81.1          .       81.1       81.1
```

Two or more comparison operators can be combined within a single **`if`** expression by using *logical operators*. Stata's logical operators are the following:

`&`	and
`¦`	or
`!`	not (~ also works)

For example, the Canadian territories (Yukon and Northwest) both have fewer than 100,000 people. To find the mean unemployment and life expectancies for the 10 Canadian provinces only, excluding both the smaller places (territories) and the largest (Canada), we might use this command:

```
. summarize unemp mlife flife if pop > 100 & pop < 20000
Variable |     Obs        Mean   Std. Dev.       Min        Max
---------+-----------------------------------------------------
   unemp |      10       12.26    4.44877          7       19.6
   mlife |      10       74.92   .6051633       73.9       75.8
   flife |      10       80.98    .586515       79.8       81.8
```

A note of caution regarding missing values: Stata shows missing values as a period, but in some operations (notably **`sort`** and **`if`**, although not in statistical calculations such as means or correlations) these same missing values are treated as if they were large positive numbers. Watch what happens if we sort places from lowest to highest unemployment rate, then ask to see places with rates higher than 15%:

```
. sort unemp

. list if unemp > 15
                       place        pop      unemp      mlife      flife
 10.    Prince Edward Island      136.1       19.1       74.8       81.3
 11.            Newfoundland      575.4       19.6       73.9       79.8
 12.   Northwest Territories       65.8          .       70.2         78
 13.                   Yukon       30.1          .       71.3       80.4
```

The two places with missing unemployment rates were included among those "greater than 15." In this instance the result is obvious, but with a larger dataset we might not notice. Suppose we were analyzing a political opinion poll. A command such as the following would tabulate the variable *vote* not only for people with ages older than 65, as intended, but also for any people whose *age* values were missing:

```
. tabulate vote if age > 65
```

Where missing values exist, we might have to deal with them explicitly as part of the **if** expression:

```
. tabulate vote if age > 65 & age != .
```

The **in** and **if** qualifiers set cases aside temporarily, so that that particular command line does not apply to them. These qualifiers have no effect on the data in memory, and the next command will apply to all cases unless it too has an **in** or **if** qualifier. To drop variables from the data in memory, use the **drop** command instead. For example, to drop *mlife* and *flife* from memory:

```
. drop mlife flife
```

We can drop observations from memory by using either **in** or **if** qualifiers. Because we earlier sorted on *unemp*, the two territories occupy the 12th and 13th positions in the data. Canada itself is 6th. One way to drop these three nonprovinces employs the **in** qualifier. **drop in 12/13** means "drop the 12th through the 13th observations."

```
. list
                       place        pop      unemp
  1.            Saskatchewan     1015.6          7
  2.                 Alberta       2747        8.4
  3.                Manitoba     1137.5        8.5
  4.                 Ontario    11100.3        9.3
  5.        British Columbia       3766        9.8
  6.                  Canada    29606.1       10.6
  7.                  Quebec     7334.2       13.2
  8.           New Brunswick      760.1       13.8
  9.             Nova Scotia      937.8       13.9
 10.    Prince Edward Island      136.1       19.1
 11.            Newfoundland      575.4       19.6
 12.                   Yukon       30.1          .
 13.   Northwest Territories       65.8          .
```

```
. drop in 12/13
(2 observations deleted)

. drop in 6
(1 observation deleted)
```

The same change could have been accomplished through an **`if`** qualifier, for example with a command that says "drop if *place* equals Canada or population is less than 100."

```
. drop if place == "Canada" | pop < 100
(3 observations deleted)
```

After dropping Canada, the territories, and the variables *mlife* and *flife*, we have the reduced dataset:

```
. list
                    place        pop      unemp
  1.         Saskatchewan     1015.6          7
  2.              Alberta       2747        8.4
  3.             Manitoba     1137.5        8.5
  4.              Ontario    11100.3        9.3
  5.     British Columbia       3766        9.8
  6.               Quebec     7334.2       13.2
  7.        New Brunswick      760.1       13.8
  8.          Nova Scotia      937.8       13.9
  9. Prince Edward Island      136.1       19.1
 10.         Newfoundland      575.4       19.6
```

We can also drop selected variables or cases through the Delete button in the spreadsheet editor.

Instead of telling Stata which variables or observations to drop, it sometimes is simpler to specify which to keep. The same reduced dataset just shown could have been obtained as follows:

```
. keep place pop unemp

. keep if place != "Canada" & pop > 100
(3 observations deleted)
```

Like any other changes to the data in memory, none of these reductions affect disk files until we **`save`** the data. At that point we will have the option of writing over the old dataset (**`save, replace`**) and thus destroying it, or just saving the newly modified dataset with a new name (by choosing Save As from the File menu, or typing a command with the form **`save`** ***`newname`***) so that both versions exist on disk.

Generating and Replacing Variables

The **generate** and **replace** commands allow us to create new variables or change the values of existing variables. For example, in Canada as in most industrialized countries, women tend to live longer than men. To analyze regional variations in this gender gap, we might retrieve dataset *canada1* and generate a new variable equal to female life expectancy (*flife*) minus male life expectancy (*mlife*). In the main part of a **generate** or **replace** statement (unlike **if** qualifiers) we use a single equals sign, meaning "Make the variable on the left side of this equation equal to the expression on the right."

```
. use canada1, clear
(Canadian data 1)

. generate dlife = flife - mlife

. label variable dlife "Female-male gap life expectancy"

. describe
Contains data from c:\data\canada1.dta
  obs:            13                           Canadian data 1
 vars:             6                           7 Nov 1996 15:12
 size:           585 (95.3% of memory free)
-----------------------------------------------------------------------
   1. place       str21  %21s                  Place name
   2. pop         float  %9.0g                 Population in 1000s 95
   3. unemp       float  %9.0g                 % all 15+ unemployed 95
   4. mlife       float  %9.0g                 Male life expectancy years
   5. flife       float  %9.0g                 Female life expectancy years
   6. dlife       float  %9.0g                 Female-male gap life expectancy
-----------------------------------------------------------------------
Sorted by:
     Note:  data has changed since last save


. list place flife mlife dlife
                       place      flife       mlife        dlife
  1.                  Canada       81.1        75.1            6
  2.            Newfoundland       79.8        73.9     5.900002
  3.    Prince Edward Island       81.3        74.8          6.5
  4.             Nova Scotia       80.4        74.2     6.200005
  5.           New Brunswick       80.6        74.8     5.799995
  6.                  Quebec       81.2        74.5     6.699997
  7.                 Ontario       81.1        75.5     5.599998
  8.                Manitoba       80.8          75     5.800003
  9.            Saskatchewan       81.8        75.2     6.600006
 10.                 Alberta       81.4        75.5     5.900002
 11.        British Columbia       81.4        75.8     5.599998
 12.                   Yukon       80.4        71.3     9.099998
 13.   Northwest Territories         78        70.2     7.800003
```

For the province of Newfoundland, the true value of *dlife* should be 79.8–73.9 = 5.9 years, but the output shows this value as 5.900002 instead. Like all computer programs, Stata stores numbers in binary form, and 5.9 has no exact binary representation. The small inaccuracies that arise from approximating decimal fractions in binary are unlikely to affect statistical calculations much because calculations are done in double precision (8 bytes per number). They do appear disconcerting in data lists, however. We can change the display format so that Stata shows only a rounded-off version. The following command specifies a fixed display format four numerals wide, with one digit to the right of the decimal:

```
. format dlife %4.1f
```

Even when the display shows 5.9, however, a command such as the following will return no observations:

```
. list if dlife == 5.9
```

This occurs because Stata considers that the value does not exactly equal 5.9. (More technically, Stata stores *dlife* values in single precision but does all calculations in double, and the single- and double-precision approximations of 5.9 are not identical.)

replace can make the same sorts of calculations as **generate**, but it changes values of an existing variable instead of creating a new variable. For example, the variable *pop* in our dataset gives population in thousands. To convert this to simple population, we just multiply (" * " means multiply) all values by 1000:

```
. replace pop = pop * 1000
```

replace can make such wholesale changes, or it can be used with **in** or **if** qualifiers to selectively edit the data. To illustrate, suppose we had questionnaire data with variables including age and year born (*born*). A command such as the following would correct one or more typos where a subject's age had been incorrectly typed as "109" instead of 19:

```
. replace age = 19 if age == 109
```

Alternatively, the following command could correct an error in the value of *age* for observation number 2053:

```
. replace age = 19 in 2053
```

For a more complicated example,

```
. replace age = 1997-born if age == . | age < 1996-born
```

This replaces values of variable *age* with 1997 minus the year of birth if *age* is missing or if the reported age is less than 1996 minus year of birth.

generate and **replace** provide tools to create categorical variables as well. We noted earlier that our Canadian dataset includes several types of observations: 2 territories, 10 provinces, and 1 country combining them all. Although **in** and **if**

qualifiers allow us to separate these, and **drop** can eliminate them from the data, it might be most convenient to have a categorical variable that indicates the observation's "type." The following example shows one way to create such a variable. We start by generating *type* as a constant, equal to 1 for each observation. Then we replace this with the value 2 for the Yukon and Northwest Territories. Finally, we replace the value with 3 for Canada. The final steps involve labeling new variable *type* and defining labels for values 1, 2, and 3.

```
. use canada1, clear
(Canadian dataset 1)

. generate type = 1

. replace type = 2 if place == "Yukon" |
      place == "Northwest Territories"
(2 real changes made)

. replace type = 3 if place == "Canada"
(1 real change made)

. label variable type "Province, territory or nation"

. label define typelbl 1 "Province" 2 "Territor" 3 "Nation"

. label values type typelbl

. list place flife mlife dlife type
                      place     flife     mlife      dlife       type
  1.                 Canada      81.1      75.1          6     Nation
  2.           Newfoundland      79.8      73.9   5.900002   Province
  3.   Prince Edward Island      81.3      74.8        6.5   Province
  4.            Nova Scotia      80.4      74.2   6.200005   Province
  5.          New Brunswick      80.6      74.8   5.799995   Province
  6.                 Quebec      81.2      74.5   6.699997   Province
  7.                Ontario      81.1      75.5   5.599998   Province
  8.               Manitoba      80.8        75   5.800003   Province
  9.           Saskatchewan      81.8      75.2   6.600006   Province
 10.                Alberta      81.4      75.5   5.900002   Province
 11.       British Columbia      81.4      75.8   5.599998   Province
 12.                  Yukon      80.4      71.3   9.099998   Territor
 13.  Northwest Territories        78      70.2   7.800003   Territor
```

As illustrated, labeling the values of a categorical variable requires two commands. The **label define** command specifies what labels go with what numbers, as in the following:

```
. label define typelbl 1 "Province" 2 "Territor" 3 "Nation"
```

The **label values** command specifies to which variable these labels apply:

```
. label values type typelbl
```

One set of labels (created through one **label define** command) can apply to any number of variables (that is, be referenced in any number of **label values** commands). Value labels are limited to eight characters or less.

generate can create new variables, and **replace** can produce new values, using any mixture of old variables, constants, random values, and expressions. For numeric variables, the following *algebraic operators* apply:

+	add
–	subtract
*	multiply
/	divide
^	raise to power

Parentheses will control the order of calculation. Without them, the ordinary rules of precedence apply. Of the algebraic operators only addition, " + ", works with string variables, where it concatenates two string values into one.

Although their purposes differ, **generate** and **replace** have similar syntax, and either can use any mathematically or logically feasible combination of Stata operators and **in** or **if** qualifiers. These commands can also employ Stata's broad array of special functions, introduced in the following section.

Using Functions

This section lists many of the functions available for use with **generate** or **replace**. For example, we could create a new variable named *loginc*, equal to the natural logarithm of income, by using the natural log function **ln** within a **generate** command:

```
. generate loginc = ln(income)
```

ln is one of Stata's *mathematical functions*. These functions are as follows:

abs(*x***)** Absolute value.

acos(*x***)** Arc-cosine returning radians. Because 360 degrees = 2π radians, **acos(***x***)*180/_pi** gives the arc-cosine returning degrees (_pi denotes the mathematical constant π).

asin(*x***)** Arc-sine returning radians.

atan(*x***)** Arc-tangent returning radians.

comb(n,k) Combinatorial function (number of possible combinations of n things taken k at a time).

cos(x) Cosine of radians. To find the cosine of y degrees:

```
. generate y = cos(x*_pi/180)
```

exp(x) Exponential (e to power).

ln(x) Natural (base e) logarithm. For any other base number B, to find the base (B) logarithm of v

```
. generate y = ln(x)/ln(B)
```

lnfact(x) Natural log of factorial. To find x factorial:

```
. generate y = round(exp(lnfact(x),1))
```

lngamma(x) Natural log of $\Gamma(x)$. To find $\Gamma(x)$:

```
. generate y = exp(lngamma(x))
```

log(x) Natural logarithm; same as **ln**(x).

log10(x) Base 10 logarithm.

mod(x,y) Modulus of x with respect to y.

sin(x) Sine of radians.

sqrt(x) Square root.

tan(x) Tangent of radians.

The following *statistical functions* exist:

Binomial(n,k,π) Probability of k or more successes in n trials, when the probability of success on a single trial is π.

binorm(h,k,ρ) Joint cumulative probability of $\Phi(h, k, \rho)$: a bivariate normal distribution with correlation ρ, cumulative over $(-\infty, h] \times (-\infty, k]$.

chiprob(df,x) Probability that a χ^2 with df degrees of freedom is equal to or larger than x.

fprob(df_1,df_2,f) Probability that an F with df_1 numerator and df_2 denominator degrees of freedom is equal to or larger than f.

gammap(a,x) Incomplete gamma function $P(a, x)$.

ibeta(a,b,x) Incomplete beta function $I_x(a, b)$.

invbinomial(*n*,*k*,*p*) — Inverse binomial; for $p \leq 0.5$, returns π (probability of success on a single trial) such that the probability of k or more successes in n trials equals p; for p > 0.5, returns π such that the probability of k or fewer successes in n trials equals $1-p$.

invchi(*df*,*p*) — Inverse of **chiprob( )**; if **chiprob**(*df*,*x*) = p, then **invchi**(*df*,*p*) = x.

invfprob(df_1,df_2,*p*) — Inverse of **fprob( )**.

invgammap(*a*,*p*) — Inverse of **gammap( )**.

invnchi(*df*,λ,*p*) — Inverse of **nchi( )**.

invnorm(*p*) — Inverse of **normprob( )**.

invt(*df*,*p*) — Inverse of **tprob( )**.

nchi(*df*,λ,*x*) — Cumulative noncentral χ^2 distribution with *df* degrees of freedom and noncentrality parameter λ.

normd(*z*) — Standard normal density.

normprob(*z*) — Probability that a standard normal variable is less than or equal to z.

npnchi(*df*,*x*,*p*) — Noncentrality parameter λ for noncentral χ^2.

tprob(*df*, *t*) — Probability that the absolute value of T with *df* degrees of freedom is greater than $|t|$.

uniform( ) — Pseudo-random number generator, returning values from a uniform distribution on the interval [0,1).

Nothing goes inside the parentheses with **uniform( )**. Optionally, we can control the pseudo-random generator's starting seed, and hence the stream of "random" numbers, by first issuing a **set seed #** command—where # could be any integer from 0 to $2^{31}-1$ inclusive. Omitting the **set seed** command corresponds to **set seed 123456789**, which will always produce the same stream of numbers.

Stata also provides *date functions* for working with date variables:

date(s_1,s_2) — Returns the elapsed date corresponding to s_1. s_1 is a string variable indicating the date, in virtually any format. Months can be spelled out, abbreviated to three characters, or given as numbers; years can include or exclude the century; blanks and punctuation are allowed. s_2 is any permutation of `m`, `d` and `[##]y` with their order defining the order that month, day, and year occur in s_1. ##, if specified, gives the century for two-digit years in s_1; the default is 19y.

day(*e***)** Returns the numeric day of the month corresponding to *e*, the elapsed date.

dow(*e***)** Returns the numeric day of the week corresponding to *e*, the elapsed date.

mdy(*m*,*d*,*y***)** Returns the elapsed date corresponding to the numeric arguments of month, day, and year.

month(*e***)** Returns the numeric month corresponding to *e*, the elapsed date.

year(*e***)** Returns the numeric year corresponding to *e*, the elapsed date.

Several useful *special functions* include the following:

autocode(*x*,*n*,*xmin*,*xmax***)** Partitions the interval from *xmin* to *xmax* into *n* equal-length intervals and returns the upper bound of the interval that contains *x*.

cond(*x*,*a*,*b***)** Returns *a* if *x* evaluates to "true" and *b* if *x* evaluates to "false."

```
. generate y = cond(inc1 > inc2, inc1, inc2)
```

creates the variable *y* as the maximum of *inc1* and *inc2* (assuming neither is missing).

group(*x***)** Creates a categorical variable that divides the data *as presently sorted* into *x* subsamples, as nearly equal-sized as possible.

int(*x***)** Returns the integer obtained by truncating (dropping fractional parts of) *x*.

max(x_1,x_2,...,x_n**)** Returns the maximum of x_1,x_2, ..., x_n. Missing values are ignored. For example, **max(3+2,1)** evaluates to 5.

min(x_1,x_2,...,x_n**)** Returns the minimum of x_1, x_2, . . . , x_n .

recode(*x*,x_1,x_2,...,x_n**)** Returns missing if *x* is missing, x_1 if $x < x_1$, otherwise x_2 if $x < x_2$, and so on.

round(*x*,*y***)** Returns *x* rounded to the nearest *y*.

sign(*x***)** Returns –1 if $x < 0$, 0 if $x = 0$, and +1 if $x > 0$ (missing if *x* is missing).

sum(*x***)** Returns the running sum of *x*, treating missing values as zero.

String functions, not described here, help when you want to manipulate and evaluate string variables. Type **help functions** for a complete list of all Stata functions, or see the *Stata Reference Manual* and *User's Guide* for more examples and details.

Multiple functions, operators, and qualifiers can be combined in one command as needed. The functions and algebraic operators just described can also be used in another way that does not create or change any dataset variables. The **display** command performs a single calculation and shows the results on screen. The following are some examples:

```
. display 2 + 3
5

. display log10(10^83)
83

. display invt(120,.95) * 34.1/sqrt(975)
2.1622305
```

display thus works as an on-screen statistical calculator.

Stata also provides another variable-creation command, **egen** ("extensions to **generate**"), which has its own set of functions to accomplish tasks not easily done by **generate**. These include such things as creating new variables representing the sums, maxima, minima, medians, interquartile ranges, standardized values, or moving averages of other variables or expressions. For example, the following creates a new variable named *zscore*, equal to the standardized (mean 0, variance 1) values of *score:*

```
. egen zscore = std(score)
```

Or the following creates new variable *price2*, equal to five-period moving averages of variable *price.*

```
. egen price2 = ma(price), t(5)
```

Consult **help egen** for a complete list of **egen** functions or the *Reference Manual* for some clarifying examples.

Converting between Numeric and String Formats

Dataset *canada2* contains one string variable, *place.* It also has a labeled categorical variable, *type.* Both seem to have nonnumerical values.

```
. use canada2, clear
(Canada data 2)
```

```
. list place type

                     place       type
  1.                Canada     Nation
  2.          Newfoundland   Province
  3.  Prince Edward Island   Province
  4.           Nova Scotia   Province
  5.         New Brunswick   Province
  6.                Quebec   Province
  7.               Ontario   Province
  8.              Manitoba   Province
  9.          Saskatchewan   Province
 10.               Alberta   Province
 11.      British Columbia   Province
 12.                 Yukon   Territor
 13. Northwest Territories   Territor
```

Beneath the labels, however, *type* remains a numeric variable, as we can see if we ask for the **nolabel** option:

```
. list place type, nolabel

                     place       type
  1.                Canada          3
  2.          Newfoundland          1
  3.  Prince Edward Island          1
  4.           Nova Scotia          1
  5.         New Brunswick          1
  6.                Quebec          1
  7.               Ontario          1
  8.              Manitoba          1
  9.          Saskatchewan          1
 10.               Alberta          1
 11.      British Columbia          1
 12.                 Yukon          2
 13. Northwest Territories          2
```

String variables can record more alphanumeric information (as many as 80 characters) than do the labels on a numeric variable (up to 8 characters). But most statistical operations and algebraic comparisons are not defined for string variables, so we might want to have both string and labeled-numeric versions of a variable in the data. The **encode** command generates a labeled-numeric variable from a string variable, using the string value's first 8 characters. The number 1 is given to the alphabetically first value of the string variable, 2 to the second, and so on. In the following example, we create a numeric variable named *place2* from the string variable *place:*

```
. encode place, gen(place2)
(note: place is str21; only first 8 characters will be used)
```

Stata reminds us that *place* values have as many as 21 characters (`str21`), but *place2* labels will have only 8.

The opposite conversion is possible too: The **decode** command generates a string variable using the labels of a labeled numeric variable. Here we create string variable *type2* from numeric variable *type:*

```
. decode type, gen(type2)
```

When listed, the new numeric variable *place2*, and the new string variable *type2*, look similar to the originals:

```
. list place place2 type type2
                       place     place2       type      type2
  1.                  Canada     Canada     Nation     Nation
  2.            Newfoundland   Newfound   Province   Province
  3.    Prince Edward Island   Prince E   Province   Province
  4.             Nova Scotia   Nova Sco   Province   Province
  5.           New Brunswick   New Brun   Province   Province
  6.                  Quebec     Quebec   Province   Province
  7.                 Ontario    Ontario   Province   Province
  8.                Manitoba   Manitoba   Province   Province
  9.            Saskatchewan   Saskatch   Province   Province
 10.                 Alberta    Alberta   Province   Province
 11.        British Columbia    British   Province   Province
 12.                   Yukon      Yukon   Territor   Territor
 13.   Northwest Territories   Northwes   Territor   Territor
```

But with the **nolabel** option, the differences become visible. Stata views *place2* and *type* basically as numbers:

```
. list place place2 type type2, nolabel
                       place     place2       type      type2
  1.                  Canada          3          3     Nation
  2.            Newfoundland          6          1   Province
  3.    Prince Edward Island         10          1   Province
  4.             Nova Scotia          8          1   Province
  5.           New Brunswick          5          1   Province
  6.                  Quebec         11          1   Province
  7.                 Ontario          9          1   Province
  8.                Manitoba          4          1   Province
  9.            Saskatchewan         12          1   Province
 10.                 Alberta          1          1   Province
 11.        British Columbia          2          1   Province
 12.                   Yukon         13          2   Territor
 13.   Northwest Territories          7          2   Territor
```

Statistical analyses such as finding means and standard deviations work only with basically numeric variables. For calculation purposes, the numeric variables' labels do not matter:

```
. summarize place place2 type type2
Variable |     Obs        Mean    Std. Dev.       Min        Max
---------+-----------------------------------------------------
   place |       0
  place2 |      13           7    3.89444          1         13
    type |      13    1.307692   .6304252          1          3
   type2 |       0
```

Occasionally we encounter a string variable where the values are all or mostly numbers. To convert these string values into their actual numerical counterparts, use the **real** function. For example, the variable *siblings* below is a string variable, although it only has one value, "4 or more," that could not be represented just as easily by a number:

```
. describe siblings
   1. siblings     str9    %9s                         Number of siblings (string)

. list
       siblings
  1.          0
  2.          1
  3.          2
  4.          3
  5. 4 or more

. generate sibs2 = real(siblings)
(1 missing value generated)
```

The new variable *sibs2* is numeric, with a missing value where *siblings* had "4 or more."

```
. list
       siblings        sibs2
  1.          0            0
  2.          1            1
  3.          2            2
  4.          3            3
  5. 4 or more             .
```

Creating New Categorical and Ordinal Variables

A previous section illustrated how to construct a categorical variable called *type* to distinguish between territories, provinces, and nation in our Canadian dataset. You can create categorical or ordinal variables in many other ways. This section gives a few examples, using the same dataset (*canada2.dta*).

type has three categories:

```
. tabulate type
Province,    |
territory or |
nation       |      Freq.     Percent        Cum.
-------------+-----------------------------------
    Province |         10       76.92       76.92
    Territor |          2       15.38       92.31
      Nation |          1        7.69      100.00
-------------+-----------------------------------
       Total |         13      100.00
```

For some purposes, such as regression analysis, we might want to re-express a polytomous (multicategory) variable as a set of dichotomies or "dummy variables," each coded 0 or 1. **tabulate** will create a set of dummy variables automatically, if we add the **generate** option. In the following example, this results in a set of dummy variables called *type1*, *type2*, and *type3* representing each of the three categories of *type:*

```
. tabulate type, generate(type)

Province,    |
territory or |
nation       |      Freq.     Percent        Cum.
-------------+-----------------------------------
    Province |         10       76.92       76.92
   Territor  |          2       15.38       92.31
      Nation |          1        7.69      100.00
-------------+-----------------------------------
       Total |         13      100.00
```

```
. describe

Contains data from c:\data\canada2.dta
  obs:            13                          Canada data 2
 vars:            10                          8 Nov 1996 12:31
 size:           676 (95.3% of memory free)
-------------------------------------------------------------------------------
  1. place       str21  %21s                  Place name
  2. pop         float  %9.0g                 Population in 1000s 95
  3. unemp       float  %9.0g                 % all 15+ unemployed 95
  4. mlife       float  %9.0g                 Male life expectancy years
  5. flife       float  %9.0g                 Female life expectancy years
  6. dlife       float  %9.0g                 Female-male gap life expectancy
  7. type        float  %9.0g      typelbl    Province, territory or nation
  8. type1       byte   %8.0g                 type==Province
  9. type2       byte   %8.0g                 type==Territor
 10. type3       byte   %8.0g                 type==Nation
-------------------------------------------------------------------------------
Sorted by:
     Note:  data has changed since last save
```

```
. list place type type1-type3

                       place       type      type1      type2      type3
  1.                  Canada     Nation          0          0          1
  2.            Newfoundland   Province          1          0          0
  3.    Prince Edward Island   Province          1          0          0
  4.             Nova Scotia   Province          1          0          0
  5.           New Brunswick   Province          1          0          0
  6.                  Quebec   Province          1          0          0
  7.                 Ontario   Province          1          0          0
  8.                Manitoba   Province          1          0          0
  9.            Saskatchewan   Province          1          0          0
 10.                 Alberta   Province          1          0          0
 11.        British Columbia   Province          1          0          0
 12.                   Yukon   Territor          0          1          0
 13.   Northwest Territories   Territor          0          1          0
```

Re-expressing categorical information as a set of dummy variables involves no loss of information; in this example *type1* through *type3* together tell us exactly as much as *type* itself does. Occasionally, however, analysts want to re-express a continuous measurement variable in categorical or ordinal form, even though this *does* result in a substantial loss of information. For example, *unemp* in *canada2.dta* gives a measure of the unemployment rate. Excluding Canada itself from the data, we see that *unemp* ranges from 7% to 19.6%, with a mean of 12.26:

```
. summarize unemp if type != 3
Variable |     Obs        Mean   Std. Dev.       Min        Max
---------+-----------------------------------------------------
   unemp |      10       12.26    4.44877          7       19.6
```

Having Canada in the data becomes a nuisance at this point, so we drop it:

```
. drop if type == 3
(1 observation deleted)
```

Two commands create a dummy variable named *unempD* with values of 0 when unemployment is below average (12.26), 1 when unemployment is equal to or above average, and missing when *unemp* is missing. In reading the second command, recall that Stata's sorting and comparison operators treat missing values as very large numbers:

```
. generate unempD = 0 if unemp < 12.26
(7 missing values generated)

. replace unempD = 1 if unemp >= 12.26 & unemp != .
(5 real changes made)
```

We might want to group the values of a measurement variable, thereby creating an ordered-category or ordinal variable. The **autocode** function (see "Using Functions" earlier in this chapter) provides automatic grouping of measurement variables. To create new ordinal variable *unempO*, which groups values of *unemp* into three equal-width groups over the interval from 5 to 20:

```
. generate unempO = autocode(unemp,3,5,20)
(2 missing values generated)
```

A list of the data shows how the new dummy (*unempD*) and ordinal (*unempO*) variables correspond to values of the original measurement variable *unemp:*

```
. list place unemp unempD unempO

                     place      unemp     unempD     unempO
  1.          Newfoundland       19.6          1         20
  2.  Prince Edward Island       19.1          1         20
  3.           Nova Scotia       13.9          1         15
  4.         New Brunswick       13.8          1         15
  5.                Quebec       13.2          1         15
  6.               Ontario        9.3          0         10
  7.              Manitoba        8.5          0         10
  8.          Saskatchewan          7          0         10
  9.               Alberta        8.4          0         10
 10.      British Columbia        9.8          0         10
 11.                 Yukon          .          .          .
 12. Northwest Territories          .          .          .
```

Both strategies just described dealt appropriately with missing values, so that Canadian places with missing values on *unemp* likewise receive missing values on the variables derived from *unemp*. Another possible approach works best if our data contain no missing values. To illustrate, we begin by dropping the Yukon and Northwest Territories:

```
. drop if unemp == .
(2 observations deleted)
```

We now can use the **group** function to create an ordinal variable not with approximately equal-width groupings, as **autocode** did, but instead with groupings of approximately equal size. We do this in two steps. First, sort the data (assuming no missing values) on the variable of interest. Second, generate a new variable using the **group(#)** function, where # indicates the number of groups desired. The example below divides our 10 Canadian provinces into 5 groups:

```
. sort unemp

. generate unempOO = group(5)

. list place unemp unempD unempO unempOO

                     place      unemp     unempD     unempO    unempOO
  1.          Saskatchewan          7          0         10          1
  2.               Alberta        8.4          0         10          1
  3.              Manitoba        8.5          0         10          2
  4.               Ontario        9.3          0         10          2
  5.      British Columbia        9.8          0         10          3
  6.                Quebec       13.2          1         15          3
  7.         New Brunswick       13.8          1         15          4
  8.           Nova Scotia       13.9          1         15          4
  9.  Prince Edward Island       19.1          1         20          5
 10.          Newfoundland       19.6          1         20          5
```

Another difference is that **autocode** assigns values equal to the upper bound of each interval, whereas **group** simply assigns 1 to the first group, 2 to the second, and so forth.

Using Explicit Subscripts with Variables

When Stata has data in memory, it also defines certain system variables that describe those data. For example, _N represents the total number of observations. _n represents the observation number: _n = 1 for the first observation, _n = 2 for the second, and so on to the last observation (_n = _N). If we issue a command such as the following, it creates a new variable, *caseID*, equal to the number of each observation as presently sorted:

```
. generate caseID = _n
```

Sorting the data another way will change each observation's value of _n, but its *caseID* value will remain unchanged. Thus, if we do sort the data another way, we can later return to the earlier order by typing

```
. sort caseID
```

We can use _n to begin an artificial dataset as well. The following commands create a new dataset with one variable: $x = 1, 2, 3, \ldots, 10{,}000$:

```
. clear

. set obs 10000
obs was 0, now 10000

. generate x = _n

. summarize

Variable |      Obs        Mean    Std. Dev.        Min         Max
---------+-----------------------------------------------------------
       x |    10000      5000.5    2886.896           1       10000
```

We can use explicit subscripts with variable names, to specify particular observation numbers. For example, the 6th observation in dataset *canada2.dta* (if we have not dropped or resorted anything) is Quebec. Consequently, *pop[6]* refers to Quebec's population, 7334 thousand:

```
. display pop[6]
7334.2002
```

Similarly, *pop[12]* is the Yukon's population:

```
. display pop[12]
30.1
```

Explicit subscripting becomes particularly useful with time series data. If we had a time series named *flow*, for instance, then either *flow* or, equivalently, *flow[_n]*

denotes the value of the _nth observation. *flow[_n–1]* denotes the previous observation, and *flow[_n+1]* denotes the next. Thus we might define a variable equal to the change in *flow* since the previous observation (that is, a "first difference")

```
. generate d1flow = flow - flow[_n-1]
```

Importing Data from Other Programs

The spreadsheet Editor provides a convenient way to enter and edit small datasets, but for larger projects we might need tools that work directly with computer files created by other programs. Such files fall in two general categories: raw data ASCII (text) files, which can be read into Stata with the appropriate Stata commands, and system (usually binary) files, which must be translated to Stata format by a special third-party program before Stata can read them.

To illustrate ASCII file methods, we return to the Canadian data of Table 2.1. Suppose that, instead of typing these data into Stata's Editor, we typed them instead into our word processor, with at least one space between each value. String values must be in double quotes if they contain any internal spaces, as does "Prince Edward Island." For other string values, quotes are optional. Most word processors allow us to save documents as ASCII (text) files, a simpler and more universal type than the word processor's usual saved-file format. We can thus create an ASCII file named *canada.raw*, that looks something like this:

```
"Canada"  29606.1  10.6  75.1  81.1
"Newfoundland"  575.4  19.6  73.9  79.8
"Prince Edward Island"  136.1  19.1  74.8  81.3
"Nova Scotia"  937.8  13.9  74.2  80.4
"New Brunswick"  760.1  13.8  74.8  80.6
"Quebec"  7334.2  13.2  74.5  81.2
"Ontario"  11100.3  9.3  75.5  81.1
"Manitoba"  1137.5  8.5  75  80.8
"Saskatchewan"  1015.6  7  75.2  81.8
"Alberta"  2747  8.4  75.5  81.4
"British Columbia"  3766  9.8  75.8  81.4
"Yukon"  30.1  .  71.3  80.4
"Northwest Territories"  65.8  .  70.2  78
```

Note the use of periods, not blanks, to indicate missing values for the Yukon and Northwest Territories. If the dataset should have five variables, then for every observation exactly five values (including periods for missing values) must exist.

infile reads into memory an ASCII file such as *canada.raw*, in which the values are separated by spaces (or commas). Its basic form is

```
. infile variable-list using filename.raw
```

With purely numeric data, the variable list could be omitted, in which case Stata assigns names *v1*, *v2*, *v3*, and so forth. On the other hand, we may want to give each variable a distinctive name. We also need to identify string variables individually. For *canada.raw*, the **infile** command might be

```
. infile str30 place pop unemp mlife flife using canada.raw,
      clear
(13 observations read)
```

The **infile** variable list specifies variables in the order they appear in the data file. The **clear** option drops any current data from memory before reading in the new file.

If any string variables exist, their names must each be preceded by a **str#** statement. **str30**, for example, informs Stata that the next-named variable (*place*) is a string variable with as many as 30 characters. Actually, none of the Canadian place names involve more than 21 characters, but we do not need to know that in advance. It is often easier to overestimate string variable lengths. Then, once data are in memory, use **compress** to ensure that no variable takes up more space than it needs. The **compress** command automatically changes all variables to their most memory-efficient storage type.

```
. compress
place was str30 now str21

. describe
Contains data
  obs:            13
 vars:             5
 size:           533 (95.4% of memory free)
-----------------------------------------------------------------------
   1. place        str21  %21s
   2. pop          float  %9.0g
   3. unemp        float  %9.0g
   4. mlife        float  %9.0g
   5. flife        float  %9.0g
-----------------------------------------------------------------------
Sorted by:
     Note:  data has changed since last save
```

We can now proceed to label variables and data as described earlier. At any point, the commands **save** ***canada3*** (or **save** ***canada3*****, replace**) would save the new dataset in Stata format, as file *canada0.dta*. The original raw-data file, *canada.raw*, remains unchanged on disk.

If our variables have short nonnumeric values (for example, "male" and "female") that we want to store as labeled numeric variables, based on their first eight characters, then adding the option **automatic** will accomplish this automatically. For example, we might read in raw survey data through this **infile** command:

```
. infile gender age income vote using survey.raw, automatic
```

Spreadsheet and database programs will often write ASCII files that have only one observation per line, with values separated by tabs or commas. To read these files into Stata, use **insheet**. Its general syntax resembles that of **infile**, with an option telling Stata whether the data are tab- or comma-delimited. For example, assuming tab-delimited data,

```
. insheet variable-list using filename.raw, tab
```

Or, assuming comma-delimited data, with the first row of the file containing variable names (also comma-delimited),

```
. insheet variable-list using filename.raw, comma names
```

With **insheet** we do not need to separately identify string variables, and the variable list could be omitted with any data (resulting in names *v1*, *v2*, *v3*, ...). Errors will result, however, if some values in our ASCII file are not separated by tabs or commas as specified in the **insheet** command.

Raw data files created by other statistical packages can be in "fixed-column" format, where the values are not necessarily delimited at all, but they do occupy predefined column positions. Both **infile** and the more specialized command **infix** permit Stata to read such files. In the command syntax itself, or (more commonly) in a "data dictionary" existing as a separate file or as the first part of the data file, we have to specify exactly how the columns should be read.

Here is a simple example. Data on four variables exist in an ASCII file named *nfresour.raw:*

```
198624087641691000
198725247430001044
198825138637481086
198925358964371140
1990    8615731195
1991    7930001262
```

These data concern natural resource production in Newfoundland. The four variables occupy fixed column positions: columns 1 through 4 are the year (1986...1991); columns 5–8 measure forestry production, in thousands of cubic meters (2408... missing); columns 9–14 measure mine production in thousands of dollars (764,169...793,000); and columns 15–18 are the consumer price index relative to 1986 (1000...1262). Notice that in fixed-column format, unlike space or tab-delimited files, blanks indicate missing values, and the raw data contain no decimal points. To read *nfresour.raw* into Stata, we specify each variable's column position:

```
. infix year 1-4 wood 5-8 mines 9-14 CPI 15-18
      using nfresour.raw, clear
(6 observations read)

. list
          year       wood       mines        CPI
  1.      1986       2408      764169       1000
  2.      1987       2524      743000       1044
  3.      1988       2513      863748       1086
  4.      1989       2535      896437       1140
  5.      1990          .      861573       1195
  6.      1991          .      793000       1262
```

More complicated fixed-column formats might require a data "dictionary." These can be straightforward, but many possible choices exist. Typing **help infix** or **help infile2** obtains brief outlines of these commands. For more examples and explanation, consult the *User's Guide* and *Reference Manuals*.

What if we need to export data from Stata to some other program? The outfile command writes ASCII files to disk. In its simplest form, a command such as the following will create a space-delimited ASCII file named *canada6.raw*, containing whatever data were in memory:

```
. outfile using canada6
```

The **infile**, **insheet**, **infix**, and **outfile** commands just described all manipulate raw data in ASCII files. It is often faster and more convenient, however, to transfer data directly between the specialized system files saved by different spreadsheet, database, or statistical programs, without going through the intermediate step of an ASCII file. Several third-party programs, notably Stat/Transfer and DBMS/COPY, perform such translations. Stat/Transfer, for example, will transfer data across formats including dBASE, Excel, GAUSS, Lotus 1-2-3, Paradox, Quattro Pro, S-Plus, SAS (transport or `.tpt` files), SPSS (export or `.xpt` files), Systat, and Stata. Transfer programs prove indispensable for researchers working in multiprogram environments or exchanging data with colleagues. See Hilbe (1996) for a comparative review of Stat/Transfer and DBMS/COPY.

Stat/Transfer is available either through Stata Corporation or from its maker, Circle Systems:

Circle Systems
1001 Fourth Avenue Plaza, Suite 3200
Seattle, WA 98154

Telephone: 206-682-3783
Fax: 206-328-4788
E-mail: stsales@circlesys.com

DBMS/COPY is available from

Conceptual Software
9660 Hillcroft, Suite 510
Houston, TX 77096

Telephone: 713-721-4200
Fax: 713-721-4298
E-mail: henry@conceptual.com

Combining Two or More Stata Files

We can combine Stata datasets in two general ways: **append** a second dataset that contains additional observations, or **merge** with a second dataset that contains new variables or values. In keeping with this chapter's Canadian theme, we will illustrate these procedures using data on Newfoundland. File *newf1* records the province's population for the years 1985 to 1989:

```
. use newf1, clear
(Newfoundland 1985-89)

. describe
Contains data from c:\data\newf1.dta
  obs:              5                             Newfoundland 1985-89
 vars:              2                             8 Nov 1996 16:56
 size:             50 (95.4% of memory free)
-------------------------------------------------------------------------
   1. year         int     %9.0g                  Year
   2. pop          float   %9.0g                  Population
-------------------------------------------------------------------------
Sorted by:

. list
          year        pop
  1.      1985     580700
  2.      1986     580200
  3.      1987     568200
  4.      1988     568000
  5.      1989     570000
```

File *newf2* has population and unemployment counts for some later years:

```
. use newf2
(Newfoundland 1990-95)

. describe
Contains data from c:\data\newf2.dta
  obs:              6                             Newfoundland 1990-95
 vars:              3                             8 Nov 1996 16:57
 size:             84 (95.4% of memory free)
-------------------------------------------------------------------------
   1. year         int     %9.0g                  Year
   2. pop          float   %9.0g                  Population
   3. jobless      float   %9.0g                  Number of people unemployed
-------------------------------------------------------------------------
Sorted by:
```

```
. list

          year        pop    jobless
  1.      1990     573400      42000
  2.      1991     573500      45000
  3.      1992     575600      49000
  4.      1993     584400      49000
  5.      1994     582400      50000
  6.      1995     575449          .
```

To combine these two datasets, with *newf2* already in memory, we use the **append** command:

```
. append using newf1

. list

          year        pop    jobless
  1.      1990     573400      42000
  2.      1991     573500      45000
  3.      1992     575600      49000
  4.      1993     584400      49000
  5.      1994     582400      50000
  6.      1995     575449          .
  7.      1985     580700          .
  8.      1986     580200          .
  9.      1987     568200          .
 10.      1988     568000          .
 11.      1989     570000          .
```

Because variable *jobless* occurs in *newf2* (1990 to 1995) but not in *newf1*, its 1985 to 1989 values are missing in the combined dataset. We can now put the observations in order from earliest to latest and save these combined data as a new file, *newf3.dta*:

```
. sort year

. list

          year        pop    jobless
  1.      1985     580700          .
  2.      1986     580200          .
  3.      1987     568200          .
  4.      1988     568000          .
  5.      1989     570000          .
  6.      1990     573400      42000
  7.      1991     573500      45000
  8.      1992     575600      49000
  9.      1993     584400      49000
 10.      1994     582400      50000
 11.      1995     575449          .

. save newf3
```

append might be compared with lengthening a sheet of paper (that is, the dataset in memory) by taping a second sheet to its bottom. **merge**, in its simplest form, corresponds to "widening" our sheet of paper by taping a second sheet to its right side, thereby adding new variables. For example, dataset *newf4.dta* contains further Newfoundland time series: numbers of births and divorces over the years 1980 to 1994. Thus it has some observations in common with our earlier dataset *newf3.dta*, as well as one variable (*year*) in common, but it also has two new variables not present in *newf3.dta*.

```
. use newf4, clear
(Newfoundland 1980-94)

. describe
Contains data from c:\data\newf4.dta
  obs:            15                          Newfoundland 1980-94
 vars:             3                          8 Nov 1996 17:03
 size:           150 (95.4% of memory free)
-------------------------------------------------------------------------
   1. year         int     %9.0g              Year
   2. births       int     %9.0g              Number of births
   3. divorces     int     %9.0g              Number of divorces
-------------------------------------------------------------------------
Sorted by:

. list
          year     births   divorces
  1.      1980      10332        555
  2.      1981      11310        569
  3.      1982       9173        625
  4.      1983       9630        711
  5.      1984       8560        590
  6.      1985       8080        561
  7.      1986       8320        610
  8.      1987       7656       1002
  9.      1988       7396        884
 10.      1989       7996        981
 11.      1990       7354        973
 12.      1991       6929        912
 13.      1992       6689        867
 14.      1993       6360        930
 15.      1994       6295        933
```

We want to merge *newf3* with *newf4*, matching observations according to *year* wherever possible. To accomplish this, both datasets must be sorted by the index variable (which in this example is *year*). We earlier issued a **sort *year*** command before saving *newf3.dta*, so we now do the same with *newf4.dta*. Then we merge the two, specifying *year* as the index variable to match:

```
. sort year

. merge year using newf3
```

```
. describe

Contains data from c:\data\newf4.dta
  obs:              16                                Newfoundland 1980-94
 vars:               6                                8 Nov 1996 17:07
 size:             304 (95.4% of memory free)
-------------------------------------------------------------------------
   1. year          int     %9.0g                     Year
   2. births        int     %9.0g                     Number of births
   3. divorces      int     %9.0g                     Number of divorces
   4. pop           float   %9.0g                     Population
   5. jobless       float   %9.0g                     Number of people unemployed
   6. _merge        byte    %8.0g
-------------------------------------------------------------------------
Sorted by:
     Note:  data has changed since last save
```

```
. list
             year     births   divorces         pop    jobless    _merge
  1.         1980      10332        555           .          .         1
  2.         1981      11310        569           .          .         1
  3.         1982       9173        625           .          .         1
  4.         1983       9630        711           .          .         1
  5.         1984       8560        590           .          .         1
  6.         1985       8080        561      580700          .         3
  7.         1986       8320        610      580200          .         3
  8.         1987       7656       1002      568200          .         3
  9.         1988       7396        884      568000          .         3
 10.         1989       7996        981      570000          .         3
 11.         1990       7354        973      573400      42000         3
 12.         1991       6929        912      573500      45000         3
 13.         1992       6689        867      575600      49000         3
 14.         1993       6360        930      584400      49000         3
 15.         1994       6295        933      582400      50000         3
 16.         1995          .          .      575449          .         2
```

In this example, we simply used **merge** to add new variables to our data, matching observations. By default, whenever the same variables are found in both datasets, those of the "master" data (the file already in memory) are retained and those of the "using" data ignored. The **merge** command has several options, however, that override this default. A command of the following form would allow any *missing values* in the master data to be replaced by corresponding nonmissing values found in the using (here, *newf5.dta*) data:

. merge *year* using *newf5*, update

Or, a command such as the following causes *any values* in the master data to be replaced by nonmissing values from the using data, if the latter are different:

. merge *year* using *newf5*, update replace

Suppose values of an index variable occur more than once in the master data; for example, suppose the year 1990 occurs twice. Then values from the using data with *year* == 1990 are matched with each occurrence of *year* == 1990 in the master data.

You can use this capability for many purposes, such as combining background data on individual patients with data on any number of separate doctor visits they made. Although **merge** makes this and many other data-management tasks straightforward, analysts should look closely at the results to be certain the command is accomplishing what they intend.

As a diagnostic aid, **merge** automatically creates a new variable called *_merge*. Unless **update** was specified, *_merge* codes have the following meanings:

1 Observation from the master dataset only.

2 Observation from the using dataset only.

3 Observation from both master and using data (using values ignored if different).

If the **update** option was specified, the codes convey what happened:

1 Observation from the master dataset only.

2 Observation from the using dataset only.

3 Observation from both, master data agrees with using.

4 Observation from both, master data updated if missing.

5 Observation from both, master data replaced if different.

Before performing another **merge** operation, it will be necessary to discard or rename this variable. For example,

```
. drop _merge
```

Or,

```
. rename _merge _merge1
```

Transposing, Reshaping, or Collapsing Data

Long after a dataset has been created, we might discover that for some analytical purposes it has the wrong organization. Fortunately, several commands facilitate drastic restructuring of datasets. We will illustrate these using data (*growth1.dta*) on recent population growth in five eastern provinces of Canada. In these data, unlike our previous examples, province names are represented by a numerical variable with eight-character labels:

```
. use growth1, clear
(Eastern Canada growth)

. describe
Contains data from c:\data\growth1.dta
  obs:             5                         Eastern Canada growth
 vars:             5                         11 Nov 1996 15:45
 size:           105 (95.1% of memory free)
-------------------------------------------------------------------------------
  1. provinc2     byte   %8.0g       provinc2  Eastern Canadian province
  2. grow92       float  %9.0g                 Pop. gain in 1000s, 1991-92
  3. grow93       float  %9.0g                 Pop. gain in 1000s, 1992-93
  4. grow94       float  %9.0g                 Pop. gain in 1000s, 1993-94
  5. grow95       float  %9.0g                 Pop. gain in 1000s, 1994-95
-------------------------------------------------------------------------------
Sorted by:

. list
      provinc2      grow92      grow93      grow94      grow95
  1. New Brun          10         2.5         2.2         2.4
  2. Newfound         4.5          .8          -3        -5.8
  3. Nova Sco        12.1         5.8         3.5         3.9
  4.  Ontario       174.9       169.1       120.9       163.9
  5.   Quebec        80.6        77.4        48.5        47.1
```

In this organization, population growth for each year is stored as a separate variable. We could analyze changes in the mean or variance of population growth from year to year. On the other hand, given this organization Stata could not readily draw a simple time plot of population growth against year, nor can Stata find the correlation between population growth in New Brunswick and Newfoundland. All the necessary information is here, but such analyses require different organizations of the data.

One simple reorganization involves transposing variables and observations. In effect, the dataset rows become its columns, and the dataset columns become its rows. This is accomplished by the **xpose** command. The option **clear** is required with this command because it always clears the present data from memory. Including the **varname** option creates an additional variable (named *_varname*) in the transposed dataset, containing original variable names as strings.

```
. xpose, clear varname

. describe
Contains data
  obs:             5
 vars:             6
 size:           160 (95.1% of memory free)
-------------------------------------------------------------------------------
  1. v1           float  %9.0g
  2. v2           float  %9.0g
  3. v3           float  %9.0g
```

```
  4. v4          float  %9.0g
  5. v5          float  %9.0g
  6. _varname    str8   %9s
----------------------------------------------------------------------
Sorted by:
     Note:  data has changed since last save
```

```
. list
             v1         v2          v3          v4          v5   _varname
  1.          1          2           3           4           5   provinc2
  2.         10        4.5        12.1       174.9        80.6     grow92
  3.        2.5         .8         5.8       169.1        77.4     grow93
  4.        2.2         -3         3.5       120.9        48.5     grow94
  5.        2.4       -5.8         3.9       163.9        47.1     grow95
```

Value labels are lost along the way, so provinces in the transposed dataset are indicated only by their numbers (1 = New Brunswick, 2 = Newfoundland, and so on). The second through last values in each column are the population gains for that province, in thousands. Thus variable *v1* has a province identification number (1, meaning New Brunswick) in its first row, and New Brunswick's population growth values for 1992 to 1995 in its second through fifth rows. We can therefore find correlations between population growth in different provinces, for instance, by typing a **correlate** command with the **in 2/5** (second through fifth observations only) qualifier:

```
. correlate v1-v5 in 2/5
(obs=4)

        |      v1         v2         v3         v4         v5
 -------+--------------------------------------------------------
     v1 |   1.0000
     v2 |   0.8058     1.0000
     v3 |   0.9742     0.8978     1.0000
     v4 |   0.5070     0.4803     0.6204     1.0000
     v5 |   0.6526     0.9362     0.8049     0.6765     1.0000
```

The strongest correlation appears between the growth of maritime provinces Newfoundland (*v2*) and Nova Scotia (*v3*): $r = .9742$. Newfoundland's growth has a much weaker correlation with that of Ontario (*v4*): $r = .4803$.

More sophisticated restructuring is possible through the **reshape** command. This command switches datasets between two basic configurations termed "wide" and "long." Dataset *growth1.dta* is initially in wide format:

```
. use growth1, clear
(Eastern Canada growth)

. list
      provinc2      grow92      grow93      grow94      grow95
  1. New Brun           10         2.5         2.2         2.4
  2. Newfound          4.5          .8          -3        -5.8
```

```
  3. Nova Sco        12.1         5.8         3.5         3.9
  4.  Ontario       174.9       169.1       120.9       163.9
  5.   Quebec        80.6        77.4        48.5        47.1
```

A sequence of four **reshape** commands switches this to long format:

```
. reshape groups year 92-95

. reshape vars grow

. reshape cons provinc2

. reshape long

. describe
Contains data from C:\temp\ST_0000f.tmp
  obs:              20                          Eastern Canada growth
 vars:               3                          11 Nov 1996 15:47
 size:             220 (95.1% of memory free)
-------------------------------------------------------------------------
   1. provinc2     byte    %8.0g      provinc2   Eastern Canadian province
   2. grow         float   %9.0g
   3. year         int     %8.0g
-------------------------------------------------------------------------
Sorted by:
     Note:  data has changed since last save


. list
      provinc2        grow       year
  1. New Brun          2.4         95
  2. Newfound         -5.8         95
  3. Nova Sco          3.9         95
  4.  Ontario        163.9         95
  5.   Quebec         47.1         95
  6. New Brun          2.2         94
  7. Newfound           -3         94
  8. Nova Sco          3.5         94
  9.  Ontario        120.9         94
 10.   Quebec         48.5         94
 11. New Brun          2.5         93
 12. Newfound           .8         93
 13. Nova Sco          5.8         93
 14.  Ontario        169.1         93
 15.   Quebec         77.4         93
 16. New Brun           10         92
 17. Newfound          4.5         92
 18. Nova Sco         12.1         92
 19.  Ontario        174.9         92
 20.   Quebec         80.6         92
```

To see more clearly what has been done, we can sort the new dataset by province and year, then save it as *growth2.dta:*

```
. sort provinc2 year

. list
      provinc2        grow       year
  1. New Brun          10         92
  2. New Brun         2.5         93
  3. New Brun         2.2         94
  4. New Brun         2.4         95
  5. Newfound         4.5         92
  6. Newfound          .8         93
  7. Newfound          -3         94
  8. Newfound        -5.8         95
  9. Nova Sco        12.1         92
 10. Nova Sco         5.8         93
 11. Nova Sco         3.5         94
 12. Nova Sco         3.9         95
 13.  Ontario       174.9         92
 14.  Ontario       169.1         93
 15.  Ontario       120.9         94
 16.  Ontario       163.9         95
 17.   Quebec        80.6         92
 18.   Quebec        77.4         93
 19.   Quebec        48.5         94
 20.   Quebec        47.1         95

. label data "Eastern Canadian growth-long"

. label variable grow "Population growth in 1000s"

. save growth2
file c:\data\growth2.dta saved
```

This sequence of **reshape** commands had the following meanings:

reshape groups — Names the single variable (*year*) that will be created when we switch from wide to long, and gives the values (92–95) it will take. These values are also the suffixes applied to the variable named in the **reshape vars** command, when the dataset is wide.

reshape vars — Names the variables (in this example, *grow*, but there might be more than one) for which there are related observations.

reshape cons — Identifies the variable that is "constant" across the related observations (in this example, the province identifier *provinc2*).

reshape long — Specifies that we are going from wide to long format. The reverse would be accomplished by typing **reshape wide** at this point.

Figure 2.1

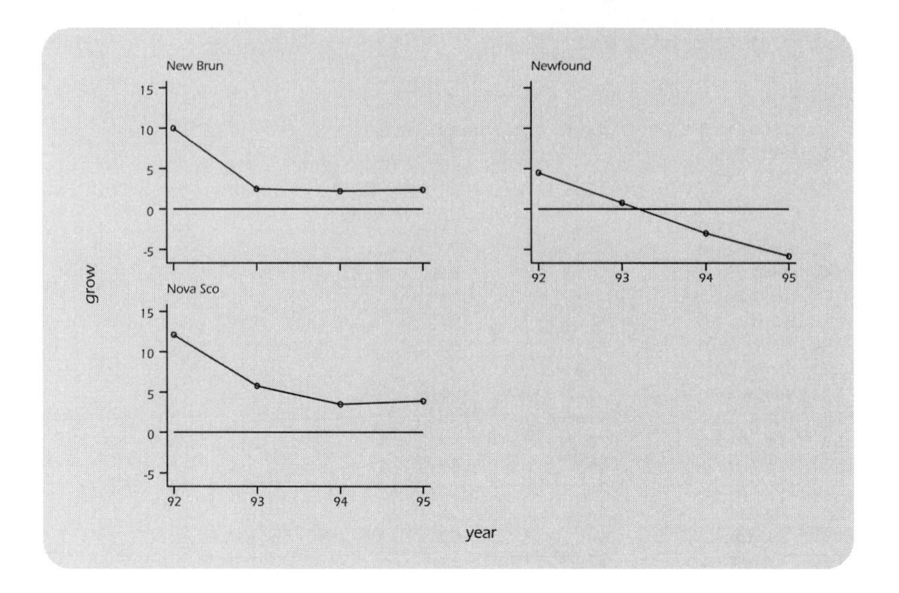

Figure 2.1 shows a possible use for the long-format dataset: With one **graph** command we can now produce a time plot comparing, for example, the population gains in New Brunswick, Newfoundland, and Nova Scotia (that is, observations for which *provinc2* < 4). The **graph** command calls for a plot of *grow* (as *y*-axis variable) against *year* (*x* axis) if *province2* < 4, with *y* and *x* axis values labeled neatly, a horizontal line at *y* = 0 (zero population growth), data points connected by line segments, and separate plots for each value of *provinc2*. Chapter 3 gives more details and examples concerning the versatile **graph** command.

```
. graph grow year if provinc2 < 4, ylabel xlabel yline(0)
      connect(l) by(provinc2)
```

The 1992 collapse of Atlantic Canada's fisheries caused economic hardship for the three provinces in Figure 2.1. Growth slowed dramatically in New Brunswick and Nova Scotia, while Newfoundland (the most fishing-dependent Atlantic province) actually lost population.

reshape works equally well in reverse, to switch data from "long" to "wide" format. Dataset *growth3.dta* serves as an example of long format:

```
. use growth3, clear

. list

      provinc2       grow       year
  1.  New Brun         10         92
  2.  New Brun        2.5         93
  3.  New Brun        2.2         94
  4.  New Brun        2.4         95
  5.  Newfound        4.5         92
```

```
 6. Newfound        .8      93
 7. Newfound        -3      94
 8. Newfound      -5.8      95
 9. Nova Sco      12.1      92
10. Nova Sco       5.8      93
11. Nova Sco       3.5      94
12. Nova Sco       3.9      95
13.  Ontario     174.9      92
14.  Ontario     169.1      93
15.  Ontario     120.9      94
16.  Ontario     163.9      95
17.   Quebec      80.6      92
18.   Quebec      77.4      93
19.   Quebec      48.5      94
20.   Quebec      47.1      95
```

To reshape this into wide format, we use the same command sequence as before, but this time conclude with **reshape, wide**:

```
. reshape groups year 92-95

. reshape vars grow

. reshape cons provinc2

. reshape wide

. list
     provinc2     grow95     grow94     grow93     grow92
 1. New Brun        2.4        2.2        2.5         10
 2. Newfound       -5.8         -3         .8        4.5
 3. Nova Sco        3.9        3.5        5.8       12.1
 4.  Ontario      163.9      120.9      169.1      174.9
 5.   Quebec       47.1       48.5       77.4       80.6
```

Many government and administrative databases are stored in long format, and the long-to-wide transformation can help you prepare these for statistical analysis. Another important tool for such data is the **collapse** command, which creates an aggregated dataset of statistics (for example, means, medians, or sums). The long *growth3* dataset has four observations for each province:

```
. use growth3, clear
(Eastern Canadian growth—long)

. list
     provinc2       grow      year
 1. New Brun          10        92
 2. New Brun         2.5        93
 3. New Brun         2.2        94
 4. New Brun         2.4        95
 5. Newfound         4.5        92
 6. Newfound          .8        93
 7. Newfound          -3        94
 8. Newfound        -5.8        95
```

```
 9. Nova Sco       12.1          92
10. Nova Sco        5.8          93
11. Nova Sco        3.5          94
12. Nova Sco        3.9          95
13.  Ontario      174.9          92
14.  Ontario      169.1          93
15.  Ontario      120.9          94
16.  Ontario      163.9          95
17.   Quebec       80.6          92
18.   Quebec       77.4          93
19.   Quebec       48.5          94
20.   Quebec       47.1          95
```

We might want to aggregate the different years into one mean growth rate for each province. In the collapsed dataset, each observation will correspond to one value of the **by( )** variable, for example, one province:

```
. collapse (mean) grow, by(provinc2)

. list
     provinc2          grow
 1. New Brun          4.275
 2. Newfound     -.8750001
 3. Nova Sco          6.325
 4.  Ontario          157.2
 5.   Quebec           63.4
```

For a slightly more complicated example, suppose we had a dataset similar to *growth3.dta* but also containing the variables *births*, *deaths*, and *income*. We want an aggregate dataset with each province's total numbers of births and deaths over these years, the mean income (to be named *meaninc*), and the median income (to be named *medinc*). If we do not specify a new variable name, as with *grow* in the previous example, or *births* and *deaths*, the collapsed variable takes on the same name as the old variable:

```
. collapse (sum) births deaths (mean) meaninc = income
      (median) medinc = income, by(provinc2)
```

In addition to a sum, mean, or median, **collapse** can create variables representing a standard deviation (**sd**), number of nonmissing observations (**count**), maximum (**max**), minimum (**min**), or any percentile (denoted by **p1**, **p2**, . . . **p99**). The default (**mean**) is calculated if we do not specify what type of statistic.

Weighting Observations

Stata understands four types of weighting:

aweight Analytical weights, used in weighted least squares (WLS) regression and similar procedures.

fweight Frequency weights, counting the number of duplicated observations. Frequency weights must be integers.

iweight Importance weights, however you define "importance."

pweight Probability or sampling weights, equal to the inverse of the probability that an observation is included because of sampling strategy.

Researchers sometimes speak of "weighted data." This might mean that the original sampling scheme selected observations in a deliberately disproportionate way, as reflected by the weights (equal to 1/probability of selection). Appropriate use of **pweight** can compensate for disproportionate sampling in certain analyses. On the other hand, "weighted data" might mean something different — an aggregate dataset, perhaps constructed from a frequency table or cross-tabulation, with one or more variables indicating how many times a particular value or combination of values occurred. In that case, we need **fweight**.

Not all types of weighting have been defined for all types of analysis. We cannot, for example, use **pweight** with the **tabulate** command, although the other three weight types are allowed. Using weights in any analysis thus requires a careful understanding of what we want weighting to accomplish in that particular analysis. The weights themselves can be any variable in the dataset.

The following small dataset (*nfschool.dta*), containing results from a survey of 1,381 rural Newfoundland high school students, illustrates a simple application of frequency weighting:

```
. describe

Contains data from c:\data\nfschool.dta
  obs:              6                        Newf.school/univer.(Seyfrit 93)
 vars:              3                        16 Nov 1996 10:19
 size:             48 (95.4% of memory free)
-------------------------------------------------------------------------------
   1. univers      byte   %8.0g      yes     Expect to attend university?
   2. year         byte   %8.0g              What year of school now?
   3. count        int    %8.0g              observed frequency
-------------------------------------------------------------------------------
Sorted by:


. list

       univers       year      count
  1.        no         10        210
  2.        no         11        260
  3.        no         12        274
  4.       yes         10        224
  5.       yes         11        235
  6.       yes         12        178
```

At first glance, the dataset seems to contain only six observations, and when we cross-tabulate whether students expect to attend a university (*univers*) by their current year in high school (*year*), we get a table with one observation per cell:

```
. tabulate univers year

Expect to   | What year of school now?
attend      |
university? |        10          11          12 |     Total
------------+-----------------------------------+-----------
         no |         1           1           1 |         3
        yes |         1           1           1 |         3
------------+-----------------------------------+-----------
      Total |         2           2           2 |         6
```

To understand these data, we need to apply frequency weights. The variable *count* gives frequencies: 210 of these students are tenth graders who said they did not expect to attend a university, 260 are eleventh graders who said no, and so on. Specifying **[fweight = *count*]** therefore obtains a cross-tabulation showing responses of all 1,381 students:

```
. tabulate univers year [fweight = count]

Expect to   | What year of school now?
attend      |
university? |       10          11          12 |      Total
------------+----------------------------------+-----------
         no |      210         260         274 |        744
        yes |      224         235         178 |        637
------------+----------------------------------+-----------
      Total |      434         495         452 |       1381
```

Carrying the analysis further, we might add options asking for a table with column percentages (**col**), no cell frequencies shown (**nof**), and a χ^2 test of independence (**chi**). This reveals that the percentage of students expecting to go to college declines with each year of high school:

```
. tab univers year [fw = count], col nof chi2

Expect to   | What year of school now?
attend      |
university? |       10          11          12 |      Total
------------+----------------------------------+-----------
         no |    48.39       52.53       60.62 |      53.87
        yes |    51.61       47.47       39.38 |      46.13
------------+----------------------------------+-----------
      Total |   100.00      100.00      100.00 |     100.00

          Pearson chi2(2) =  13.8967   Pr = 0.001
```

A special set of commands allow you to analyze more complex survey data with probability weights, clustering, or stratification. Type **help svy** for a list and description of these commands.

Using Random Variables and Random Sampling

The pseudo-random number function **uniform( )** lies at the heart of Stata's considerable abilities to generate random data, or to sample randomly from the data at hand. The *User's Guide* provides a technical description of this 32-bit pseudo-random generator, which is more sophisticated than the one used by earlier versions of Stata. If we presently have data in memory, then a command such as the following creates a new variable named *randnum*, having apparently random 16-digit values between 0 and 1, for each case in the data:

```
. generate randnum = uniform()
```

Alternatively, we might create a random dataset from scratch. Suppose we want to start a new dataset containing 10 random values. We first clear any other data from memory (if they were valuable, **save** them first). Next set the number of observations desired for the new dataset. Explicitly setting the seed number ensures that someone else (the reader, for instance) can reproduce the same "random" results. Finally, we generate our random variable:

```
. clear

. set obs 10
obs was 0, now 10

. set seed 12345

. generate randnum = uniform()

. list

       randnum
  1.   .309106
  2.  .6852276
  3.  .1277815
  4.  .5617244
  5.  .3134516
  6.  .5047374
  7.  .7232868
  8.  .4176817
  9.  .6768828
 10.  .3657581
```

In combination with Stata's algebraic, statistical, and special functions, **uniform()** can simulate values sampled from a variety of theoretical distributions. If we want *newvar* sampled from a uniform distribution over [0,428) instead of the usual [0,1):

```
. generate newvar = 428 * uniform()
```

These will still be 16-digit values. Perhaps we want only integers from 1 to 428 (inclusive):

```
. generate newvar = 1 + int(428 * uniform())
```

To simulate 1,000 throws of a six-sided die:

```
. clear

. set obs 1000
obs was 0, now 1000

. generate die = 1 + int(6 * uniform())

. tabulate die

        die |      Freq.     Percent        Cum.
------------+-----------------------------------
          1 |        171       17.10       17.10
          2 |        164       16.40       33.50
          3 |        150       15.00       48.50
          4 |        170       17.00       65.50
          5 |        169       16.90       82.40
          6 |        176       17.60      100.00
------------+-----------------------------------
      Total |       1000      100.00
```

We might theoretically expect 16.67% ones, 16.67% twos, and so on, but in any one sample like these 1,000 "throws," the observed percentages will vary randomly around their expected values.

Simulate 1000 throws of a pair of six-sided dice:

```
. generate dice = 2 + int(6 * uniform()) + int(6 * uniform())

. tabulate dice

       dice |      Freq.     Percent        Cum.
------------+-----------------------------------
          2 |         26        2.60        2.60
          3 |         62        6.20        8.80
          4 |         78        7.80       16.60
          5 |        120       12.00       28.60
          6 |        153       15.30       43.90
          7 |        149       14.90       58.80
          8 |        146       14.60       73.40
          9 |         96        9.60       83.00
         10 |         88        8.80       91.80
         11 |         53        5.30       97.10
         12 |         29        2.90      100.00
------------+-----------------------------------
      Total |       1000      100.00
```

A standard normal distribution has mean $\mu = 0$ and standard deviation = 1. We can generate a variable theoretically following a standard normal distribution by applying the inverse-normal function **invnorm** to **uniform()**:

```
. generate z = invnorm(uniform())
```

To obtain values from a normal distribution with $\mu = 500$ and $\sigma = 75$:

```
. generate x = 500 + 75 * invnorm(uniform())
```

Two standard normal variables with theoretical correlation $\rho = 0.8$:

```
. generate z1 = invnorm(uniform())

. generate z2 = z1 * 0.8 + invnorm(uniform() * sqrt(1 - 0.8^2)
```

The 0.8 in this example could be replaced with any other positive or negative correlation desired.

After they are generated, variables can be moved to another mean and standard deviation without affecting their correlation. For example, to simulate *z1* ~ *N*(100,15) and *z2* ~ *N*(500,75), with correlation $\rho = 0.8$, first **generate** *z1* and z2 as above. Then change their means and standard deviations:

```
. replace z1 = 100 + 15 * z1

. replace z2 = 500 + 75 * z2
```

Random-variable generation simulates random sampling from populations having the specified parameters. Sample means, correlations, and so on will not exactly equal these parameters.

If *x* follows a normal distribution, $w = e^x$ follows a lognormal distribution. To form a lognormal variable *w* based on a standard normal *x:*

```
. generate w = exp(invnorm(uniform())
```

To form a lognormal variable *w* based on an N(100,15) distribution:

```
. generate w = exp(100 + 15 * invnorm(uniform())
```

Taking logarithms, of course, normalizes a lognormal variable.

To simulate *y* values drawn randomly from an exponential distribution with mean and standard deviation $\mu = \sigma = 3$:

```
. generate y = -3 * ln(uniform())
```

For other means and standard deviations, substitute other values for 3.

X1 follows a χ^2 distribution with one degree of freedom, which is the same as a squared standard normal:

```
. generate x1 = (invnorm(uniform())^2
```

By similar logic, *X2* follows a χ^2 with two degrees of freedom:

```
. generate X2 = (invnorm(uniform()))^2 + invnorm(uniform()))^2
```

Other statistical distributions including *t* and *F* can be simulated along the same lines. In addition, programs have been written for Stata to generate random samples following distributions such as binomial, Poisson, gamma, and inverse Gaussian.

The command **sample** makes unobtrusive use of the **uniform** random generator to obtain random samples of the data in memory. For example the following command discards all but a 10% random sample of the original data:

```
. sample 10
```

When we add an **in** or **if** qualifier, **sample** applies only to those observations meeting our criteria. For example,

```
. sample 10 if age < 26
```

This would leave us with a 10% sample of those observations with *age* less than 26, plus 100% of the original observations with *age* ≥ 26.

We could use **uniform** directly for sampling in many ways. To put the data in random order, as for a lottery,

```
. generate randnum = uniform()

. sort randnum
```

Typing the following would then leave us with a random sample of $n = 15$ observations in memory:

```
. keep in 1/15
```

To create a marker variable named *chosen*, equal to 1 for a randomly selected subsample and "missing" otherwise, with each observation having a 20% chance of selection:

```
. generate chosen = 1 if uniform() < .2
```

The sections in Chapter 13 on bootstrapping and Monte Carlo simulations provide further examples of random sampling and random variable generation.

Writing Programs for Data Management

Data management on larger projects can involve repetitive or error-prone tasks that are best handled by writing specialized Stata programs. Chapter 13 addresses programming in more detail, but you can begin by writing simple programs that consist of nothing more than a sequence of Stata commands, typed and saved as an ASCII file using your favorite word processor or text editor. For example, we might create a file named *canada.do*, which contains the commands for reading in a raw data file named *canada.raw*, then labeling the dataset and its variables, compressing it, and saving in

Stata format. The commands in this file are identical to those seen earlier when we went through the example step by step:

```
infile str30 place pop unemp mlife flife using canada.raw
label data "Canada dataset 1"
label variable pop "Population in 1000s 95"
label variable unemp "% all 15+ unemployed 95"
label variable mlife "Male life expectancy years"
label variable flife "Female life expectancy years"
compress
save canada1, replace
```

Once this *canada.do* file has been written and saved, simply typing the following command causes Stata to read the file and run each command in turn:

```
. do canada
```

In Stata-speak, such batch-mode programs are termed "do-files" and saved with a `.do` extension. More elaborate programs (defined through "automatic do" or `.ado` files) can be stored in memory and can call other programs in turn — opening worlds of possibility for adventurous analysts.

Stata ordinarily interprets the end of a command line as the end of that command. This is reasonable on screen, where the line can be arbitrarily long, but does not work as well when we are typing commands in a text file. To avoid line-length problems, we use the **#delimit** command, which can set some other character as the end-of-command delimiter. In the following example, we make a semicolon the delimiter; then type two long commands that do not end until a semicolon appears; then finally reset the delimiter to its usual value, a carriage return (`cr`):

```
#delimit ;
infile str30 place pop unemp mlife flife births deaths
    marriage medinc mededuc using newcan.raw;
order place pop births deaths marriage medinc mededuc
    unemp mlife flife;
#delimit cr
```

Stata also normally pauses each time the Results window becomes full of information and waits to proceed until we press any key. Instead of waiting indefinitely, we can ask Stata to wait a certain number of seconds before scrolling on. The following command, typed in the Command window or as part of a program, tells Stata to wait only one second before scrolling:

```
. set more 1
```

This is convenient if our program produces much screen output we don't want to see, or if it is writing to a log file that we will examine later. Typing the following returns to the usual mode of waiting indefinitely for keyboard input before scrolling:

```
. set more 0
```

3

Graphs

The **graph** command forms the core of Stata's graphics. It offers seven basic styles:

hist	Histograms (default if only one variable is listed)
oneway	One-way scatterplots
twoway	Two-way scatterplots (default if two or more variables are listed)
matrix	Scatterplot matrix
box	Boxplots
bar	Bar charts
star	Star charts
pie	Pie charts

With most of these styles we can accept simple default versions or choose among many further options that control details of their appearance.

In addition to **graph**, Stata has numerous specialized graphing commands, including symmetry plots, quantile plots, and quality control charts. Other graphs used in connection with particular statistical models are described in later chapters such as those about regression diagnostics, nonlinear regression, and survival analysis.

Example Commands

. graph ***y***

Draws basic unadorned histogram of variable *y* (same as **graph** ***y*****, hist**).

```
. graph y, bin(9) norm ylabel xlabel border
```

Draws histogram of *y* with 9 vertical bars (default is 5), normal curve, automatically-selected *y* and *x* axis values labeled, and a border drawn around.

```
. sort x
. graph y, by(x) ylabel xlabel total
```

In one figure, draws separate histograms of *y* for each value of *x*. Also includes a "total" histogram showing distribution of *y* across all values of *x*.

```
. graph y time, connect(l) sort ylabel xlabel(1900,1940,1980)
```

Plots *y* against *time*. Connects the points with line segments. Labels the *y* axis automatically, but labels the *x* axis specifically at points 1900, 1940, and 1980.

```
. graph y1 y2 time, connect(ll) sort ylabel xlabel rlabel
  rescale
```

Plots *y1* and *y2* against *time*, connects with line segments, and rescales the right-hand axis to match the units of *y2*.

```
. graph y x
```

Displays a basic two-variable scatterplot of *y* against *x*. Is equivalent to **graph** ***y x*****, twoway**. Add the option **border** to include a border.

```
. graph y x, oneway twoway box ylabel xlabel
```

Constructs scatterplot of *y* versus *x*, with automatically labeled axes. In the margins, shows one-way scatterplots and boxplots of *y* and *x*.

```
. sort x2
. graph y x1, by(x2)
```

In one figure, draws separate *y* versus *x1* scatterplots for each value of *x2*.

```
. graph y x1 [iweight = x2]
```

Draws a scatterplot of *y* versus *x1*, with symbols sized in proportion to *x2*.

```
. graph y1 y2 y3, box ylabel
```

Constructs boxplots of variables *y1*, *y2* and *y3*, with labeled vertical axis.

```
. sort x
. graph y, box by(x) ylabel
```

Constructs boxplots of *y* for each value of *x*.

```
. graph w x y z, pie
```

Draws one pie chart with slices indicating the relative amounts of variables *w*, *x*, *y*, and *z*. The variables must have similar units.

. graph *w x y z*, **star**

Draws a star chart for each observation in the data, showing that observation's relative values on variables *w*, *x*, y, and *z*. The variables need not have similar units.

. graph *w x y*, **bar**

Shows the sums of variables *w*, *x*, and *y* as side-by-side bars in a bar chart.

. sort *z*

. graph *w x y*, **bar by(***z***) means**

Constructs bar chart showing the means of *w*, *x*, and *y* for each value of *z*.

. qnorm *y*

Draws a quantile-normal plot (normal probability plot) showing quantiles of *y* versus corresponding quantiles of a normal distribution.

. rchart *x1 x2 x3 x4 x5*, **connect(l)**

Constructs a quality-control R chart graphing the range of values represented by variables *x1* through *x5*.

Histograms

Typing simply **graph** ***varname*** produces a histogram, Stata's default one-variable display. For examples we turn to *states90.dta*, which contains selected environment and education measures on the 50 U.S. states plus the District of Columbia (data originally from the League of Conservation Voters 1991; National Center for Education Statistics 1992, 1993; World Resources Institute 1993).

```
. use states90
(U.S. states data 1990-91)

. describe
Contains data from c:\data\states90.dta
  obs:            51                             U.S. states data 1990-91
 vars:            21                             15 Dec 1996 12:10
 size:         4,080 (96.4% of memory free)
-------------------------------------------------------------------------------
   1. state        str20  %20s                   State
   2. region       byte   %9.0g     region       Geographical region
   3. pop          float  %9.0g                  1990 population
   4. area         float  %9.0g                  Land area, square miles
   5. density      float  %7.2f                  People per square mile
   6. metro        float  %5.1f                  Metropolitan area population, %
   7. waste        float  %5.2f                  Per capita solid waste, tons
   8. energy       int    %8.0g                  Per capita energy consumed, Btu
   9. miles        int    %8.0g                  Per capita miles driven/year
  10. toxic        float  %5.2f                  Per capita toxics released, lbs
```

```
 11. green        float  %5.2f              Per capita greenhouse gas, tons
 12. house        byte   %8.0g              House '91 environ. voting, %
 13. senate       byte   %8.0g              Senate '91 environ. voting, %
 14. csat         int    %9.0g              Mean composite SAT score
 15. vsat         int    %8.0g              Mean verbal SAT score
 16. msat         int    %8.0g              Mean math SAT score
 17. percent      byte   %9.0g              % HS graduates taking SAT
 18. expense      int    %9.0g              Per pupil expenditures prim&sec
 19. income       long   %10.0g             Median household income
 20. high         float  %9.0g              % over 25 w/HS diploma
 21. college      float  %9.0g              % over 25 w/bachelor's degree +
---------------------------------------------------------------------------
Sorted by:  state
```

Figure 3.1 shows a simple histogram of *college*, the percentage of a state's over-25 population with a bachelor's degree or higher. It was produced by the following command:

```
. graph college, title(Figure 3.1)
```

Menu options allow us to print whatever graph is on screen, save it to disk, or cut and paste it into another program such as a word processor. Figure 3.2 illustrates two ways to enhance a graph.

1. **ylabel** and **xlabel** options ask for round-number labeling on the *y* and *x* axes (compare with the default labeling seen in Figure 3.1).

2. We can add as many as eight titles, two each at the top, bottom, left, and right. The first bottom title (called **title**) appears in a larger font. The **title( )** option was used to include figure numbers for Figures 3.1 and 3.2.

Figure 3.1

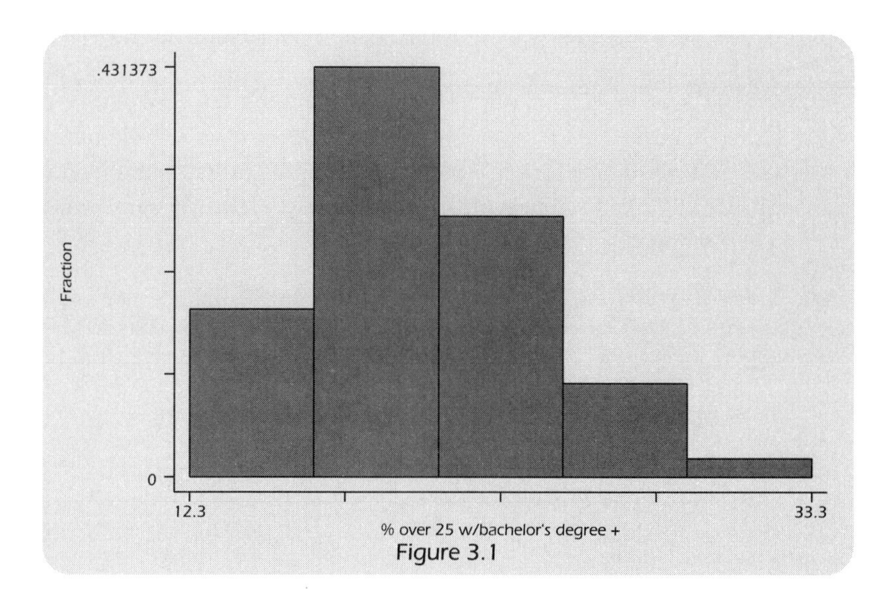

Figure 3.2

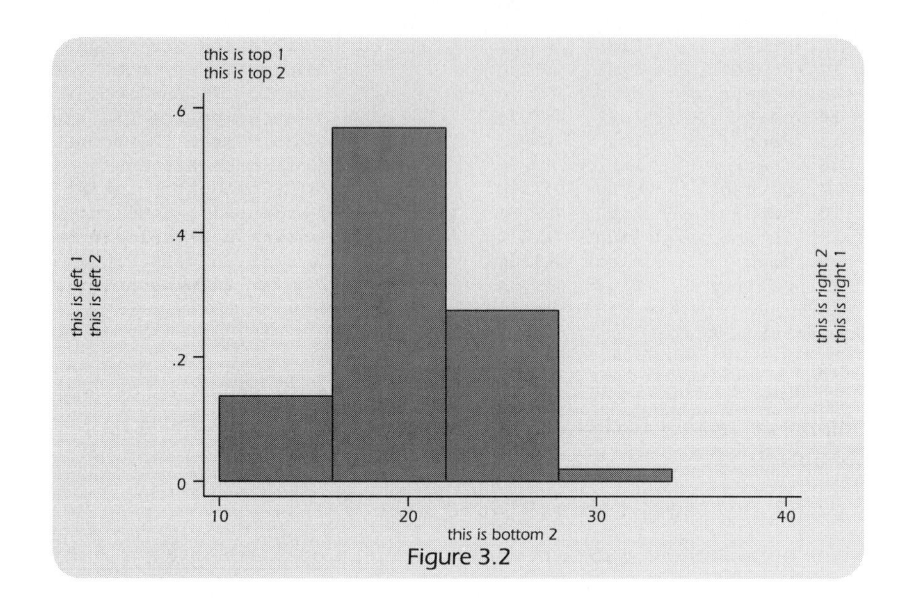


```
. graph college, ylabel xlabel title(Figure 3.2)
      b2(this is bottom 2) l1(this is left 1)
      l2(this is left 2) t1(this is top 1)
      t2(this is top 2) r1(this is right 1)
      r2(this is right 2)
```

Some further **graph** options include the following:

bin(#) Draws the histogram with # vertical bars. The default is **bin(5)**.

noaxis Suppresses the lines for *x* and *y* axes.

norm Draws a normal curve over the histogram, based on sample mean and standard deviation. We can specify other parameters; **norm(100,15)**, for example, would draw a normal curve with $\mu = 100$ and $\sigma = 15$.

ylabel Uses automatically chosen round numbers to label the *y* axis. We can ask for certain labels specifically; **ylabel(0,2,4,6)** calls for labels at *y* = 0, 2, 4, and 6.

xlabel Same as **ylabel**, applied to the *x* axis.

ytick( ) Places *y*-axis tick marks at the specified locations. **ytick(1,3,5,7)** calls for ticks at *y* = 1, 3, 5, and 7.

yline( )	Draws horizontal lines across the graph at specified heights. **yline(0,5)** calls for lines at $y = 0$ and at $y = 5$.
xtick( )	Same as **ytick( )**, applied to the x axis.
xline( )	Same as **yline( )**, applied to the x axis.
freq	Labels the histogram's vertical axis with frequencies rather than proportions.

Figure 3.3 illustrates these options.

```
. graph college, norm border xlabel(12,16,20,24,28,32)
      xtick(14,18,22,26,30,34) bin(11) freq
      ylabel(0,2,4,6,8,10,12) ytick(1,3,5,7,9,11,13)
```

Figure 3.3

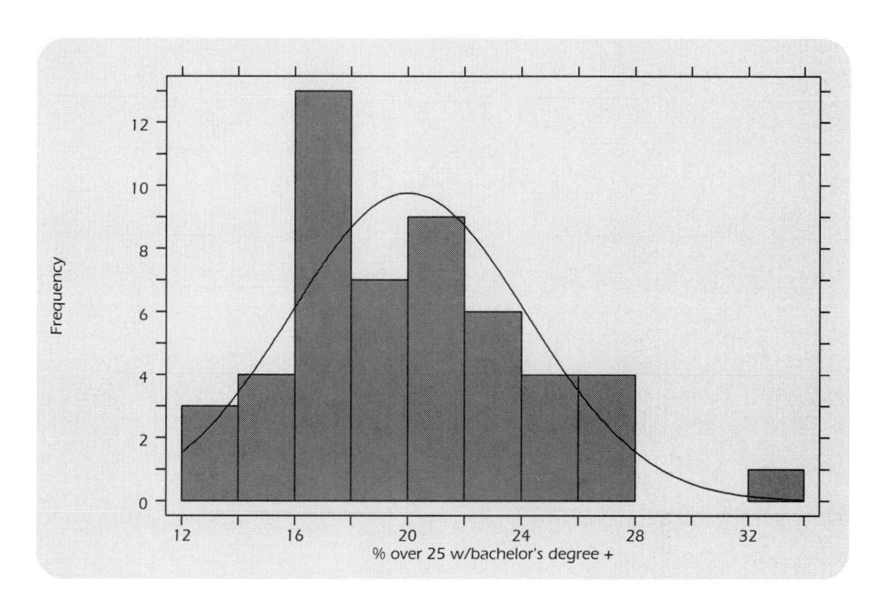

Suppose we want to see how the distribution of *college* (percent of state's adult population with college degrees) varies by region (Figure 3.4):

```
. sort region
```

```
. graph college, by(region) ylabel xlabel bin(4)
```

The **by** option obtains a separate histogram for each value of the categorical variable *region*. Note that we first had to sort the data by *region*. To get a whole-sample histogram in addition to separate regional histograms, add the **total** option. In Figure 3.5 we also overlay normal curves based not on the sample mean and standard deviation but on the more outlier-resistant median and pseudo-standard deviation (obtained via **lv *college***; see Chapter 4):

Figure 3.4

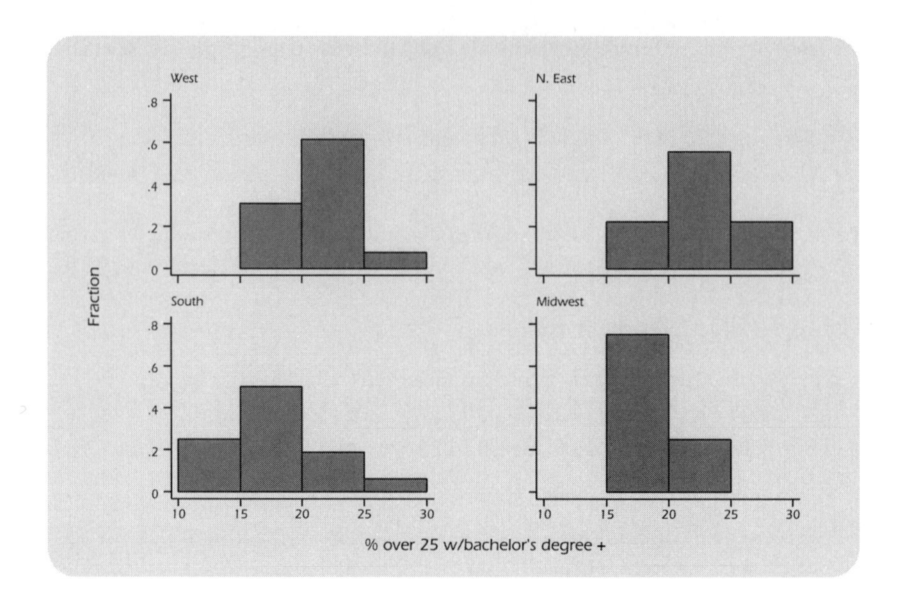


```
. graph college, by(region) ylabel xlabel bin(4) total
       norm(20.02,4.17)
```

Figure 3.5

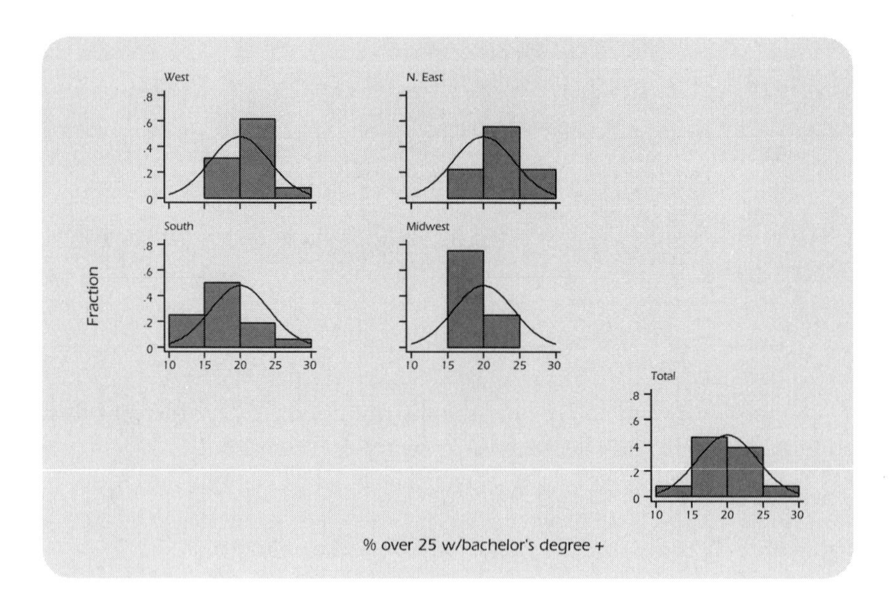

Some general points regarding the **graph** command:

1. Its syntax has the form:

```
. graph variable-list, option1 option2...
```

We can list the desired options in any order, as long as they all come after the single comma.

2. Except for **freq**, **bin( )** and **norm**, the options illustrated in Figures 3.2 through 3.5 work with other types of graphs besides histograms. For example, **ylabel** also labels *y*-axis values in a scatterplot or a boxplot. **by(*varname*)** can produce a display containing separate one or two-way scatterplots, boxplots, bars, or pie charts for each value of *varname.*

3. A good way to get exactly the graph you want is to build it in steps. Begin with a simple version, view that, and then add options a few at a time, building on previous commands retrieved from the Review window. Save only your final version.

Scatterplots 1: Time Plots

Stata does not have a special command for time plots because they are essentially just one kind of scatterplot. If **graph** is followed by two or more variable names, Stata produces a scatterplot unless we tell it otherwise. When the data comprise a series and the variable listed last in the *graph* command is a measure of elapsed time, our scatterplot amounts to a time plot. For example, we can construct a time plot showing the worldwide catch of blue whales using data in *whales.dta* (Figure 3.6):

```
. use whales
(World Whale Catch 1920-85)

. describe
Contains data from c:\data\whales.dta
  obs:            19                          World Whale Catch 1920-85
 vars:             4                          3 Apr 1996 18:58
 size:           228 (96.5% of memory free)
-------------------------------------------------------------------------------
   1. year          int     %8.0g             Year
   2. blue          int     %8.0g             Blue Whales
   3. fin           int     %8.0g             Fin Whales
   4. sei           int     %8.0g             Sei Whales
-------------------------------------------------------------------------------
Sorted by:  year

. graph blue year, connect(l) sort ylabel xlabel
```

With two variables listed, the first-named variable (*blue*) defines the scatter plot's vertical or *y* axis, and the second (*year*) its horizontal or *x* axis. **connect(l)** instructs

Figure 3.6

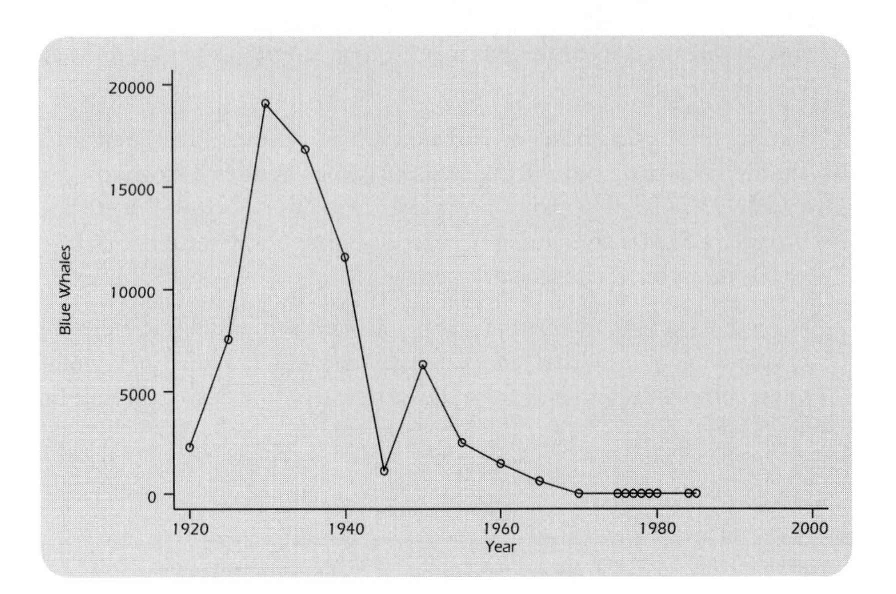

Stata to connect the plotted points with line segments, thus producing the time plot seen in Figure 3.6. (Read carefully: That is a lower-case letter **l**, standing for "line segment," inside the parentheses.) The **sort** option with **graph** temporarily sorts the data so that points are connected in *x*-axis order.

We could include as many as 20 *y* variables, or 20 time series, in one graph. Here is an example with 3 series, adding fin and sei whale catches to our blue whale graph. It also has better labels for the *x* axis (Figure 3.7):

```
. graph blue fin sei year, connect(lll) sort ylabel
      xlabel(1920,1930,1940,1950,1960,1970,1980)
```

connect(lll) means to connect all three *y* variables with line segments in Figure 3.7. The **connect( )** option allows points to be connected in a number of different ways in any scatterplot or time plot:

connect(l) Connects the points with line segments. Unless the data are already in order by *x*, you might want to specify **connect(l) sort**.

connect(L) Connects with line segments, as long as *x* keeps ascending.

connect(m) Connects cross-medians of several vertical bands. The **bands( )** option specifies how many bands. For example **connect(m) bands(6)** calls for dividing the data into six equal-width bands. The default is **bands(200)**.

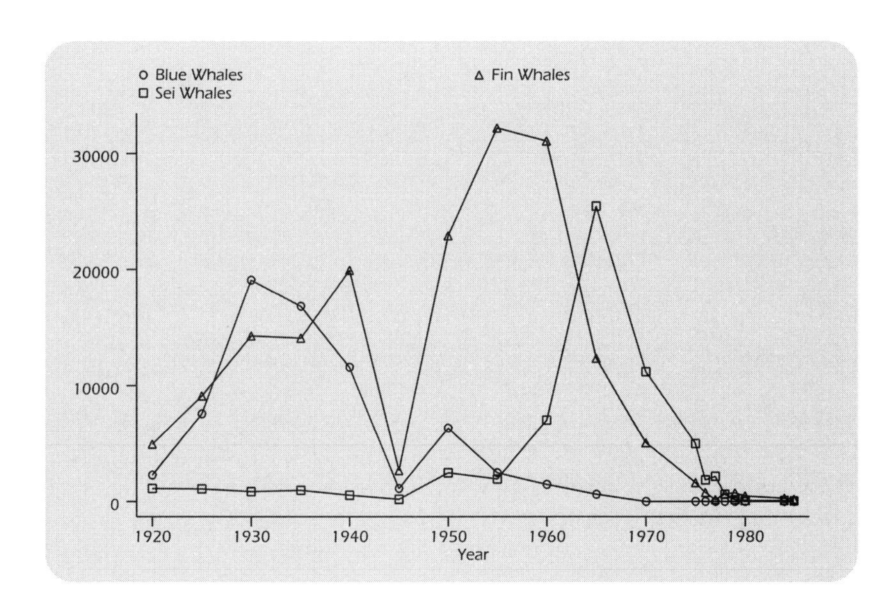

Figure 3.7

connect(s) Connects cross-medians using smooth curves (cubic splines). As with **connect(m)**, we can also specify **bands( )**. We can also specify **density( )**, the number of points to be calculated on the cubic spline. The default is **density(5)**.

connect(.) Does not connect points. This is helpful if we have two or more *y* variables. For example, **connect(.l)** does not connect the first *y* variable but connects the second with line segments.

connect(J) Connects in steps, like stairs.

connect(||) Connects a pair of variables with vertical bars (as in high-low plots).

connect(II) Same as **connect(||)**, but with bars capped like capital I's (as in error-bar plots).

Note that the similar-appearing letter **l**, vertical line |, and capital **I** mean quite different things to **graph**.

Another important **graph** option, **symbol( )**, controls the plotting symbols used in scatterplots:

symbol(O) Large circle

symbol(T) Large triangle

symbol(S) Large square

symbol(o)	Small circle
symbol(p)	Small plus sign
symbol(d)	Small diamond
symbol(.)	Dot
symbol(i)	Invisible
symbol([_n])	Observation number (note inside square brackets)
symbol([*varname*])	Values of variable *varname* (note square brackets)

We can specify as many plotting symbols as we have *y* variables. For example, **symbol(Sd[*varname*])** requests the first-named *y* plotted as large squares, the second as small diamonds, and the third using values of *varname*, which could be numbers, labels or strings. When we leave out the **symbol( )** option, **graph** assigns symbols to *y* variables in the order listed. For example, in Figure 3.7, the first-named variable *blue* is plotted with large circles, *fin* with large triangles, and *sei* with large squares. The defaults thus worked as if we had typed

```
. graph blue fin sei year, connect(lll) symbol(OTS)
```

Scatterplots 2: Smoothing Time Plots

The whale-catch variables changed gradually, but many time series exhibit rapid up and down fluctuations that make it difficult to discern underlying patterns. Smoothing such series aids analysis by breaking the data into two parts, one that varies smoothly and a second containing all the leftover rapid changes:

data = smooth + rough

Dataset *nhwater.dta* has data on daily water consumption for the town of Milford, New Hampshire, during the first half of 1983:

```
Contains data from c:\data\nhwater.dta
  obs:           212                          Milford NH water 1/83-7/83
 vars:             4                          15 Dec 1996 12:22
 size:         2,120 (96.5% of memory free)
-------------------------------------------------------------------------------
  1. month        byte   %9.0g                Month
  2. day          byte   %9.0g                Date
  3. year         int    %9.0g                Year
  4. water        int    %9.0g                Water use in 1000 gallons
-------------------------------------------------------------------------------
Sorted by:
```

Before graphing, we need to convert the month, day, and year information into a single numerical index of time. Stata's **mdy( )** function does this, creating an elapsed-date variable (here named *date*) indicating the number of days since January 1, 1960:

```
. generate date = mdy(month,day,year)

. list in 1/5

        month      day      year     water      date
  1.        1        1      1983       520      8401
  2.        1        2      1983       600      8402
  3.        1        3      1983       610      8403
  4.        1        4      1983       590      8404
  5.        1        5      1983       620      8405
```

The January 1, 1960, reference date is an arbitrary default, and we can easily rescale *date* so that its first value is "1". A graph of *water* against *date* shows much day-to-day variation (Figure 3.8):

```
. replace date = date - 8400
(212 real changes made)

. sort date

. label variable date "Date: 1 = January 1, 1983"

. graph water date, connect(l) symbol(.) xlabel
      ylabel(400,500,600,700,800,900)
```

The **smooth** command performs outlier-resistant nonlinear smoothing, employing methods and a terminology detailed by Velleman and Hoaglin (1981) and

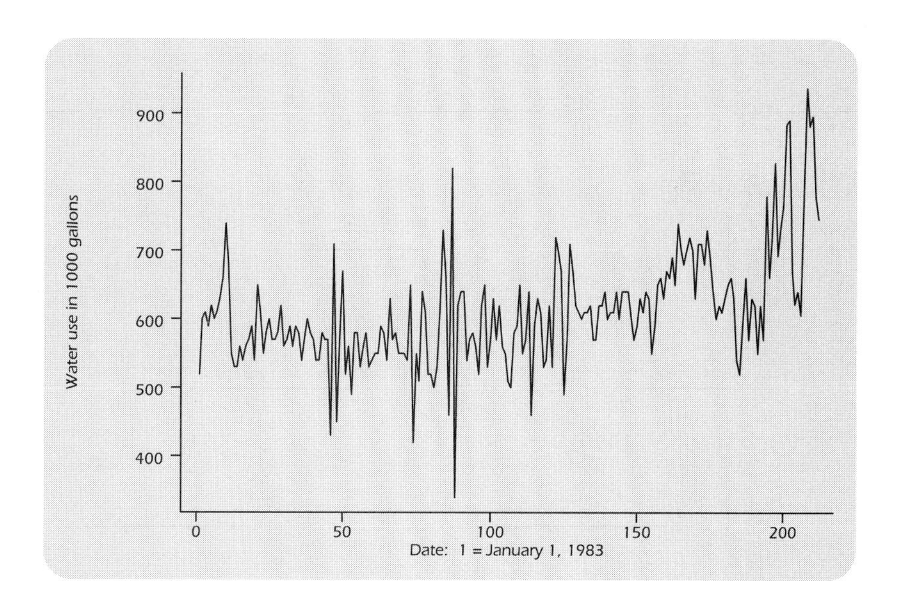

Figure 3-8

Velleman (1982). For example, the following command creates a new variable named *water1*, holding the values of *water* after smoothing by running medians of span 3:

```
. smooth 3 water, gen(water1)
```

Compound smoothers using running medians of different spans, in combination with hanning (¼, ½, ¼ weighted moving averages of span three) and other techniques can be specified in Velleman's original notation. One compound smoother that Velleman recommends is called "4253h,twice." Here we apply it to *water:*

```
. smooth 4253h,twice water, generate(water2)

. graph water2 date, connect(l) symbol(.) xlabel
      ylabel(400,500,600,700,800,900)
```

Compare Figure 3.9 with Figure 3.8 to see the effects of 4253h,twice smoothing. Sometimes our goal in smoothing is to look for patterns in such smoothed plots. With these particular data, however, the "rough" or residuals after smoothing actually hold more interest. We can calculate the rough as the difference between data and smooth and graph the results in another time plot (Figure 3.10):

```
. generate rough = water - water2

. label variable rough "Residuals from 4253h,twice"

. graph rough date, connect(l) symbol(.) xlabel
      ylabel(-200,-100,0,100,200)
```

The wildest fluctuations in Figure 3.10 occur around dates 86 to 88 (March 27 to 29). Water use abruptly dropped, rose again, and then dropped even further before

Figure 3.9

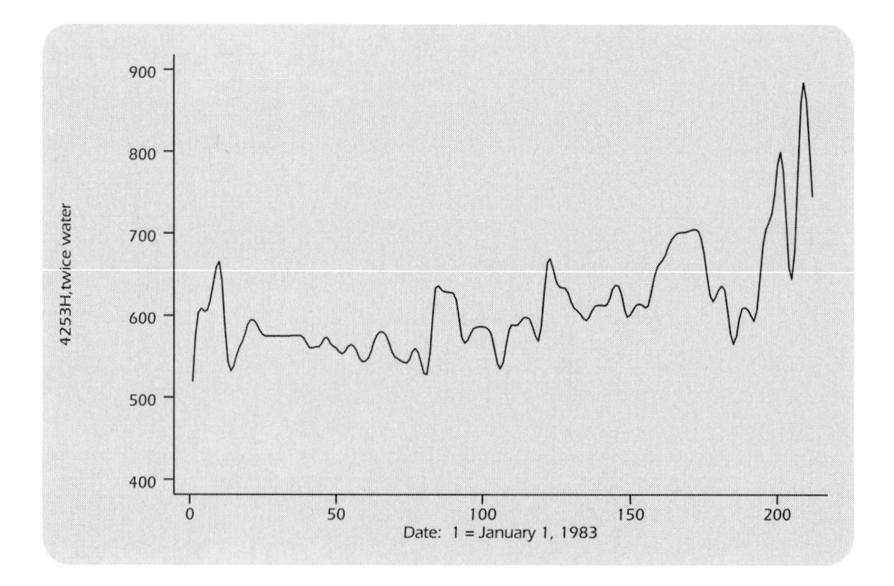

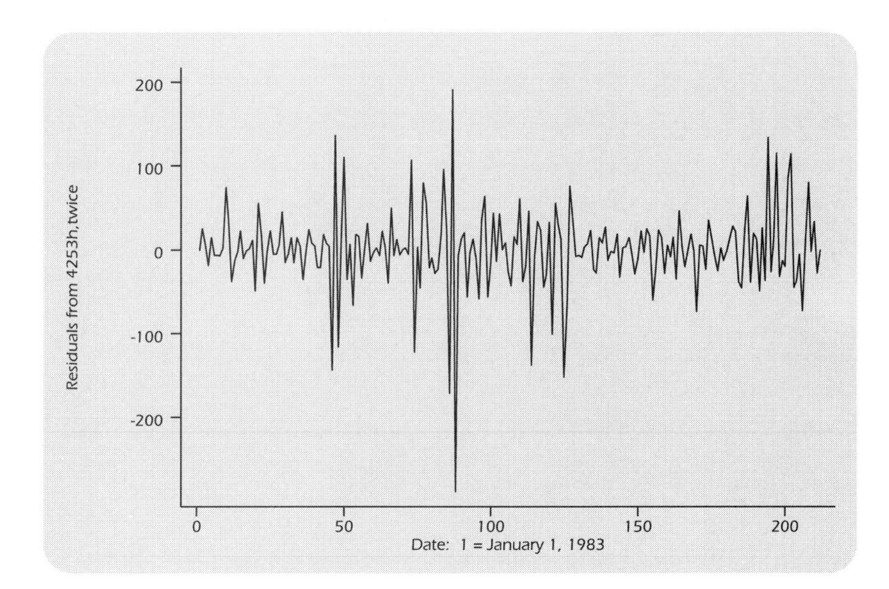

Figure 3.10

returning to more usual levels. On these days local newspapers carried stories that hazardous chemical wastes had been discovered in one of the wells that supplied the town's water. Initial reports alarmed people, but they were reassured after the questionable well was taken off line.

The smoothing techniques described in this section require variables with no missing values and make the most sense when observations are equally spaced in time. For data with missing values or uneven spacing in time, a regression-based technique such as lowess regression (Chapter 8) might work better.

Scatterplots 3

Scatterplots have many other applications besides time series. They are Stata's most versatile style of graph. A basic two-way scatterplot requires only **graph** followed by two variable names for the *y* and *x* axis, respectively. To illustrate, we return to the data on U.S. states, clearing any previous dataset from memory:

```
. use states90, clear
(U.S. states data 1990-91)
```

Figure 3.11 shows a scatterplot of *waste* (per capita solid wastes generated) against *metro* (percent population living in metropolitan areas).

Figure 3.11

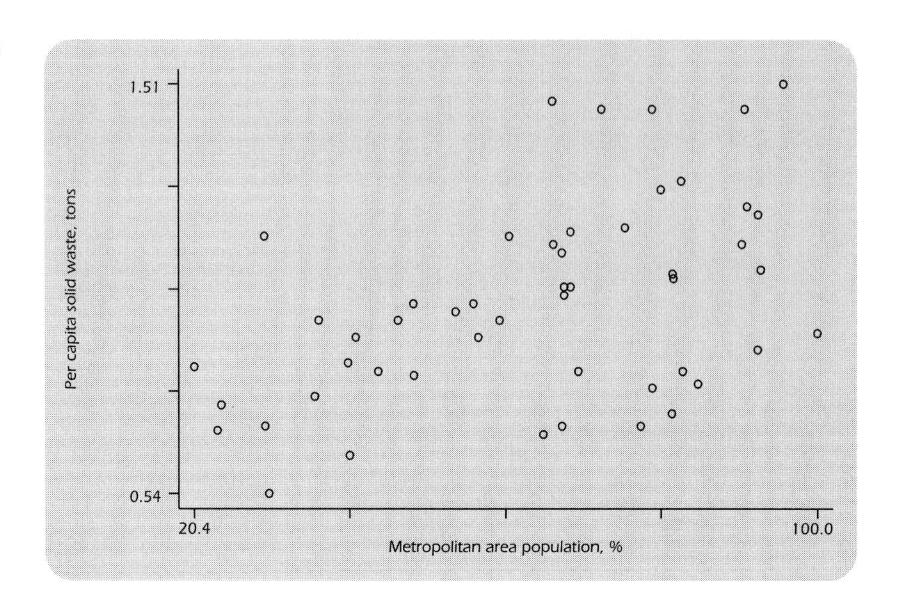


```
. graph waste metro
```

We could readily add options to connect points or to control the titles, labeling, or symbols. Scatterplots also permit "importance weighting," in which the area of each plotting symbol becomes proportional to a weighting variable. For example, importance-weighting by state population (*pop*) and labeling axes produces Figure 3.12:

```
. graph waste metro [iweight = pop], ylabel xlabel
```

Figure 3.12

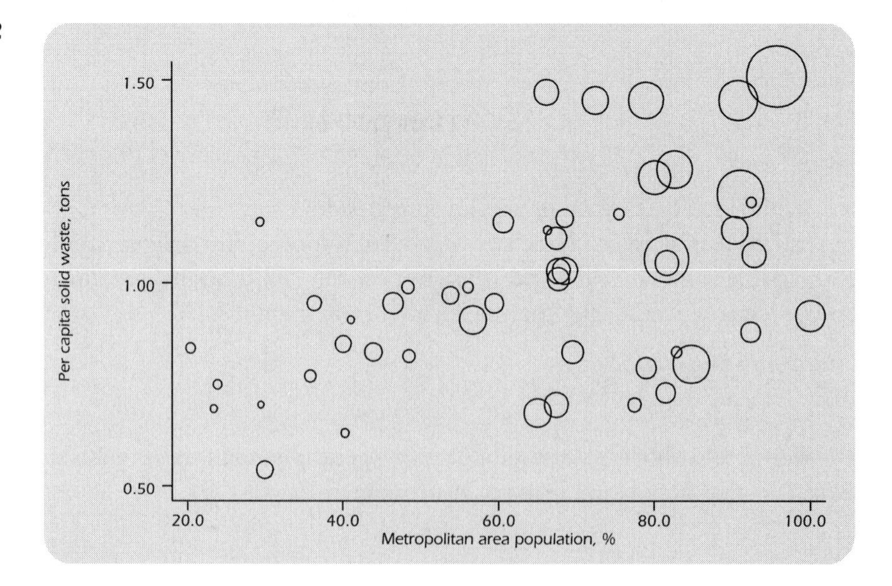

Figure 3.13

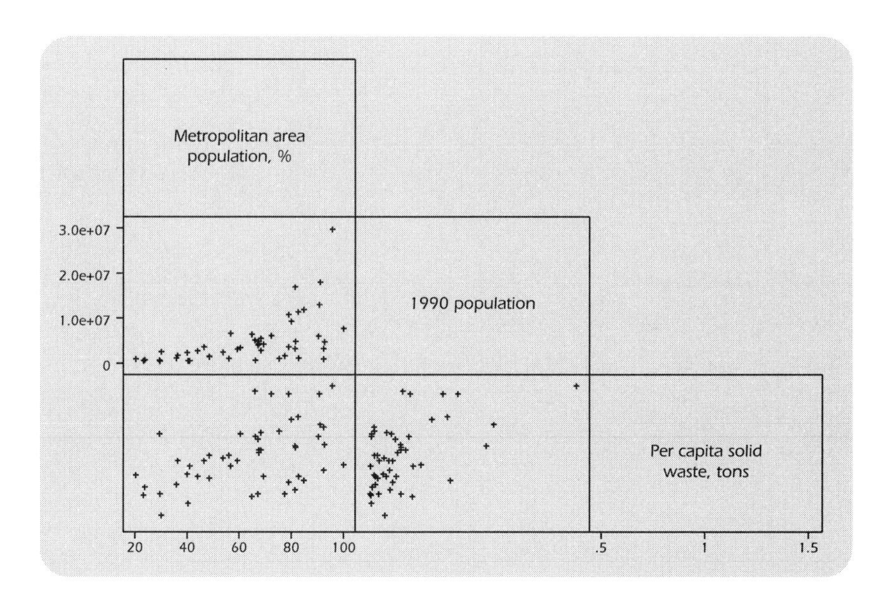

Plots of all possible pairs from a variable list appear in a scatterplot matrix (Figure 3.13):

```
. graph metro pop waste, matrix label symbol(p) half
```

The **half** option displays only the lower triangular half of a scatterplot matrix. If our data include one "dependent" variable and a number of "independent" variables, it makes sense to list the dependent variable last—as done here with *waste*. Then the bottom row of the matrix consists of a series of dependent-versus-independent variable plots. The option **label** calls for round-number labeling of axes; **ylabel** and **xlabel** do not work with scatterplot matrices.

Two-variable scatterplots can have boxplots or oneway scatterplots in their margins, as seen in Figure 3.14. The options **twoway oneway box** invoke all three styles at once. Marginal plots display the individual distributions of each variable, at the same time the main scatterplot shows their joint distribution. The marginal plots assist in noticing outliers and skewness.

```
. graph waste metro, twoway oneway box ylabel xlabel
```

Boxplots and One-Way Scatterplots

Boxplots convey information about center, spread, symmetry, and outliers at a glance. To obtain a single boxplot, type a command of the form:

```
. graph y, box
```

Figure 3.14

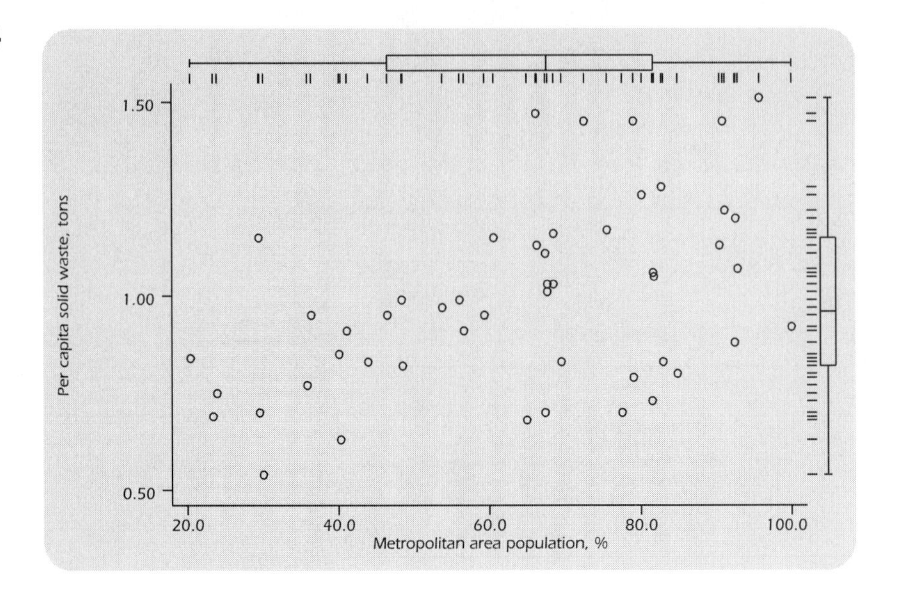

If they have roughly similar scales, we can visually compare the distributions of several different variables through commands of the form:

```
. graph w x y z, box
```

One of the most common applications of boxplots involves comparing the distribution of one variable across categories of a second. To illustrate, Figure 3.15 compares the distribution of *college* (percent adults with college degrees) across states of four U.S. regions:

```
. sort region
```

```
. graph college, box by(region) ylabel yline(19.3)
```

The **yline(19.3)** option drew a horizontal line at $y = 19.3$, the 50-state median (found via **summarize *college*, detail** ; see Chapter 4). In a boxplot, the box extends from approximate first to third quartiles, a distance called the interquartile range (IQR). Outliers, data points more than 1.5IQR beyond the first or third quartile, are plotted individually in a boxplot — usually as large circles, although we can control what symbol gets used for outliers by specifying the **symbol( )** option. No outliers appear among the four distributions in Figure 3.15. Stata's boxplots define quartiles in the same manner as **summarize, detail**. This is not the same approximation used to calculate "fourths" for letter-value displays, **lv** (Chapter 4). See Frigge, Hoaglin, and Iglewicz (1989) and Hamilton (1992b) for more about quartile approximations and their role in identifying outliers.

Figure 3.15

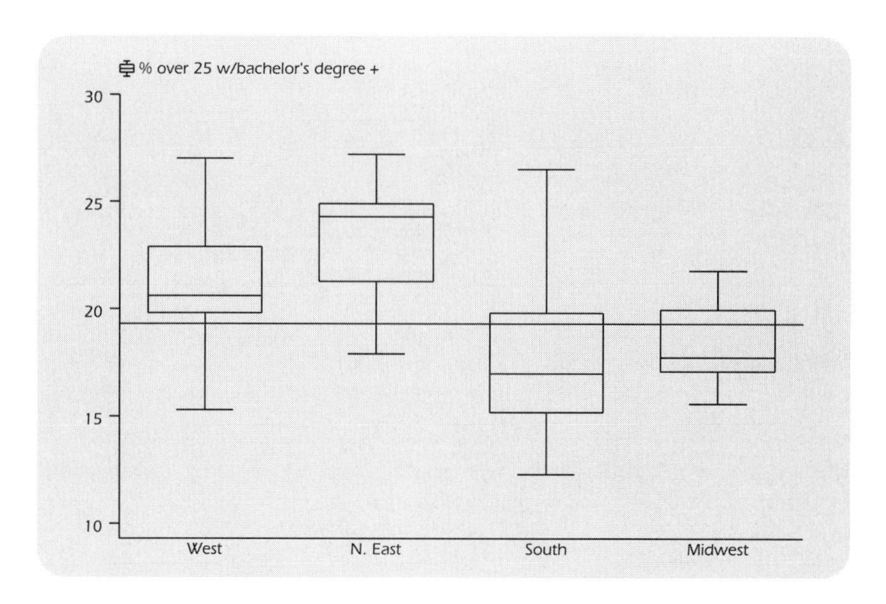

One-way scatterplots, Stata's simplest style of graph, also display variable distributions. They work best for continuous variables that have few or no tied values. Figure 3.16 shows a one-way scatterplot of *college.*

```
. graph college, oneway
```

Figure 3.16

We earlier saw one-way scatterplots combined with boxplots and a two-way scatterplot (Figure 3.14). Combining just boxplots and one-way scatterplots also makes a useful display for comparing distributions of several different variables:

```
. graph w x y z, oneway box
```

Alternatively, we could compare the distribution of one variable across categories of a second. Figure 3.17 does this, giving a slightly different view of the same information presented in Figure 3.15. The percentage of adults with a college degree tends to be highest among Northeastern states and lower in the South.

```
. sort region
```

```
. graph college, oneway box by(region)
```

Figure 3.17

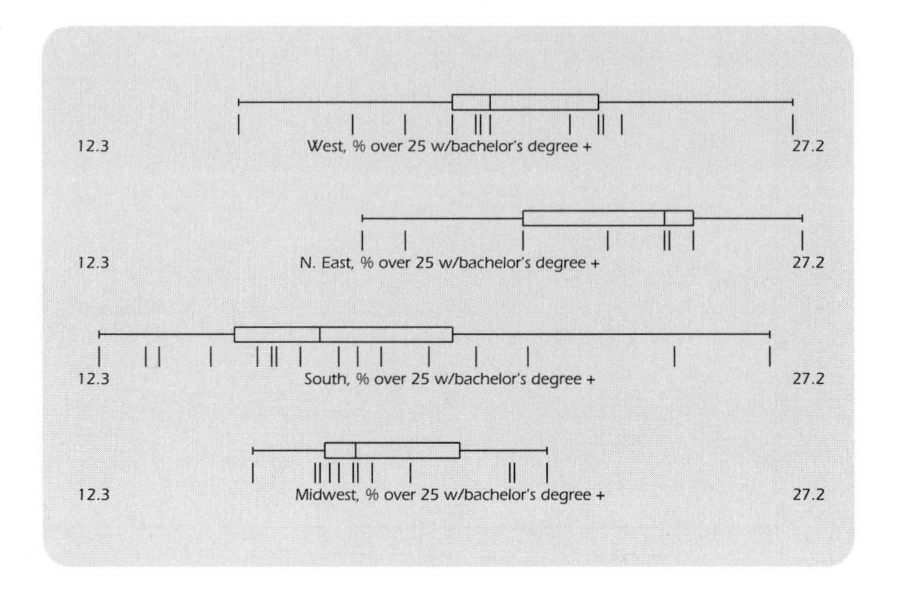

Pie Charts and Star Charts

Pie charts are key tools of "presentation graphics," although they have less value for analytical work. Stata's basic pie chart command has the form:

```
. graph w x y z, pie
```

where the variables *w*, *x*, *y*, and *z* all measure quantities of something, in similar units (for example, all are in dollars, or hours, or people).

Dataset *akethnic.dta*, on the ethnic composition of Alaska, provides an illustration. Alaska's indigenous Native population divides into three broad cultural and linguistic groups: Aleut, Indian (including Athabaska, Tlingit, and Haida) and Eskimo (Yupik and Inupiat). The variables *aleut*, *indian*, *eskimo*, and *nonnativ* are population counts for each of these groups, taken from the 1990 U.S. Census. This dataset contains only three observations, representing three types or sizes of communities: cities of 10,000 people or more; towns of 1,000 to 10,000 people; and villages with fewer than 1,000 people.

```
Contains data from c;\data\akethnic.dta
  obs:             3                                  Alaska ethnicity 1990
 vars:             7                                  15 Dec 1996 12:36
 size:            63 (96.3% of memory free)
-------------------------------------------------------------------------------
   1. comtype      byte   %8.0g       popcat      Community type (size)
   2. pop          float  %9.0g                   Population
   3. n            int    %8.0g                   number of communities
   4. aleut        int    %8.0g                   Aleuts
```

```
   5. indian       int     %8.0g                     Indians
   6. eskimo       int     %8.0g                     Eskimos
   7. nonnativ     float   %9.0g                     Non-Natives
-------------------------------------------------------------------------------
Sorted by:
```

The majority of the state's population is non-Native, as clearly seen in a pie chart (Figure 3.18; like bar charts, Stata's pie charts look better in color).

```
. graph aleut indian eskimo nonnativ, pie
```

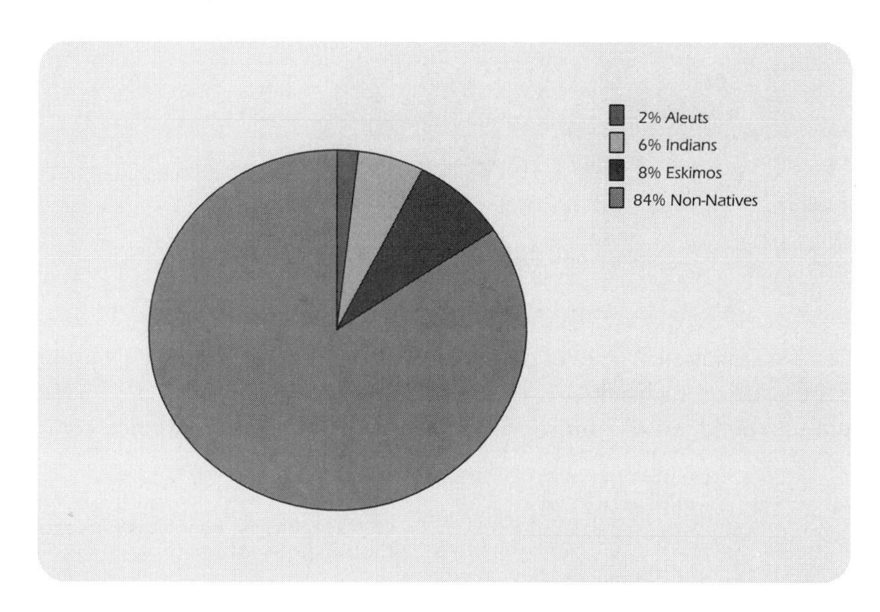

Figure 3.18

Non-Natives constitute the dominant group in Figure 3.18, but if we draw separate pies for each type of community, new details emerge (Figure 3.19). Although Natives are only a small fraction of the population in Alaskan cities, they constitute the majority among those living in villages. This gives Alaskan villages a quite different character from Alaskan cities.

```
. sort comtype
. graph aleut indian eskimo nonnativ, pie by(comtype) total
```

In contrast with the ubiquitous pie chart, star charts are relatively uncommon. They belong to a family of methods (which includes "Chernof faces") meant to compare observations visually of a number of variables at once. A basic star chart command has the form:

```
. graph w x y z, star label(name)
```

Unlike pie charts, star charts do not require that the variables be in similar units. `by(varname)` options do not work with star charts. Instead, we automatically get

Figure 3.19

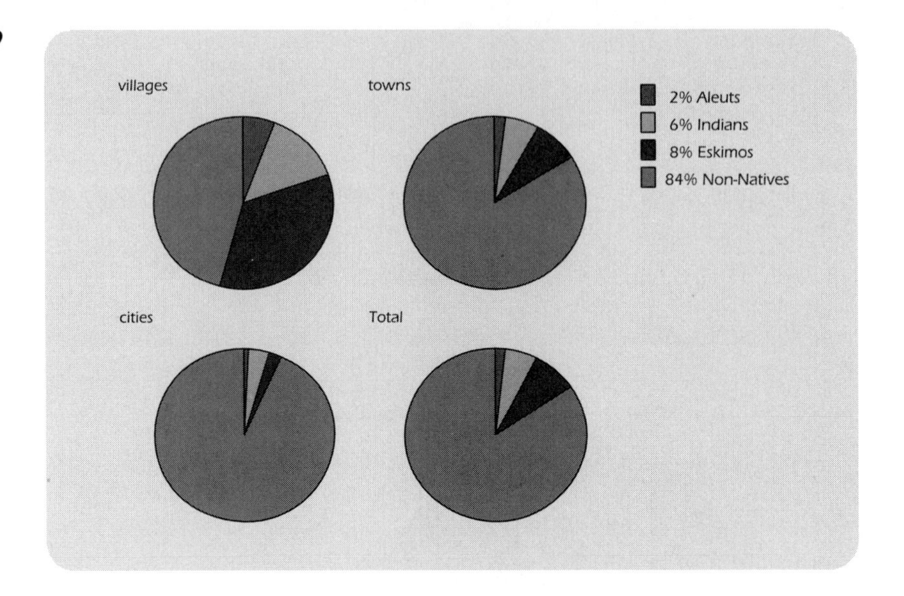

one star for each observation. Values of the variable specified in the **label( )** option can provide the labels for each star.

We could use star charts to compare a number of technically similar microcomputers (file *micro.dta*).

```
Contains data from c:\data\micro.dta
  obs:            51                            PC Magazine's benchmark tests
 vars:             9                            15 Dec 1996 12:38
 size:         1,887 (96.2% of memory free)
-------------------------------------------------------------------------------
   1. make         byte   %8.0g       mk        Manufacturer
   2. model        float  %9.0g       md        Model
   3. chip         float  %9.0g                 CPU chip type
   4. disk         float  %9.0g                 DOS disk access time
   5. clock        float  %9.0g                 Clock speed in MHZ
   6. nop          float  %9.0g                 No operation loop
   7. mix          float  %9.0g                 CPU instruction mix
   8. fp           float  %9.0g                 Floating point test
   9. ram          float  %9.0g                 RAM read/write test
-------------------------------------------------------------------------------
Sorted by:  chip
```

The variables *disk, clock, nop, mix, fp,* and *ram* measure different aspects of computing speed. Most are indices, and higher values mean a slower computer. Variable *make* stores the computer brand names. Figure 3.20 shows a star chart for the 17 computers that use an 80386 chip and do not have missing values on the memory read/write test (**if *chip* == 80386 & *ram* != .**).

```
. graph disk clock nop mix fp ram if chip == 80386 &
      ram != ., star label(make)
```

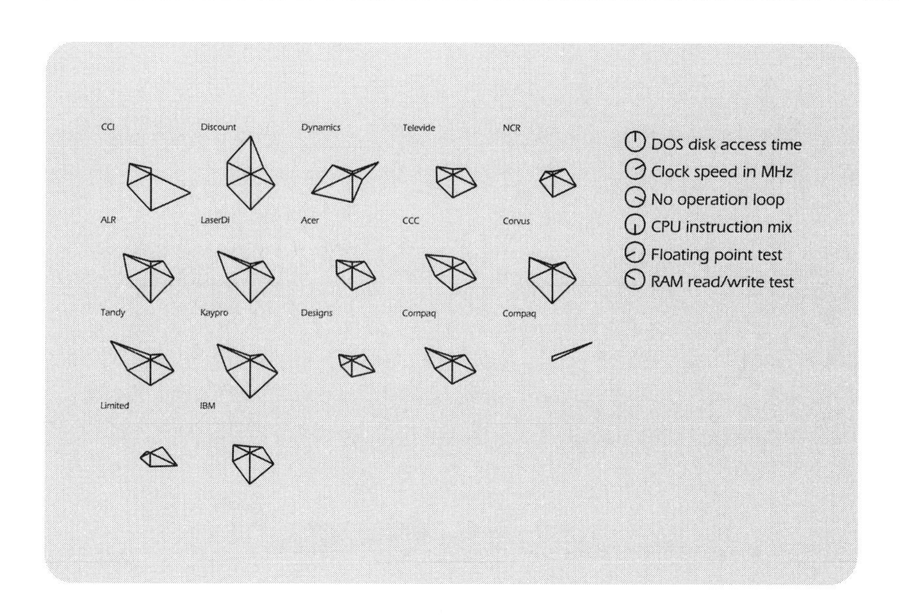

Figure 3.20

A long line segment in the *disk* direction indicates the computer has a high value on this variable. That is, its disk access is slow. The same holds for the other five variables: The slower the computer is, the longer the corresponding line segment is. We cannot read the variables' actual values from a star chart. Instead we scan for general patterns. For example, the second Compaq and the Limited stand out as fast computers overall; the Discount and Laser appear generally slow. Several computers have quite similar performance profiles, such as the Laser and Kaypro, or the Televideo and Acer. Note that we compare these 17 computers on six variables at once; the chart contains a lot of information.

Bar Charts

Bar charts provide simple yet versatile tools for depicting the distributions of categorical or mixed categorical and measurement variables. Two key options are the following:

`means` Draws bar heights to reflect variable means (default is variable sums).

`stack` Stacks the bars for listed variables on top of each other instead of placing them side by side.

Figure 3.21

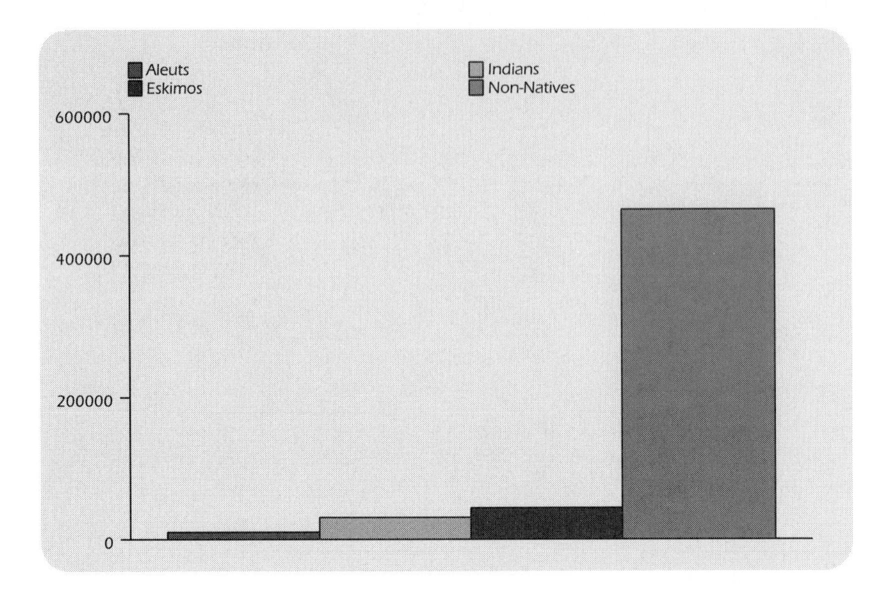

Returning to the Alaskan ethnicity dataset *akethnic.dta*, we can display the same information as the earlier pie charts but this time in bar chart form. Figure 3.21 shows the state's overall ethnic composition:

```
. graph aleut indian eskimo nonnativ, bar ylabel
```

Figure 3.22 breaks the population down by community type (*comtype*), and stacks bars for different ethnic groups on top of each other. Without the **stack** option, we would have four side-by-side bars for each community type. The pie charts in

Figure 3.22

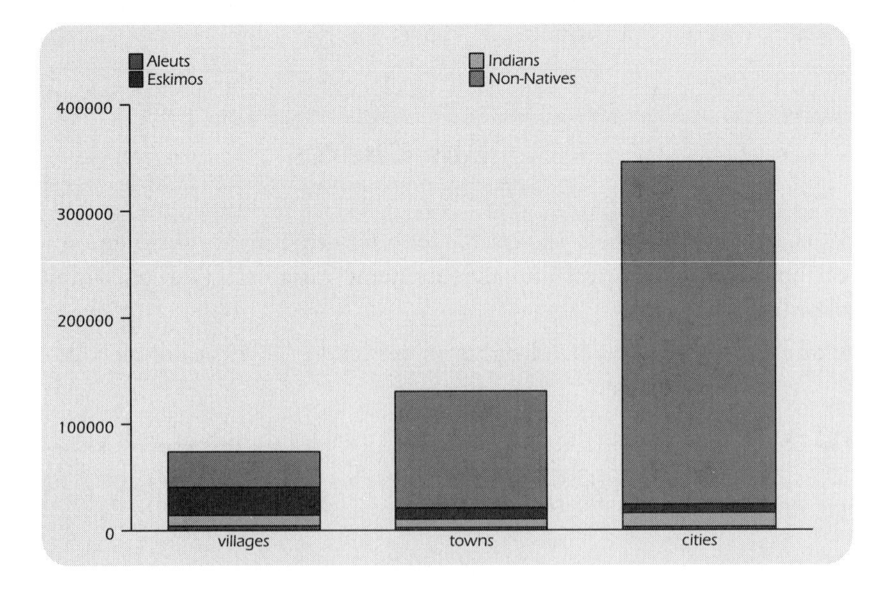

Figure 3.19 depict the relative sizes (percentages) of different ethnic groups, whereas the bars in Figure 3.22 depict their absolute sizes (sums).

```
. sort comtype

. graph aleut indian eskimo nonnativ, bar ylabel stack
        by(comtype)
```

Bar charts can display means instead of sums. For example, Figure 3.23 charts the means of median household income in U.S. states by geographical region (from *states90.dta*). The **yline(33957)** option includes a horizontal line indicating the overall mean, $33,957, found via **summarize *income***.

```
. sort region

. graph income, bar means ylabel yline(33957) by(region)
```

Stata's bar charts work most naturally with measurement variables, but they will also show categorical-variable distributions. Forty statistics students surveyed for dataset *stats.dta* indicated their political party preference as follows:

```
. tabulate party

Political   |
party       |
preference  |      Freq.     Percent        Cum.
------------+-----------------------------------
   Democrat |         17       42.50       42.50
   Republic |          7       17.50       60.00
   Independ |         16       40.00      100.00
------------+-----------------------------------
      Total |         40      100.00
```

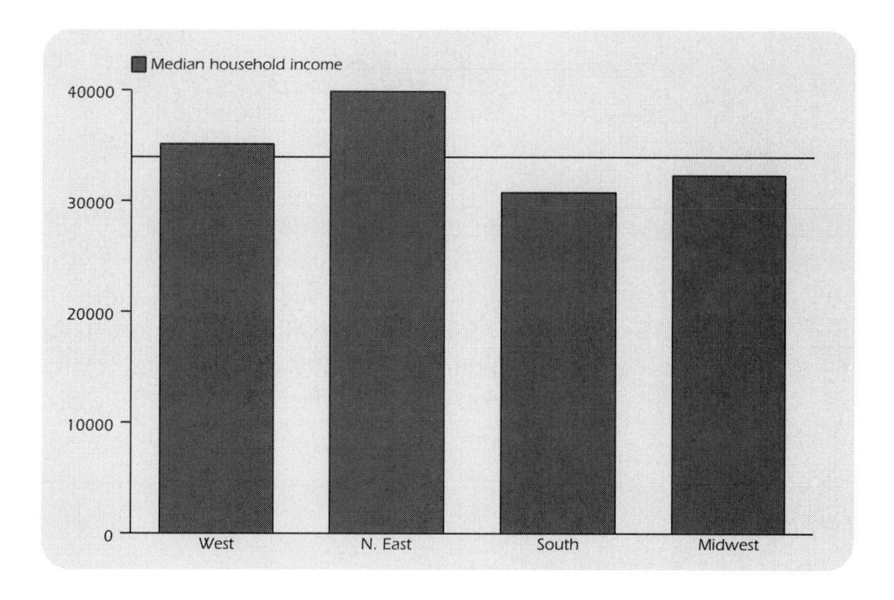

Figure 3.23

graph, bar displays either sums or means, but neither summary ordinarily makes sense with a purely categorical variable such as *party*. To graph the distribution of one categorical variable therefore requires an extra step:

1. Create a new "variable" (if one does not already exist) equal to one for every observation in the data:

```
. generate count = 1
```

2. Create a bar chart showing sums of this new "variable," by values of the categorical variable of interest (Figure 3.24). Because *count* equals 1 for each observation, its sum within a category equals the frequency of that category:

```
. sort party
. graph count, bar by(party) ylabel
```

Figure 3.24

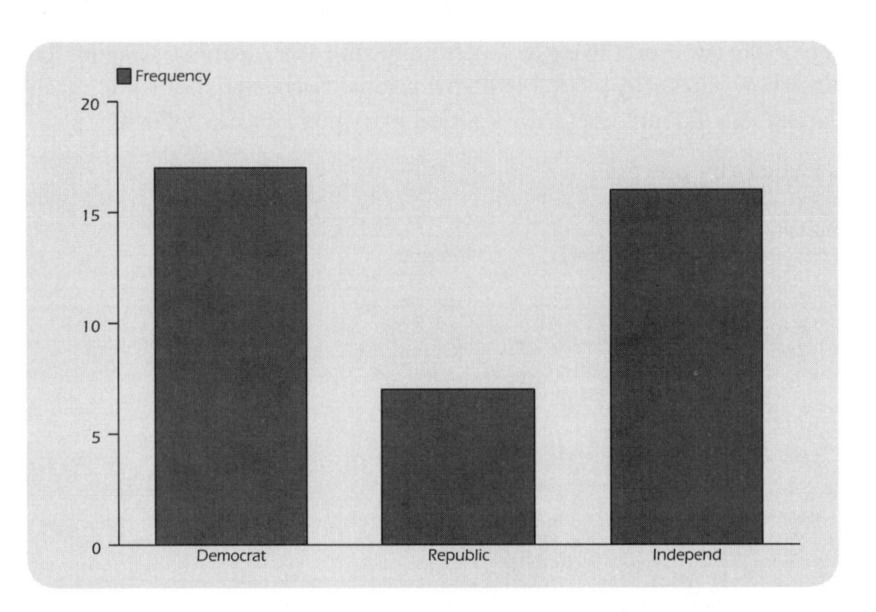

A similar trick allows us to graph the relation between two categorical variables such as *party* and *gender* in one bar chart. *gender* exists in *stats.dta* as a dummy variable, coded 0 = "male" and 1 = "female." For this graph we will want a new indicator coded 0 = "male" and 100 = "female." Also, for reasons that will soon be apparent, we label this new indicator "Percent female," although really it is just an indicator with values of 0 or 100. We then use the new variable *female* as the basis for our bar chart. The result, Figure 3.25, shows the percent female within each category of political party preference.

```
. generate female = 0
. replace female = 100 if gender == 1
(31 real changes made)
```

Figure 3.25

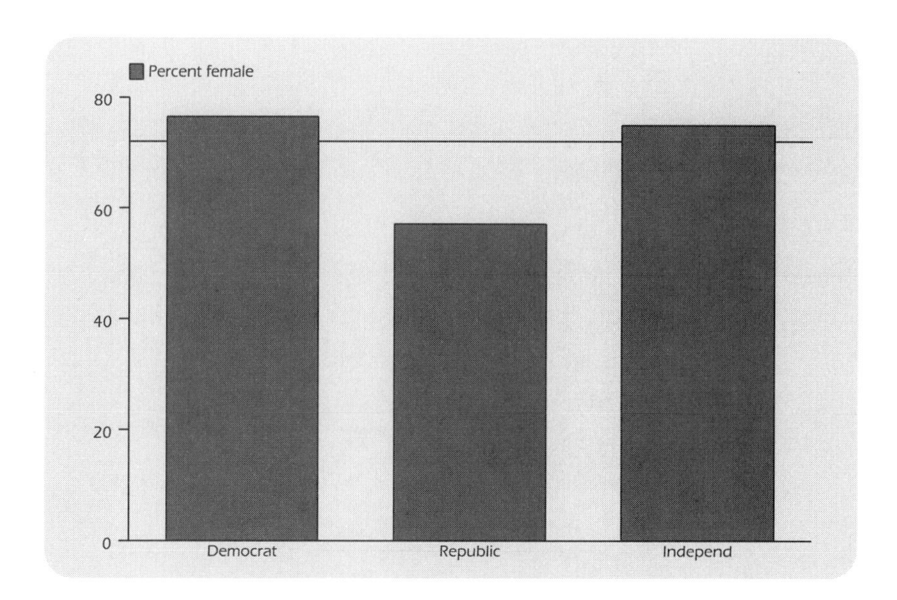


```
. label variable female "Percent female"
. graph female, bar by(party) means ylabel yline(72)
```

Note that we used the **means** option to form Figure 3.25. Means are needed here because the mean of a {0,100} dichotomy really does equal the percentage coded 100, that is, the percent female. (Similarly, the mean of a {0,1} dummy variable equals the proportion coded 1.)

Symmetry and Quantile Plots

Boxplots and histograms summarize measurement variable distributions, hiding individual data points to clarify overall patterns. Symmetry and quantile plots, on the other hand, include points for every observation in a distribution. They are harder to read than summary graphs but convey more detailed information.

A histogram of per-capita energy consumption in the 50 U.S. states appears in Figure 3.26. The skewed distribution includes a handful of very high-consumption states, which happen to be oil producers. Bar heights indicate the fraction of observations with each range of values. For example, about .35 (35%) of these states had *energy* values between 200 and 300.

```
. graph energy, xlabel ylabel bin(8) norm
```

Figure 3.27 depicts this distribution as a symmetry plot. It plots the distance of the *i*th observation above the median (vertical) against the distance of the *i*th

Figure 3.26

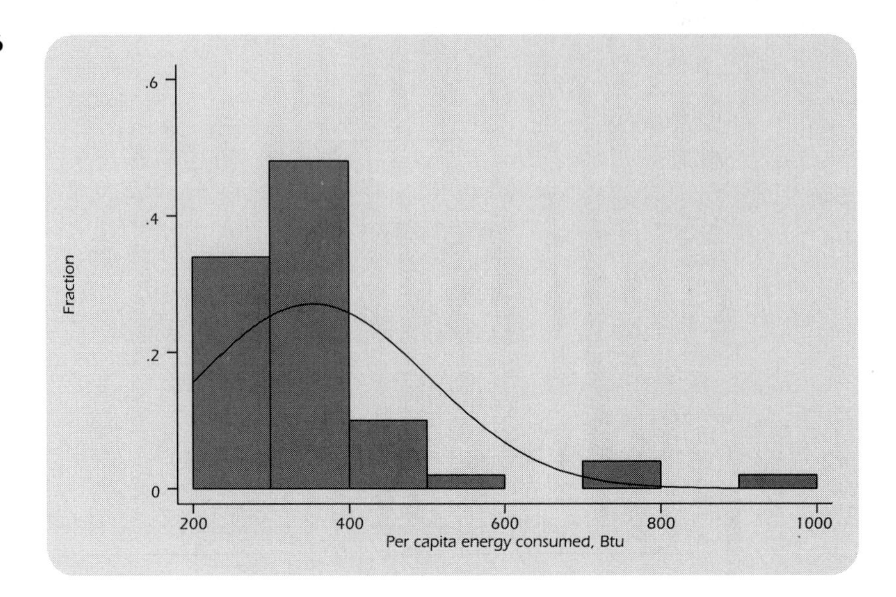

Figure 3.27

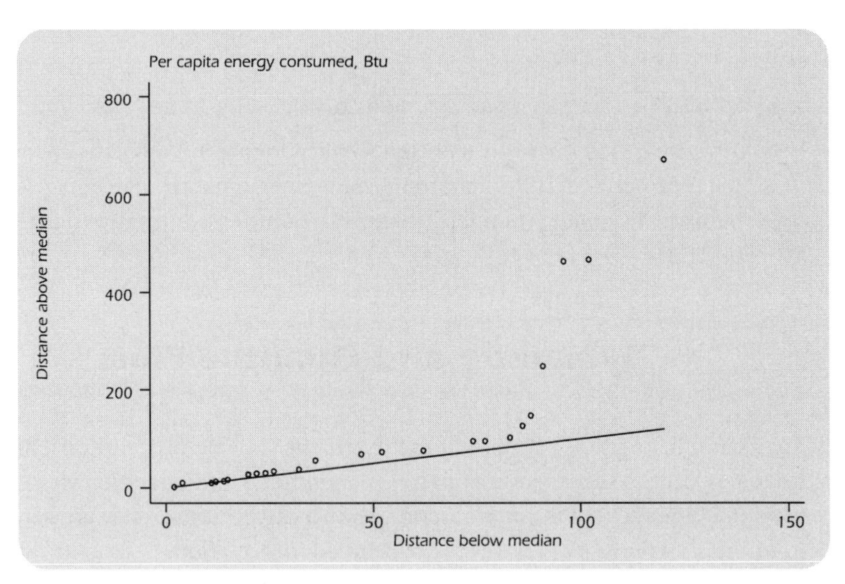

observation below the median. All points would lie on the diagonal line if this distribution were symmetrical. Instead we see that distances above the median grow steadily larger than corresponding distances below the median, a symptom of positive skew. Unlike Figure 3.26, Figure 3.27 also reveals that the energy-consumption distribution is approximately symmetrical near its center.

```
. symplot energy, xlabel ylabel
```

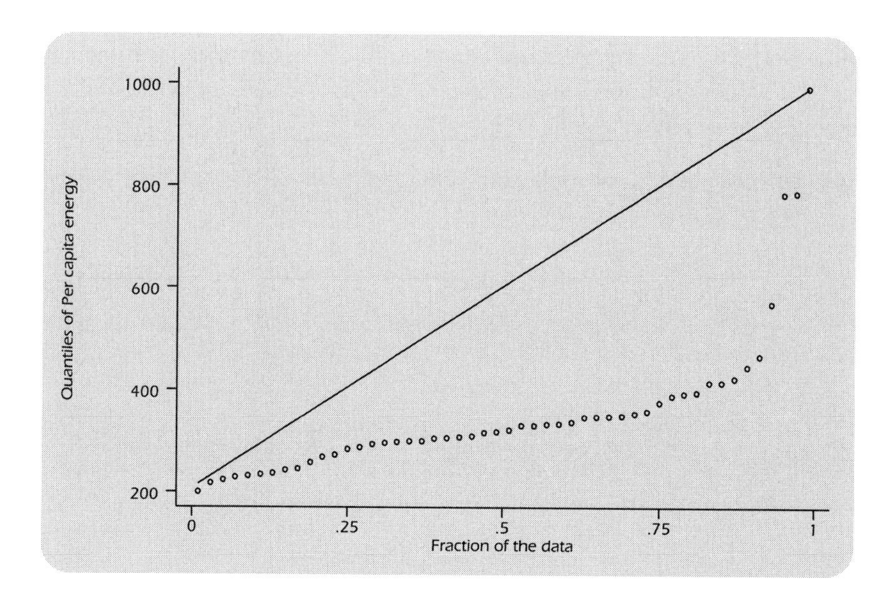

Figure 3.28

Quantiles are values below which a certain fraction of the data lie. For example, a .3 quantile is that value higher than 30% of the data. If we sort *n* observations in ascending order, the *i*th value forms the $(i- .5)/n$ quantile. To calculate quantiles of variable *energy:*

```
. drop if energy == .
. sort energy
. generate quantile = (_n - .5)/_N
```

_n and _N are Stata system variables, always unobtrusively present when there are data in memory. _n represents the current observation number, and _N represents the total number of observations. Quantile plots automatically calculate what fraction of the observations lie below each data value and display the results graphically as in Figure 3.28. Quantile plots provide a graphic reference for someone who does not have the original data at hand. From well-labeled quantile plots, we can estimate order statistics such as median (.5 quantile) or quartiles (.25 and .75 quantiles). The IQR equals the rise between .25 and .75 quantiles. We could also read a quantile plot to estimate the fraction of observations falling below a given value.

```
. quantile energy, ylabel
```

Quantile-normal plots, also called normal probability plots, compare quantiles of a variable's distribution with quantiles of a theoretical normal distribution that has the same mean and standard deviation. Quantile-normal plots allow visual inspection for departures from normality in every part of a distribution, which can help guide decisions regarding normality assumptions and efforts to find a normalizing transformation. Figure 3.29, a quantile-normal plot of *energy*, confirms the severe positive skew that we had already observed.

```
. qnorm energy, ylabel xlabel
```

Figure 3.29

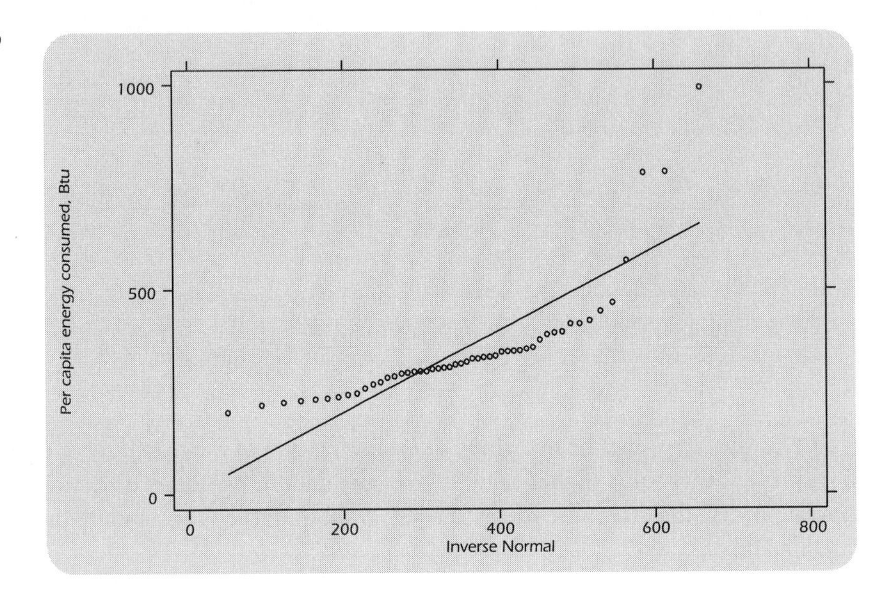

Quantile-quantile plots resemble quantile-normal plots, but they compare quantiles (ordered data points) of two empirical distributions instead of comparing one empirical distribution with a theoretical normal distribution. Figure 3.30 shows a quantile-quantile plot of the mean math SAT score versus the mean verbal SAT score in 50 states and the District of Columbia, from *states90.dta*. If the two distributions were identical, we would see points along the diagonal line. Instead, data points form a straight line roughly parallel to the diagonal, indicating that the two variables have different means but similar shapes and standard deviations.

```
. qqplot msat vsat, ylabel xlabel
```

Regression with Graphics (Hamilton 1992a) includes an introduction to reading quantile-based plots. Chambers and associates (1983) provide more details. Related Stata commands include **pnorm** (standard normal probability plot), **pchi** (chi-square probability plot) and **qchi** (quantile–chi-square plot).

Figure 3.30

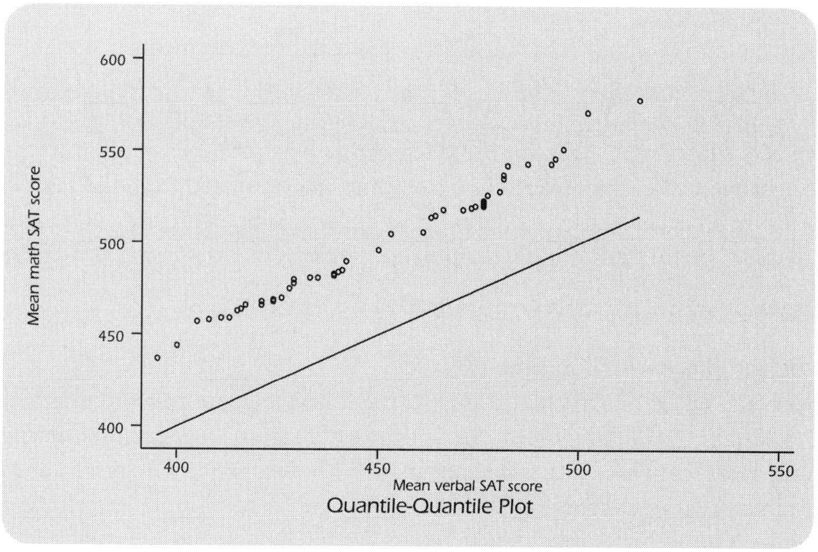

Quantile-Quantile Plot

Quality Control Charts

You can use quality control charts to monitor output from a repetitive process such as industrial production. Stata offers four basic types: c chart, p chart, R chart, and $\overline{x}$ chart. A fifth type, called Shewhart after the inventor of these methods, consists of vertically-aligned $\overline{x}$ and R charts. Iman (1994) provides a brief introduction to R and $\overline{x}$ charts, including the tables used in calculating their control limits. The *Stata Reference Manual* gives the command details and formulas used by Stata. Basic outlines of these commands are as follows:

. cchart ***defects unit***

Constructs a c chart with the number of nonconformities or defects (*defects*) graphed against the unit number (*unit*). Upper and lower control limits, based on the assumption that number of nonconformities per unit follows a Poisson distribution, appear as horizontal lines in the chart. Observations with values outside these limits are said to be "out of control."

. pchart ***rejects unit ssize***

Constructs a p chart with the proportion of items rejected (*rejects / ssize*) graphed against the unit number (*unit*). Upper and lower control limit lines derive from a normal approximation, taking sample size (*ssize*) into account. If *ssize* varies across units, the control limits will vary too, unless we add the option **stabilize**.

. rchart ***x1 x2 x3 x4 x5*****, connect(l)**

Constructs an R (range) chart using the replicated measurements in variables *x1* through *x5*—that is, in this example, five replications per sample. Graphs the range within each sample against the sample number, and (optionally) connects successive ranges with line segments. Horizontal lines indicate the mean range and control limits. Control limits are estimated from the sample size if the process standard deviation is unknown. When σ is known, we can include this information in the command. For example, assuming σ = 10:

```
. rchart x1 x2 x3 x4 x5, connect(l) std(10)
```

. xchart ***x1 x2 x3 x4 x5*****, connect(l)**

Constructs an $\overline{x}$ (mean) chart using the replicated measurements in variables *x1* through *x5*. Graphs the mean within each sample against the sample number and connects successive means with line segments. The mean range is estimated from the mean of sample means, and control limits are estimated from sample size, unless we override these defaults. For example, if we know that the process actually has μ = 50 and σ =10,

```
. xchart x1 x2 x3 x4 x5, connect(l) mean(50) std(10)
```

Alternatively, we could specify particular upper and lower control limits:

```
. xchart x1 x2 x3 x4 x5, connect(l) mean(50) lower(40)
     upper(60)
```

. shewhart ***x1 x2 x3 x4 x5*****, mean(50) std(10)**

In one figure, vertically aligns an $\overline{x}$ chart with an R chart. **shewhart** serves as a rough analytical aid; better-looking displays of this sort can be constructed by saving an $\overline{x}$ and an R chart separately, then combining them using the Stata Graphics Editor.

To illustrate a p chart, we turn to the quality inspection data in *qual1.dta:*

```
Contains data from c:\data\qual1.dta
  obs:            16                          Quality control example 1
 vars:             3                          15 Dec 1996 16:33
 size:           112 (96.5% of memory free)
-------------------------------------------------------------------------------
   1. day          byte    %9.0g              Day sampled
   2. ssize        byte    %9.0g              Number of units sampled
   3. rejects      byte    %9.0g              Number of units rejected
-------------------------------------------------------------------------------
Sorted by:
```

```
. list in 1/5

          day       ssize     rejects
  1.       58          53          10
  2.        7          53          12
  3.       26          52          12
  4.       21          52          10
  5.        6          51          10
```

Note that sample size varies from unit to unit, and the units (days) are not in order. **pchart** handles these complications automatically, creating the graph with changing control limits seen in Figure 3.31. (For constant control limits despite changing sample sizes, add the **stabilize** option.)

```
. pchart rejects day ssize, ylabel xlabel
```

Dataset *qual2.dta,* borrowed from Iman (1994:662), illustrates **rchart** and **xchart**. Variables *x1* through *x4* represent repeated measurements from an industrial production process; 25 units with four replications each form the dataset.

```
Contains data from c:\data\qual2.dta
  obs:            25                          Quality control (Iman 1994:662)
 vars:             4                          15 Dec 1996 16:35
 size:           500 (96.5% of memory free)
-----------------------------------------------------------------------------
  1. x1            float  %9.0g
  2. x2            float  %9.0g
  3. x3            float  %9.0g
  4. x4            float  %9.0g
-----------------------------------------------------------------------------
Sorted by:
```

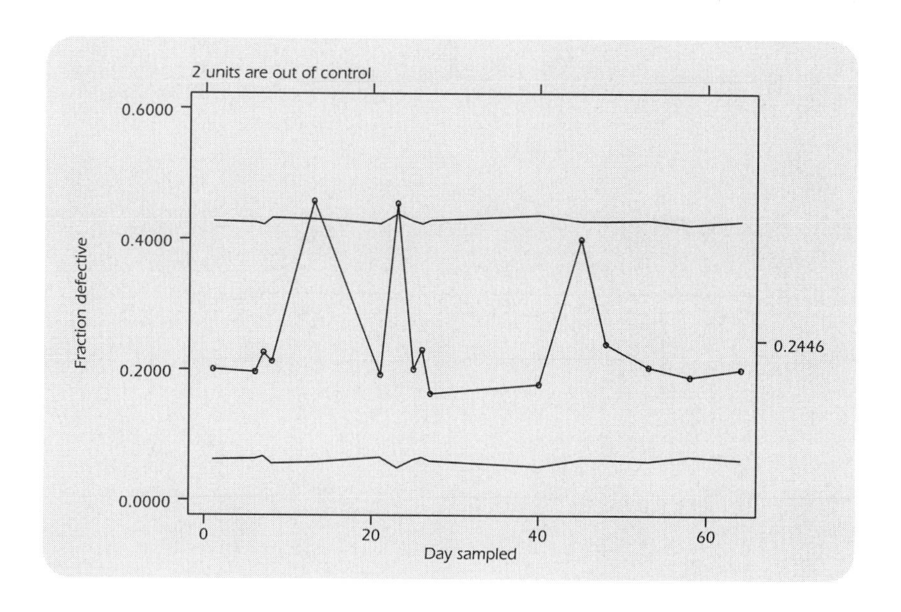

Figure 3.31

```
. list in 1/5

            x1         x2         x3         x4
  1.       4.6          2          4        3.6
  2.       6.7        3.8        5.1        4.7
  3.       4.6        4.3        4.5        3.9
  4.       4.9          6        4.8        5.7
  5.       7.6        6.9        2.5        4.7
```

Figure 3.32, an R chart, graphs variation in the process range over the 25 units. **rchart** informs us that one unit's range is "out of control."

```
. rchart x1 x2 x3 x4, connect(l)
```

Figure 3.32

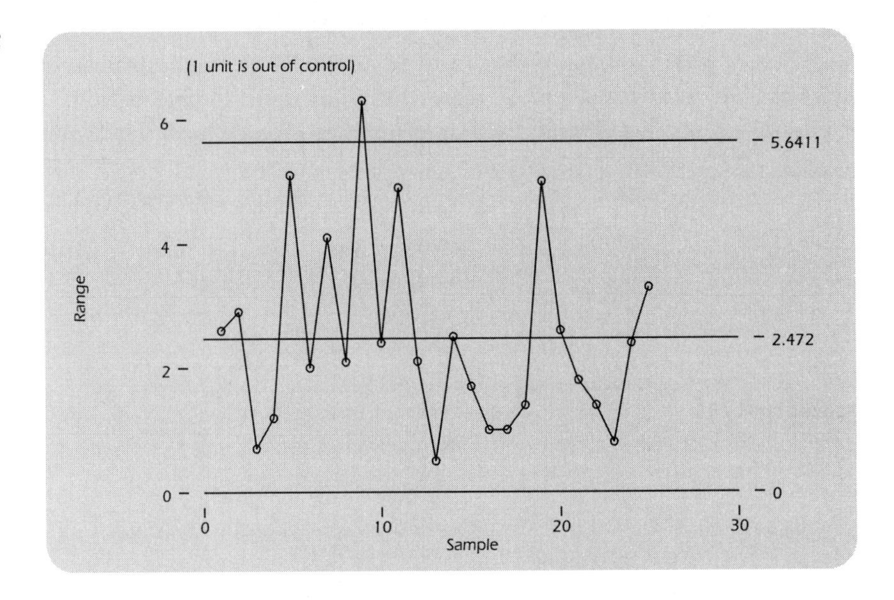

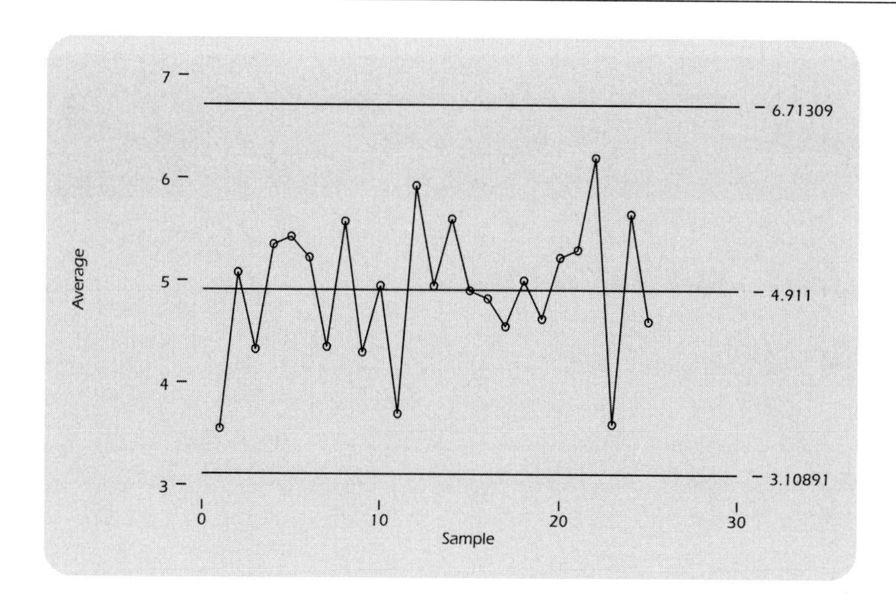

Figure 3.33

Figure 3.33, an $\overline{x}$ chart, shows variation in the process mean. None of these 25 means falls outside the control limits.

```
. xchart x1 x2 x3 x4, connect(l)
```

4

Tables and Summary Statistics

The **summarize** command obtains descriptive statistics such as medians, means, and standard deviations of measurement variables. **tabulate** obtains frequency distribution tables, cross-tabulations, and assorted tests or measures of association. **tabulate** can also construct one- or two-way tables of means and standard deviations. A more general table-making command, **table**, produces as many as seven-way tables in which the cells contain statistics such as frequencies, sums, means, or medians.

This chapter also describes further one-variable methods including normality tests, transformations, and displays for exploratory data analysis (EDA). Information about tables of particular interest to epidemiologists, not described in this chapter, can be seen by typing **help epitab**.

Example Commands

. summarize ***y1 y2 y3***

Calculates basic summary statistics (means, standard deviations, and numbers of observations) for the variables listed.

. summarize ***y1 y2 y3*****, detail**

Obtains detailed summary statistics including percentiles, median, mean, standard deviation, variance, skewness, and kurtosis.

`. summarize y1 if x1 > 3 & x2 != .`

Finds summary statistics on *y1* using only those observations for which variable *x1* is greater than 3 and *x2* is not missing.

`. summarize y1 [fweight = w], detail`

Calculates detailed summary statistics on *y1* using the frequency weights in variable *w*.

`. tabulate x1`

Displays a frequency distribution table including percentages for all nonmissing values of variable *x1*.

`. tab1 x1 x2 x3 x4`

Displays a series of frequency distribution tables, one for each of the variables listed.

`. tabulate x1 x2`

Displays a two-variable cross-tabulation with *x1* as row variable, *x2* as the columns.

`. tabulate x1 x2, chi2 nof column`

Produces a cross-tabulation and Pearson chi-square test of independence. Does not show cell frequencies, but instead gives the column percentages in each cell.

`. tabulate x1 x2, missing row all`

Produces a cross-tabulation that includes missing values in the table and in the calculation of percentages. Calculates "all" available statistics (Pearson and likelihood chi-squares, Cramer's *V*, Goodman and Kruskal's gamma, and Kendall's τ_b).

`. tab2 x1 x2 x3 x4`

Performs all possible two-way cross-tabulations of the listed variables.

`. tabulate x1, summ(y)`

Produces a one-way table showing the mean, standard deviation, and frequency of *y* values within each category of *x1*.

`. tabulate x1 x2, summ(y) means`

Produces a two-way table showing the mean of *y* at each combination of *x1* and *x2* values.

`. sort x3`

`. by x3: tabulate x1 x2, exact`

Creates a three-way cross-tabulation, with subtables for *x1* (row) by *x2* (column) at each value of *x3*. Calculates Fisher's exact test for each subtable. **by *varname*:** works with many Stata commands, but must always be preceded by **sort *varname*** (unless the data are sorted already).

`. table y x1 x3, by(x4 x5) contents(freq)`

Creates a five-way cross-tabulation, of *y1* (row) by *x2* (column) by *x3* (super-column), by *x4* (superrow 1) by *x5* (superrow 2). Cells contain frequencies.

`. table x1 x2, contents(mean y1 sum y2)`

Creates a two-way table of *x1* (row) by *x2* (column). Cells contain the mean of *y1* and the sum of *y2*.

Summary Statistics for Measurement Variables

Dataset *vttown.dta* contains information from residents of a town in Vermont. This survey was conducted soon after routine state testing had detected trace amounts of toxic chemicals in the town's water supply. Higher concentrations were found in several private wells and near the town's public schools. Worried citizens held meetings to discuss possible solutions to this problem.

```
Contains data from c:\data\vttown.dta
  obs:         153                             VT town survey (Hamilton 85)
 vars:           7                             16 Dec 1996 09:36
 size:       1,683 (96.5% of memory free)
-------------------------------------------------------------------------------
  1. gender      byte   %8.0g     sexlbl     Respondent's gender
  2. lived       byte   %8.0g                Years lived in town
  3. kids        byte   %8.0g     kidlbl     Have children <19 in town?
  4. educ        byte   %8.0g                Highest year school completed
  5. meetings    byte   %8.0g     kidlbl     Attended 2 or more meetings
  6. contam      byte   %8.0g     contamlb   Believe own property/water cont
  7. school      byte   %8.0g     close      School closing opinion
-------------------------------------------------------------------------------
Sorted by:  gender
```

To find the mean and standard deviation of the variable *lived* (years the respondent had lived in town), type

```
. summarize lived

Variable |     Obs        Mean   Std. Dev.       Min        Max
---------+-----------------------------------------------------
   lived |     153    19.26797   16.95466          1         81
```

This table also provides the number of nonmissing observations and the variable's minimum and maximum values. If we had typed simply the following command with no variable list, we would obtain means and standard deviations for every numerical variable in the dataset:

```
. summarize
```

To see more detailed summary statistics,

```
. summarize lived, detail

                      Years lived in town
-------------------------------------------------------------
      Percentiles      Smallest
 1%            1              1
 5%            2              1
10%            3              1       Obs                 153
25%            5              1       Sum of Wgt.         153
50%           15                      Mean           19.26797
                        Largest       Std. Dev.      16.95466
75%           29             65
90%           42             65       Variance       287.4606
95%           55             68       Skewness       1.208804
99%           68             81       Kurtosis       4.025642
```

This **summarize, detail** output includes basic statistics plus the following:

Percentiles: Notably the first quartile (25th percentile), median (50th percentile) and third quartile (75th percentile). Because many samples do not divide evenly into quarters, or other standard divisions, these percentiles are approximations.

Four smallest and four largest values, where outliers might show up.

Sum of weights: Stata understands four types of weights: analytical weights (**aweight**), frequency weights (**fweight**), importance weights (**iweight**), and sampling weights (**pweight**). Different procedures allow, and make sense with, different kinds of weights. **summarize**, for example, permits only **aweight** or **fweight.** Type **help weights** for explanations.

Variance: Standard deviation squared (more properly, standard deviation equals the square root of variance).

Skewness: The direction and degree of asymmetry. A perfectly symmetrical distribution has skewness = 0. Positive skew (heavier right tail) results in skewness > 0; negative skew (heavier left tail) results in skewness < 0.

Kurtosis: Tail weight. A normal (Gaussian) distribution is symmetrical and has kurtosis = 3. If a symmetrical distribution has heavier-than-normal tails (that is, is sharply peaked), it will have kurtosis > 3. Kurtosis < 3 indicates lighter-than-normal tails.

The statistics produced by **summarize** describe the sample at hand. We might also want to make inferences about the population, for example by constructing a confidence interval for the mean of *lived*:

```
. ci lived, level(99)
Variable |      Obs        Mean     Std. Err.       [99% Conf. Interval]
---------+-------------------------------------------------------------------
   lived |      153    19.26797    1.370703        15.69241    22.84354
```

Based on this sample, we could be 99% confident that the population mean lies somewhere in the interval from 15.69 to 22.84 years. Here we used a **level( )** option to specify a 99% confidence interval. If we leave out this option, **ci** defaults to a 95% confidence interval. Other options allow **ci** to calculate exact confidence intervals for variables that have binomial or Poisson distributions. Type **help ci** for details.

Exploratory Data Analysis

Statistician John Tukey invented a toolkit of methods for analyzing data in an exploratory and skeptical way without making unneeded assumptions (see Tukey 1977; also Hoaglin, Mosteller, and Tukey 1983, 1985). Boxplots, introduced in Chapter 3, are one of Tukey's best-known innovations. Another is the stem-and-leaf display, in which initial digits form the "stems" and following digits for each observation make up the "leaves":

```
. stem lived
Stem-and-leaf plot for lived (Years lived in town)

  0* | 1111111222222333333344444444
  0. | 55555555555566666666777889999
  1* | 0000001122223333334
  1. | 55555567788899
  2* | 000000111112224444
  2. | 56778899
  3* | 00000124
  3. | 5555666789
  4* | 0012
  4. | 59
  5* | 00134
  5. | 556
  6* |
  6. | 5558
  7* |
  7. |
  8* | 1
```

stem automatically chose a double-stem version here, where 1* denotes first digits of 1, second digits of 0 through 4 (that is, respondents who had lived in town 10 through 14 years). `1.` denotes first digits of 1, second digits of 5 to 9 (15 through 19 years). We can control the number of lines per initial digit with the **lines( )** option. For example, a five-stem version (where the `1*` stem holds leaves of 0 through 1, `1t` holds leaves of 2 through 3, `1f` holds leaves of 4 through 5, `1s` holds leaves of 6 through 7, and `1.` holds leaves of 8 through 9) could be obtained by typing

```
. stem lived, lines(5)
```

Type **help stem** for information about other options.

Letter-value displays use order statistics to dissect a distribution:

```
. lv lived

#      153              Years lived in town
                  --------------------------------
M       77    |               15              |       spread  pseudosigma
F       39    |        5      17        29    |           24      17.9731
E       20    |        3      21        39    |           36     15.86391
D       10.5  |        2      27        52    |           50     16.62351
C        5.5  |        1   30.75      60.5    |         59.5     16.26523
B        3    |        1      33        65    |           64     15.15955
A        2    |        1    34.5        68    |           67     14.59762
Z        1.5  |        1   37.75      74.5    |         73.5     15.14113
         1    |        1      41        81    |           80     15.32737
              |                               |
              |                               |      # below      # above
inner fence   |      -31                65    |            0            5
outer fence   |      -67               101    |            0            0
```

M denotes the median, and *F* the "fourths" (quartiles, using a different approximation than **summarize, detail** does). *E, D, C, . . .* denote cutoff points such that roughly 1/8, 1/16, 1/32, . . . of the distribution remains outside in the tails. The second column of numbers gives the "depth," or distance from nearest extreme, for each letter value. Within the center box, the middle column gives "midsummaries," which are averages of the two letter values. If midsummaries drift away from the median, as they do for *lived*, that tells us that the distribution becomes progressively more skewed as we move farther out into the tails. The "spreads" are differences between pairs of letter values. The spread between *F*'s, 29–25=24, equals approximate interquartile range, for instance. Finally, the right-hand column of "pseudosigmas" estimates what the standard deviation should be if these letter values described a Gaussian population. The *F* pseudosigma, sometimes called a "pseudo standard deviation" (*PSD*), provides a simple check for approximate normality in symmetrical distributions:

1. Comparing mean with median diagnoses overall skew:

mean > median	positive skew
mean = median	symmetry
mean < median	negative skew

2. If mean and median indicate symmetry, comparing standard deviation with *PSD* evaluates tail normality (resemblance to a Gaussian curve):

standard deviation > *PSD*	heavier-than-normal tails
standard deviation = *PSD*	normal tails
standard deviation < *PSD*	lighter-than-normal tails

Let F_1 and F_3 denote 1st and 3rd fourths (approximate 25th and 75th percentiles). Then the interquartile range, *IQR*, equals $F_3 - F_1$, and $PSD = IQR/1.349$.

`lv` also identifies mild and severe outliers. We call an x value a "mild outlier" when it lies outside the inner fence, but not outside the outer fence:

$$F_1 - 3IQR \leq x < F_1 - 1.5IQR \quad \text{or} \quad F_3 + 1.5IQR < x \leq F_3 + 3IQR$$

x is a "severe outlier" if it lies outside the outer fence:

$$x < F_1 - 3IQR \quad \text{or} \quad x > F_3 + 3IQR$$

`lv` gives these cutoffs and the number of outliers of each type. Severe outliers, values beyond the outer fences, occur sparsely (about two per million) in normal populations. Monte Carlo simulations suggest that the presence of any severe outliers in samples of $n = 15$ to about 20,000 should be sufficient evidence to reject a normality hypothesis at $\alpha = .05$ (Hamilton 1992b). Furthermore, severe outliers create problems for many kinds of statistical techniques.

`summarize`, **`stem`**, and **`lv`** all reveal that *lived* has a positively skewed sample distribution, not at all resembling a theoretical normal curve. The next section introduces a more formal normality test and a transformation method that can reduce a variable's skew.

Normality Tests and Box-Cox Transformations

Many statistical procedures work best when applied to variables that follow normal (Gaussian) distributions. Chapter 3 presented several graphical ways to check whether sample distributions appear approximately normal, and the previous section considered exploratory numerical methods.

A skewness-kurtosis test, which uses the skewness and kurtosis statistics shown by **`summarize, detail`**, can more formally evaluate the null hypothesis that the sample came from a normally distributed population:

```
. sktest lived

                   Skewness/Kurtosis tests for Normality
                                                 ------- joint -------
  Variable |  Pr(Skewness)   Pr(Kurtosis)  adj chi-sq(2)  Pr(chi-sq)
-----------+------------------------------------------------------
     lived |       0.000         0.028            24.79         0.0000
```

sktest here rejects normality: *lived* appears significantly nonnormal in skewness ($P = .000$), kurtosis ($P = .028$), and in both statistics considered jointly ($P = .0000$). Stata rounds off displayed probabilities to three or four decimals; "$P = 0.0000$" really means $P < .00005$.

Nonlinear transformations such as square roots and logarithms are often employed to change distributions' shapes, with the aim of making skewed distributions more symmetrical and hence more nearly normal. Transformations might also help linearize relations between variables. Box-Cox transformation provides a formal method for accomplishing both goals. Let y be the dependent variable, which must have only positive values (add a suitable constant first if necessary to achieve this). $x1$, $x2$ and so forth represent any independent variables. Assuming normal errors (ε), we estimate a linear model:

$$y^{(\lambda)} = ß_0 + ß_1\, x1 + ß_2\, x2 + \ldots + \varepsilon$$

where $y^{(\lambda)}$ is a nonlinear transformation of y, from the general family:

$$y^{(\lambda)} = \{y^{\lambda} - 1\} / \lambda \qquad \lambda > 0 \text{ or } \lambda < 0$$

$$y^{(\lambda)} = \ln(y) \qquad \lambda = 0$$

Note that $\lambda = 1$ implies a linear transformation, or no change in shape.

Stata's **boxcox** procedure uses maximum-likelihood methods to estimate the optimal value for λ. In the one-variable case:

$$y^{(\lambda)} = ß_0 + \varepsilon$$

this amounts to estimating λ such that $y^{(\lambda)}$ most resembles a normal (Gaussian) distribution.

For example, to find the Box-Cox transformation that comes closest to normalizing variable *lived*:

```
. boxcox lived, nolog level(95) gen(newlive)
Transform:   (lived^L-1)/L

                  L       [95% Conf. Interval]     Log Likelihood
             --------------------------------------------------
              0.2136       0.0717      0.3698        -387.59998

Test:   L == -1      chi2(1) =   283.56    Pr>chi2 =   0.0000
        L ==  0      chi2(1) =     8.61    Pr>chi2 =   0.0033
        L ==  1      chi2(1) =    89.94    Pr>chi2 =   0.0000
```

The options **nolog level(95)** tell Stata not to show the iterations log (list of λ and log likelihoods at each step of the estimation process) but to include a 95% confidence interval for λ. **boxcox** confidence intervals are approximate at best, but we receive some encouragement here because the interval rejects the hypothesis that the best transformation is no transformation ($\lambda = 1$):

$$.0717 < \lambda < .3698$$

The **gen( )** option generates a new variable, here named *newlive*, holding transformed values of *lived:*

$$newlive = (lived^{.2136} - 1)/.2136$$

Figures 4.1 and 4.2 illustrate the effect of this Box-Cox transformation. Figure 4.1 shows the original distribution of *lived*, years lived in town, with its obvious positive skew.

```
. graph lived, bin(9) ylabel xlabel(0,20,40,60,80)
      xtick(10,30,50,70,90) xscale(0,90) t1(raw data)
```

Figure 4.2 shows the distribution of *newlive,* the new variable generated by **boxcox**. *newlive*'s distribution more closely resembles a theoretical normal curve. The resemblance is far from perfect, but **sktest** (not shown) indicates that with *newlive* we can no longer reject the normality hypothesis.

```
. graph newlive, bin(8) ylabel xlabel norm t1(transformed data)
```

boxcox offers advanced options for transforming distributions that are already symmetrical and for tracking or controlling the iterative fitting procedure. Consult **help boxcox** and the *Stata Reference Manual* for details.

Figure 4.1

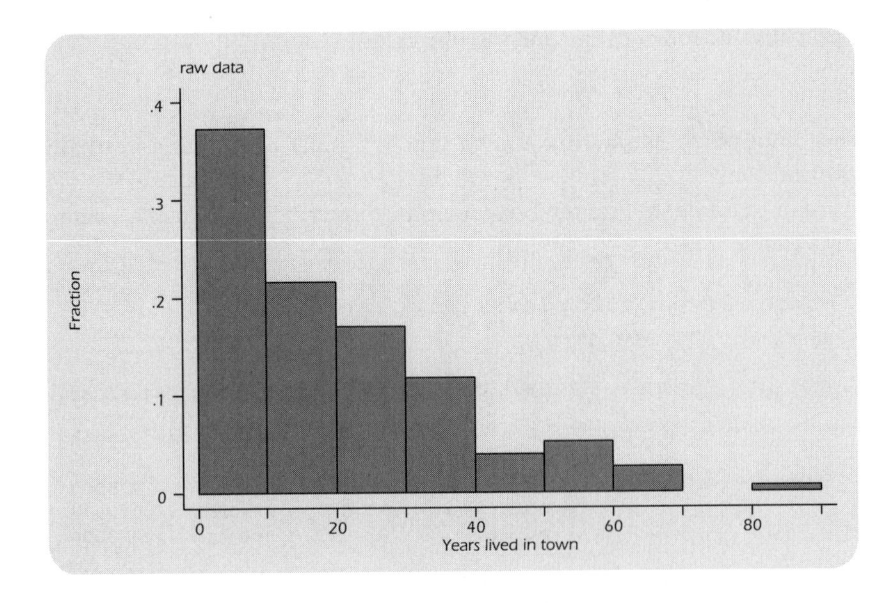

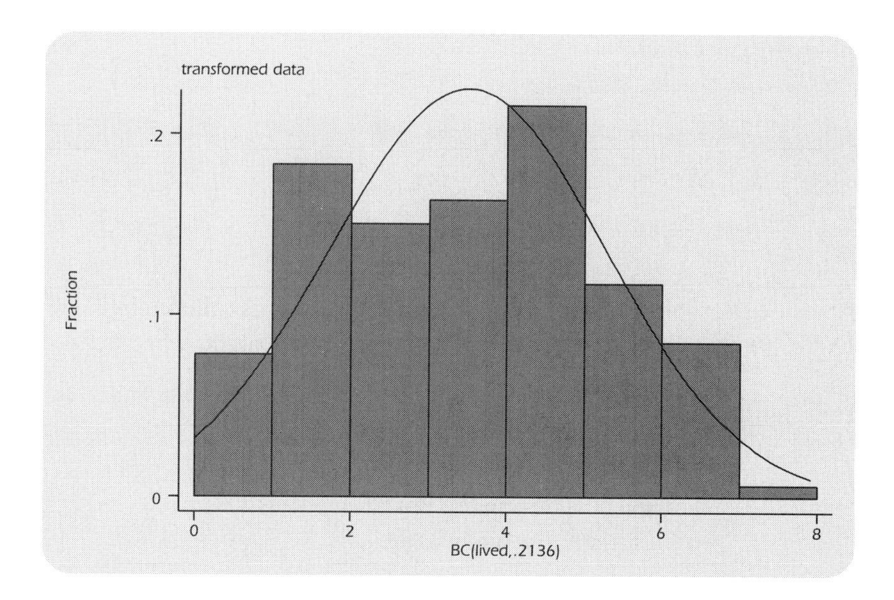

Figure 4.2

Frequency Tables and Two-Way Cross-Tabulations

The methods previously described apply to measurement variables. Categorical variables require other approaches, such as tabulation. To find the percentage of respondents attending two or more meetings concerning the pollution problem, **tabulate** the categorical variable *meetings:*

```
. tabulate meetings

Attended 2  |
or more     |
meetings    |      Freq.     Percent        Cum.
------------+-----------------------------------
         no |        106       69.28       69.28
        yes |         47       30.72      100.00
------------+-----------------------------------
      Total |        153      100.00
```

tabulate can produce frequency distributions for variables that have many values—as many as 500 with Small Stata or 3,000 with Intercooled Stata. To construct a more manageable frequency distribution table, however, you might first group values by applying **generate** with its **recode** or **autocode** options.

tabulate followed by two variable names creates a two-way cross-tabulation. For example, here is a cross-tabulation of *meetings* by *kids* (whether respondent has children younger than 19 living in town):

```
. tabulate meetings kids

Attended 2 | Have children <19 in town?
or more    |
meetings   |        no         yes |     Total
-----------+----------------------+----------
        no |        52          54 |       106
       yes |        11          36 |        47
-----------+----------------------+----------
     Total |        63          90 |       153
```

The first-named variable forms the rows, and the second forms columns in the resulting table. We see that only 11 of these 153 people were nonparents who attended the meetings.

tabulate has a number of options useful with frequency tables:

cell Shows total percentages for each cell.

chi2 Pearson χ^2 (chi-square) test of hypothesis that row and column variables are independent.

column Shows column percentages for each cell.

exact Fisher's exact test of the independence hypothesis. Superior to **chi2** if the table contains thin cells with low expected frequencies. Often too slow to be practical in large tables, however.

gamma Goodman and Kruskal's γ (gamma), with its asymptotic standard error (ASE). Measures association between ordinal variables, based on the number of concordant and discordant pairs (ignoring ties). $-1 \leq \gamma \leq 1$.

generate(*new*) Creates a set of dummy variables named *new1, new2,* and so on to represent the values of the tabulated variable.

lrchi2 Likelihood-ratio χ^2 (chi-square) test of independence hypothesis. Not obtainable if table contains any empty cells.

missing Includes "missing" as one row and/or column of the table.

nolabel Shows numerical values rather than value labels of labeled numeric variables.

row Shows row percentages for each cell.

nofreq Does not show cell frequencies.

taub Kendall's τ_b (tau-b), with its ASE. Measures association between ordinal variables, similar to **gamma** but uses a correction for ties. $-1 \leq \tau_b \leq 1$.

`V` — Cramers *V*; measures association for nominal variables. In 2×2 tables, $1 \le V \le 1$. In larger tables, $0 \le V \le 1$.

`all` — Equivalent to the options **`chi2 lrchi2 gamma taub V`**. Note that not all of these options might be equally appropriate for a given table. **`gamma`** and **`taub`** assume that both variables have ordered categories, whereas **`chi2`**, **`lrchi2`**, and **`V`** do not.

These same options work (where appropriate) with the related **`tabi`**, **`tab1`**, and **`tab2`** commands, discussed later. For example, to get column percentages (because the column variable, *kids*, is the independent variable in this analysis) and a χ^2 test:

```
. tabulate meetings kids, column chi2

Attended 2 | Have children <19 in town?
or more    |
meetings   |         no        yes |     Total
-----------+----------------------+ ---------
        no |         52         54 |       106
           |      82.54      60.00 |     69.28
-----------+----------------------+ ---------
       yes |         11         36 |        47
           |      17.46      40.00 |     30.72
-----------+----------------------+ ---------
     Total |         63         90 |       153
           |     100.00     100.00 |    100.00

         Pearson chi2(1) =    8.8464   Pr = 0.003
```

Forty percent of the respondents with children attended meetings, compared with only 17.46% of the respondents without children. This association is statistically significant ($P = .003$). Occasionally we might analyze a cross-tabulation without retrieving the original raw data. A special command, **`tabi`** ("immediate" tabulation), accomplishes this. Type the cell frequencies on the command line, with table rows separated by " \ ". For illustration, here is how **`tabi`** could reproduce the previous chi square analysis, given only the four cell frequencies:

```
. tabi 52 54 \ 11 36, column chi2

           | col
       row|          1          2 |     Total
-----------+----------------------+ ---------
         1 |         52         54 |       106
           |      82.54      60.00 |     69.28
-----------+----------------------+ ---------
         2 |         11         36 |        47
           |      17.46      40.00 |     30.72
-----------+----------------------+ ---------
     Total|          63         90 |       153
           |     100.00     100.00 |    100.00

          Pearson chi2(1) =    8.8464   Pr = 0.003
```

Unlike **tabulate**, **tabi** does not require or refer to any data in memory. By adding the **replace** option, however, we can ask **tabi** to replace whatever data are in memory with the new cross-tabulation.

Multiple Tables and Multi-Way Cross-Tabulations

With surveys and other large datasets, we might need frequency distributions for many different variables. Instead of asking for each table separately, for example by typing **tabulate *meetings***, then **tabulate *gender*,** and then **tabulate *kids*,** we could simply use another specialized command, **tab1**:

```
. tab1 meetings gender kids
```

Or, to produce one-way frequency tables for each variable in this dataset (the maximum is 30 at one time):

```
. tab1 gender-school
```

Similarly, **tab2** creates multiple two-way tables. For example, the following command cross-tabulates every two-way combination of the listed variables:

```
. tab2 meetings gender kids
```

Returning to the ordinary **tabulate** command: with a **by** prefix, we can use this to form multi-way contingency tables. Here is a three-way cross-tabulation of *meetings* by *kids* by *contam* (respondent believes his or her own property or water contaminated):

```
. sort contam

. by contam: tabulate meetings kids, column

-> contam=        no
Attended 2 | Have children <19 in town?
or more    |
meetings   |        no        yes |     Total
-----------+----------------------+----------
        no |        42         44 |        86
           |     91.30      68.75 |     78.18
-----------+----------------------+----------
       yes |         4         20 |        24
           |      8.70      31.25 |     21.82
-----------+----------------------+----------
     Total |        46         64 |       110
           |    100.00     100.00 |    100.00
```

```
-> contam=      yes
Attended 2 | Have children <19 in town?
or more    |
meetings   |        no        yes |     Total
-----------+----------------------+----------
        no |        10         10 |        20
           |     58.82      38.46 |     46.51
-----------+----------------------+----------
       yes |         7         16 |        23
           |     41.18      61.54 |     53.49
-----------+----------------------+----------
     Total |        17         26 |        43
           |    100.00     100.00 |    100.00
```

The "parenthood effect" shows up regardless of whether respondents believed their own property had been affected.

This approach can be extended to tabulations of greater complexity. For example, to get a four-way cross-tabulation of *gender* by *contam* by *meetings* by *kids* (results not shown):

```
. sort gender contam

. by gender contam: tabulate meetings kids, column
```

An easier way to produce multi-way tables, if we do not need percentages or statistical tests, is through Stata's general table-making command, **table**. This versatile command has many options, only a few of which are illustrated here. To construct a simple frequency table of *meetings:*

```
. table meetings, contents(freq)

----------+-----------
Attended  |
2 or more |
meetings  |      Freq.
----------+-----------
       no |        106
      yes |         47
----------+-----------
```

For a two-way frequency table or cross-tabulation:

```
. table meetings kids, contents(freq)

----------+-----------
          |   Have
          | children
Attended  |  <19 in
2 or more |  town?
meetings  |  no   yes
----------+-----------
       no |  52    54
      yes |  11    36
----------+-----------
```

If we specify a third categorical variable, it forms the "supercolumns" of a three-way table:

```
. table meetings kids contam, contents(freq)

----------+-------------------------
          |       Believe own
          | property/water cont and
          |   Have children <19 in
Attended  |          town?
2 or more | -- no --      --yes ---
meetings  |  no    yes       no   yes
----------+-------------------------
       no |  42     44       10    10
      yes |   4     20        7    16
----------+-------------------------
```

More complicated tables require the **by( )** option, which allows as many as four "superrow" variables. **table** thus can produce up to seven-way tables: one row, one column, one supercolumn, and four superrows. Here is a four-way example:

```
. table meetings kids contam, contents(freq) by(gender)

-----------+ -----------------------
Respondent's|       Believe own
gender      | property/water cont and
and         |   Have children <19 in
Attended    |          town?
2 or more   |-- no --      --yes --
meetings    |  no   yes       no   yes
-----------+ -----------------------
male        |
         no |  18    18        3     3
        yes |   2     7        3     6
-----------+ -----------------------
female      |
         no |  24    26        7     7
        yes |   2    13        4    10
-----------+ -----------------------
```

The **contents( )** option of **table** specifies what statistics the table's cells contain:

contents(freq)	Frequency
contents(mean *varname*)	Mean of *varname*
contents(sd *varname*)	Standard deviation of *varname*
contents(sum *varname*)	Sum of *varname*
contents(rawsum *varname*)	Sums ignoring optionally specified weight
contents(count *varname*)	Count of nonmissing observations of *varname*

contents(n *varname*)	Same as **count**
contents(max *varname*)	Maximum of *varname*
contents(min *varname*)	Minimum of *varname*
contents(median *varname*)	Median of *varname*
contents(iqr *varname*)	Interquartile range (IQR) of *varname*
contents(p1 *varname*)	1st percentile of *varname*
contents(p2 *varname*)	2nd percentile of *varname* (so forth to **p99**)

The next section illustrates several more of these options.

Model fitting and hypothesis testing in multi-way cross-tabulations require more advanced methods such as logit regression, described in Chapter 10.

Tables of Means

tabulate readily produces tables of means and standard deviations within categories of the tabulated variable. For example, to form a one-way table with means of *lived* within each category of *meetings:*

```
. tabulate meetings, summ(lived)
Attended 2  |      Summary of Years lived in town
or more     |
meetings    |        Mean   Std. Dev.       Freq.
------------+------------------------------------
         no |   21.509434   17.743809         106
        yes |   14.212766   13.911109          47
------------+------------------------------------
      Total |   19.267974   16.954663         153
```

Meetings attenders appear to be relative newcomers, averaging 14.2 years in town compared with 21.5 years among those who did not attend.

We can also use **tabulate** to form a two-way table of means:

```
. tabulate meetings kids, summ(lived) means
                   Means of Years lived in town

Attended 2 | Have children <19 in town?
or more    |
meetings   |        no        yes |     Total
-----------+----------------------+----------
        no | 28.307692  14.962963 | 21.509434
       yes | 23.363636  11.416667 | 14.212766
-----------+----------------------+----------
     Total | 27.444444  13.544444 | 19.267974
```

Both parents and nonparents among the meeting attenders tend to have lived fewer years in town, so the newcomer/oldtimer division noticed in the previous table is not a spurious result of parenthood.

The **means** option called for a tabulation containing only means. Otherwise we get a bulkier table including means, standard deviations, and frequencies in each cell. Chapter 5 describes statistical tests for hypotheses about subgroup means.

Although it performs no tests, **table** nicely constructs up to seven-way tables containing means, standard deviations, sums, medians, or other statistics (see the option list in previous section). Here is a one-way table showing means of *lived*, within categories of *meetings*:

```
. table meetings, contents(mean lived)

---------+------------
Attended |
2 or more |
meetings | mean(lived)
---------+------------
      no |   21.50943
     yes |   14.21277
---------+------------
```

A two-way table of means is a straightforward extension:

```
. table meetings kids, contents(mean lived)

---------+-------------------
Attended | Have children <19
2 or more |      in town?
meetings |       no        yes
---------+-------------------
      no | 28.30769  14.96296
     yes | 23.36364  11.41667
---------+-------------------
```

Table cells can contain more than one statistic. Suppose we wanted a two-way table with both means and medians of the variable *lived:*

```
. table meetings kids, contents(mean lived median lived)

---------+-------------------
Attended | Have children <19
2 or more |      in town?
meetings |       no        yes
---------+-------------------
      no | 28.30769  14.96296
         |     27.5      12.5
         |
     yes | 23.36364  11.41667
         |       21         6
---------+-------------------
```

The cell contents could be means, medians, sums, and so on for two or more different variables.

Using Frequency Weights

summarize, **tabulate**, **table**, and related commands can be used with frequency weights that indicate the number of replicated observations. For example, file *sextab2.dta* contains results from a British survey of sexual behavior (Johnson et al., 1992). It apparently has 48 observations:

```
. use sextab2, clear

. describe
Contains data from c:\data\sextab2.dta
  obs:            48                          British sex survey (Johnson 92)
 vars:             4                          16 Dec 1996 10:06
 size:           432 (95.8% of memory free)
-------------------------------------------------------------------------------
   1. age          byte   %8.0g     age        Age
   2. gender       byte   %8.0g     gender     Gender
   3. lifepart     byte   %8.0g     partners   # heterosex partners lifetime
   4. count        int    %8.0g                Number of individuals
-------------------------------------------------------------------------------
Sorted by:  age  lifepart  gender
```

One variable, *count*, indicates the number of individuals with each combination of characteristics, so this small dataset actually contains information from more than 18,000 respondents. For example, 405 respondents were male, ages 16 to 24, and reported having no heterosexual partners so far in their lives:

```
. list in 1/5
          age    gender   lifepart      count
  1.    16-24      male       none        405
  2.    16-24    female       none        465
  3.    16-24      male        one        323
  4.    16-24    female        one        606
  5.    16-24      male        two        194
```

We use *count* as a frequency weight to create a cross-tabulation of *lifepart* by *gender:*

```
. tabulate lifepart gender [fweight = count]
# heterosex| Gender
partners   |
lifetime   |       male     female |      Total
-----------+----------------------+----------
      none |        544        586 |       1130
       one |       1734       4146 |       5880
       two |        887       1777 |       2664
       3-4 |       1542       1908 |       3450
       5-9 |       1630       1364 |       2994
       10+ |       2048        708 |       2756
-----------+----------------------+----------
     Total |       8385      10489 |      18874
```

The usual **tabulate** options work as expected with frequency weights. Here is the same table showing column percentages instead of frequencies:

```
. tabulate lifepart gender [fweight = count], column nof

# heterosex| Gender
partners   |
lifetime   |       male     female |      Total
-----------+----------------------+----------
      none |       6.49       5.59 |       5.99
       one |      20.68      39.53 |      31.15
       two |      10.58      16.94 |      14.11
       3-4 |      18.39      18.19 |      18.28
       5-9 |      19.44      13.00 |      15.86
       10+ |      24.42       6.75 |      14.60
-----------+----------------------+----------
     Total |     100.00     100.00 |     100.00
```

Other types of weights such as probability or analytical weights do not work with **tabulate**, because their meanings are unclear regarding its principal options.

A different application of frequency weights can be demonstrated with **summarize**. File *college1.dta* contains information on a random sample consisting of 11 U.S. colleges, from *Barron's Compact Guide to Colleges* (1992).

```
. use c:\data\college1
(Colleges sample 1 (Barron's 92))

. describe
Contains data from c:\data\college1.dta
  obs:            11                      Colleges sample 1 (Barron's 92)
 vars:             5                      16 Dec 1996 10:08
 size:           429 (95.8% of memory free)
-------------------------------------------------------------------------------
  1. school         str28  %28s           College or university
  2. enroll         int    %8.0g          Full-time students 1991
  3. pctmale        byte   %8.0g          Percent male 1991
  4. msat           int    %8.0g          Average math SAT
  5. vsat           int    %8.0g          Average verbal SAT
-------------------------------------------------------------------------------
Sorted by:
```

The variables include *msat,* the mean math Scholastic Aptitude Test score at each of the 11 schools:

```
. list school enroll msat

                             school     enroll       msat
  1.               Brown University       5550        680
  2.                    U. Scranton       3821        554
  3.  U. North Carolina/Asheville         2035        540
  4.              Claremont College        849        660
  5.              DePaul University       6197        547
  6.          Thomas Aquinas College        201        570
  7.               Davidson College       1543        640
```

```
  8.        U. Michigan/Dearborn      3541       485
  9.        Mass. College of Art       961       482
 10.             Oberlin College      2765       640
 11.         American University      5228       587
```

We can easily find the mean *msat* value among these 11 schools:

```
. summarize msat

Variable |     Obs        Mean   Std. Dev.       Min        Max
---------+-----------------------------------------------------
    msat |      11    580.4545   67.63189        482        680
```

This summary table gives each school's mean math SAT score the same weight. DePaul University, however, has 30 times as many students as Thomas Aquinas College. To take the different enrollments into account we could weight by *enroll:*

```
. summarize msat [fweight = enroll]

Variable |     Obs        Mean   Std. Dev.       Min        Max
---------+-----------------------------------------------------
    msat |   32691     583.064   63.10665        482        680
```

The enrollment-weighted mean, unlike the unweighted mean, is equivalent to the mean for the 32,691 students at these colleges (assuming they all took the SAT). Note, however, that we could not say the same thing about the standard deviation, minimum, or maximum: Apart from the mean, most individual-level statistics cannot be calculated simply by weighting data that already are aggregated. Thus we need to use weights with caution; they might make sense in the context of one particular analysis, but seldom do so for the dataset as a whole when many different kinds of analysis are needed.

5

ANOVA and Other Comparison Methods

Analysis of variance (ANOVA) encompasses a set of methods for testing hypotheses about differences between means. Its applications range from simple analyses where we compare the means of *y* across categories of *x*, to more complicated situations with two or more *x* variables. *t* tests for hypotheses regarding a single mean (one-sample) or a pair of means (two-sample) correspond to elementary forms of ANOVA.

Rank-based "nonparametric" tests, including sign, Mann-Whitney, and Kruskal-Wallis, take a different approach to comparing distributions. These tests make weaker assumptions about measurement, distribution shape, and spread, and so remain valid under a wider range of conditions than ANOVA and its "parametric" relatives.

Careful analysts sometimes use parametric and nonparametric tests together, checking to see whether both point toward the same conclusions. Further troubleshooting is called for when the parametric and nonparametric tests disagree.

Example Commands

`. ttest y = 23.4`

Performs a one-sample *t* test of the null hypothesis that the population mean of *y* equals 23.4.

`. ttest y1 = y2`

Performs a *t* test of the null hypothesis that the population mean of *y1* equals that of *y2*. The default form of this command assumes data are paired. With unpaired data (*y1* and *y2* are measured from two independent sets of observations), add the option **`unpaired`**.

`. ttest` ***`y`*** **`, by(`*****`x`*****`) unequal`**

Performs a two-sample *t* test of the null hypothesis that the population mean of *y* is the same for both categories of dichotomous variable *x*. Does not assume that the populations have equal variances. (Without the **`unequal`** option, **`ttest`** does assume equal variances.)

`. oneway` ***`y x`***

Performs a one-way analysis of variance (ANOVA), testing for differences among the means of *y* across categories of *x*. The same analysis, with a different output table, is produced by **`anova`** ***`y x`***.

`. oneway` ***`y x`*** **`, tabulate scheffe`**

Performs one-way ANOVA, including a table of sample means and Scheffé multiple-comparison tests in the output.

`. anova` ***`y x1 x2`***

Performs two-way ANOVA, testing for differences among the means of *y* across categories of *x1* and *x2*.

`. anova` ***`y x1 x2 x1*x2`***

Performs a two-way factorial ANOVA, including both the main and interaction effects of categorical variables *x1* and *x2*.

`. anova` ***`y x1 x2 x3 x4`*** **`, continuous(`*****`x3 x4`*****`) regress`**

Performs analysis of covariance (ANOCOVA) with four independent variables, two of them (*x1* and *x2*) categorical and two of them (*x3* and *x4*) measurements. Shows the results in the form of a regression table instead of the default ANOVA table.

`. serrbar` ***`ybar se x`*** **`, scale(2)`**

Constructs a standard-error-bar plot from a dataset of means. Variable *ybar* holds the group means of *y*, *se* the standard errors, and *x* the values of categorical variable *x*. The dataset contains one "observation" for each value of *x*. **`scale(2)`** asks for bars extending to ±2 standard errors around each mean (default is ±1 standard error).

`. signrank` ***`y1`*** **`=`** ***`y2`***

Performs a Wilcoxon matched-pairs signed-rank test for the equality of the rank distributions of *y1* and *y2*. We could test whether the median of *y1* differs from a constant such as 23.4 by typing the command **`signrank`** ***`y1`*** **`= 23.4`**.

. signtest *y1* **=** *y2*

Tests the equality of the medians of *y1* and *y2* (assuming matched data, that is, both variables measured on the same sample of observations). Typing **signtest** *y1* **= 23.4** would perform a sign test of the null hypothesis that the median of *y1* equals 23.4.

. ranksum *y*, **by(***x***)**

Performs a Wilcoxon rank-sum test (also known as a Mann-Whitney *U* test) of the null hypothesis that *y* has identical rank distributions for both categories of dichotomous variable *x*. If we can assume that both rank distributions possess the same shape, this amounts to a test for whether the two medians of *y* are equal.

. kwallis *y*, **by(***x***)**

Performs a Kruskal-Wallis test of the null hypothesis that *y* has identical rank distributions across the *k* categories of *x* ($k > 2$).

One-Sample Tests

One-sample *t* tests have two seemingly different applications:

1. Testing whether a sample mean $\bar{y}$ differs significantly from an hypothesized value μ_0.
2. Testing whether the means of y_1 and y_2, two variables measured over the same set of observations, differ significantly from each other. This is equivalent to testing whether the mean of a "difference score" variable created by subtracting y_1 from y_2 equals zero.

We use essentially the same formulas for either application, although the second starts with information on two variables instead of one.

The data in *writing.dta* were collected to evaluate a college writing course that employed microcomputers for word processing (Nash and Schwartz, 1987). Measures such as the number of sentences completed in timed writing were collected both before and after students took the course. The researchers wanted to know whether the postcourse measures showed improvement.

```
Contains data from c:\data\writing.dta
  obs:            24                          Nash and Schwartz (1987)
 vars:             9                          17 Dec 1996 08:59
 size:           312 (95.5% of memory free)
-------------------------------------------------------------------------------
   1. id          byte   %8.0g       slbl     Student ID
   2. preS        byte   %8.0g                # of sentences (pre-test)
   3. preP        byte   %8.0g                # of paragraphs (pre-test)
   4. preC        byte   %8.0g                Coherence scale 0-2 (pre-test)
   5. preE        byte   %8.0g                Evidence scale 0-6 (pre-test)
   6. postS       byte   %8.0g                # of sentences (post-test)
   7. postP       byte   %8.0g                # of paragraphs (post-test)
   8. postC       byte   %8.0g                Coherence scale 0-2 (post-test)
   9. postE       byte   %8.0g                Evidence scale 0-6 (post-test)
-------------------------------------------------------------------------------
Sorted by:
```

Suppose we knew that students in previous years were able to complete an average of 10 sentences. Before examining whether the students in *writing.dta* improved during the course, we might want to learn whether at the start of the course they were essentially like earlier students—that is, whether their pre-test (*preS*) mean differs significantly from the mean of previous students (10). To perform a one-sample *t* test of $H_0{:}\mu = 10$:

```
. ttest preS = 10

One-sample t test                                  Number of obs =       24

------------------------------------------------------------------------------
Variable |       Mean    Std. Err.        t      P>|t|     [95% Conf. Interval]
---------+--------------------------------------------------------------------
    preS |   10.79167    .9402034     11.478   0.0000     8.846708    12.73663
------------------------------------------------------------------------------
Degrees of freedom: 23

                          Ho: mean(preS) = 10

     Ha: mean < 10           Ha: mean ~= 10            Ha: mean > 10
       t =   0.8420           t =   0.8420              t =   0.8420
   P < t =   0.7958       P > |t| =   0.4084          P > t =   0.2042
```

The notation `Prob < t` means "the probability of a lower value of *t*"—that is, the one-tail test probability. The two-tail probability of a greater absolute *t* appears as `P > |t|` = .4084. Because this probability is high, we have no reason to reject $H_0{:}\mu = 10$. Note also that **ttest** automatically provides a 95% confidence interval for the mean. We could get a different confidence interval, such as 99%, by adding a **level(99)** option to this command.

A nonparametric counterpart, the sign test, employs the binomial distribution to test hypotheses about single medians. For example, we could test whether the median of *preS* equals 10. **signtest** gives us no reason to reject that null hypothesis either:

```
. signtest preS = 10
Sign test

    sign |    observed    expected
---------+------------------------
positive |          12          11
negative |          10          11
    zero |           2           2
---------+------------------------
     all |          24          24

One-sided tests:
Ho: median of preS - 10 = 0 vs. Ha: median of preS - 10 > 0
    Pr(#positive >= 12)
    = Binomial(n = 22, x >= 12, p = 0.5) =  0.4159
Ho: median of preS - 10 = 0 vs. Ha: median of preS - 10 < 0
    Pr(#negative >= 10)
    = Binomial(n = 22, x >= 10, p = 0.5) =  0.7383

Two-sided test:
Ho: median of preS - 10 = 0 vs. Ha: median of preS - 10 ~= 0
    Pr(#positive >= 12 or #negative >= 12)
    = min(1, 2*Binomial(n = 22, x >= 12, p = 0.5)) =  0.8318
```

signtest includes right-tail, left-tail, and two-tail probabilities. (Unlike the symmetrical *t* distributions used by **ttest**, the binomial distributions used by **signtest** have different left and right-tail probabilities.) In this example only the two-tail probability matters, because we were testing whether the *writing.dta* students "differ" from their predecessors.

Next, we can test for improvement during the course by testing the null hypothesis that the mean number of sentences completed before and after the course (that is, the means of *preS* and *postS*) are equal. The **ttest** command accomplishes this as well, finding a significant improvement:

```
. ttest postS = preS
Paired t test                                    Number of obs =        24

------------------------------------------------------------------------------
Variable |      Mean    Std. Err.        t      P>|t|     [95% Conf. Interval]
---------+--------------------------------------------------------------------
   postS |    26.375    1.693779    15.5717    0.0000     22.87115    29.87885
    preS |  10.79167    .9402034     11.478    0.0000     8.846708    12.73663
---------+--------------------------------------------------------------------
    diff |  15.58333    1.383019    11.2676    0.0000     12.72234    18.44433
------------------------------------------------------------------------------
Degrees of freedom: 23

                          Ho: mean diff = 0

    Ha: diff < 0            Ha: diff ~= 0             Ha: diff > 0
       t = 11.268              t = 11.268               t = 11.268
   P < t = 1.0000        P > |t| = 0.0000           P > t = 0.0000
```

Because we expect "improvement," not just "difference" between the *preS* and *postS* means, a one-tail test is appropriate. The displayed two-tail probability rounds off four decimal places to zero ("$P = .0000$" really means $P < .00005$). Students' mean sentence completion does significantly improve. Based on this sample, we are 95% confident that it improves by between 12.7 and 18.4 sentences.

t tests assume that variables follow a normal distribution. This assumption is usually not critical because the tests are moderately robust. When nonnormality involves severe outliers, however, or occurs in small samples, we might be safer turning to medians instead of means and employing a nonparametric test that does not assume normality. The Wilcoxon signed-rank test, for example, assumes only that the distributions are symmetrical and continuous. Applying a signed-rank test to these data yields essentially the same conclusion as **ttest**, that students' sentence completion significantly improved. Because both tests agree on this conclusion, we can assert it with more assurance.

```
. signrank postS = preS
Wilcoxon signed-rank test

    sign |      obs   sum ranks    expected
---------+---------------------------------
positive |       24         300         150
negative |        0           0         150
    zero |        0           0           0
---------+---------------------------------
     all |       24         300         300

unadjusted variance      1225.00
adjustment for ties        -1.62
adjustment for zeros        0.00
                        --------
adjusted variance        1223.38

Ho: postS = preS
              z =   4.289
    Prob > |z| =   0.0000
```

Two-Sample Tests

The remainder of this chapter draws examples from a survey of college undergraduates from Ward and Ault (1990) (*student1.dta*):

```
Contains data from c:\data\student1.dta
  obs:          243                           Student survey (Ward 1990)
 vars:           19                           20 Apr 1997 14:07
 size:        6,561 (95.3% of memory free)
-------------------------------------------------------------------------------
   1. id           int    %8.0g               Student ID
   2. year         byte   %8.0g       v1      Year in college
   3. age          byte   %8.0g               Age at last birthday
```

```
  4. gender      byte   %9.0g        s         Gender (male)
  5. major       byte   %8.0g                  Student major
  6. relig       byte   %8.0g        v4        Religious preference
  7. drink       byte   %9.0g                  33-point drinking scale
  8. gpa         float  %9.0g                  Grade Point Average
  9. grades      byte   %8.0g        grades    Guessed grades this semester
 10. belong      byte   %8.0g        belong    Belong to fraternity/sorority
 11. live        byte   %8.0g        v10       Where do you live?
 12. miles       byte   %8.0g                  How many miles from campus?
 13. study       byte   %8.0g                  Avg. hours/week studying
 14. athlete     byte   %8.0g        yes       Are you a varsity athlete?
 15. employed    byte   %8.0g        yes       Are you employed
 16. allnight    byte   %8.0g        allnight  How often study all night?
 17. ditch       byte   %8.0g        times     How many class/month ditched?
 18. hsdrink     byte   %9.0g                  High school drinking scale
 19. aggress     byte   %9.0g                  Aggressive behavior scale
-------------------------------------------------------------------------------
Sorted by:  year
```

About 19% of these students belong to a fraternity or sorority:

```
. tabulate belong

Belong to   |
fraternity/ |
sorority    |      Freq.     Percent        Cum.
------------+-----------------------------------
     member |         47       19.34       19.34
   nonmembe |        196       80.66      100.00
------------+-----------------------------------
      Total |        243      100.00
```

Another variable, *drink*, measures how often and heavily a student drinks alcohol, on a 33-point scale. Campus rumors might lead one to suspect that fraternity and sorority members tend to differ from other students in their drinking behavior. Boxplots comparing the *drink* values reported by fraternity and sorority members and nonmembers (Figure 5.1) appear consistent with these rumors.

The **ttest** command, used earlier for one-sample and paired-difference tests, can perform two-sample tests as well. In this application its general syntax is **ttest** ***measurement*, by(*categorical*)**. For example,

```
. ttest drink, by(belong)

Two-sample t test with equal variances         member: Number of obs =       47
                                             nonmembe: Number of obs =      196

------------------------------------------------------------------------------
Variable |      Mean    Std. Err.         t     P>|t|     [95% Conf. Interval]
---------+--------------------------------------------------------------------
  member |   24.7234    .7124518    34.7019    0.0000     23.28931     26.1575
nonmembe |   17.7602    .4575013      38.82    0.0000     16.85792    18.66249
---------+--------------------------------------------------------------------
    diff |    6.9632    .9978608    6.97813    0.0000     4.997558    8.928842
------------------------------------------------------------------------------
Degrees of freedom: 241
```

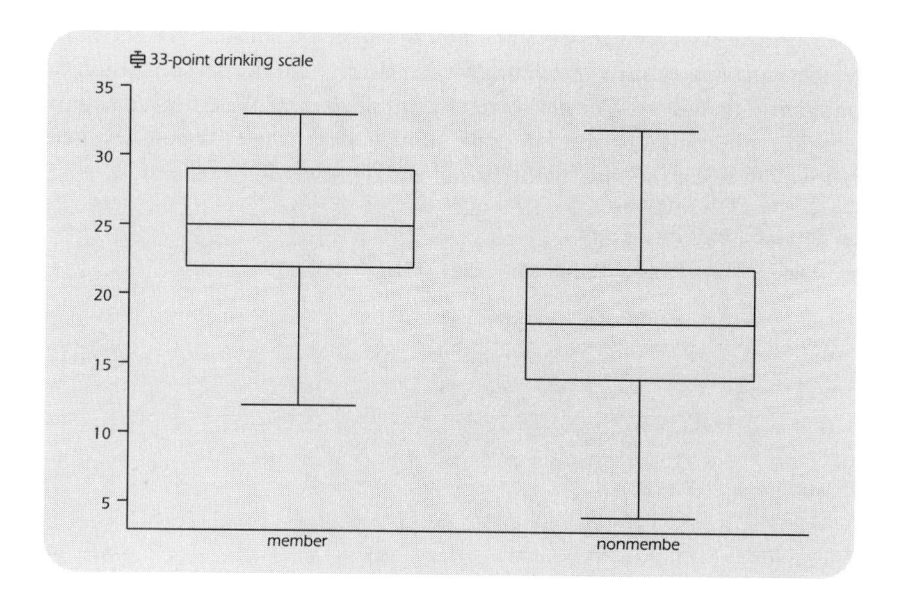

Figure 5.1

```
           Ho: mean(member) - mean(nonmembe) = diff = 0

     Ha: diff < 0            Ha: diff ~= 0             Ha: diff > 0
       t =   6.9781            t =   6.9781              t =   6.9781
   P < t =   1.0000      P > |t| =   0.0000          P > t =   0.0000
```

As the output notes, this *t* test rests on an equal-variances assumption. But the fraternity and sorority members' sample standard deviation appears somewhat lower—they are more alike than nonmembers in their reported drinking behavior. To perform a similar test but without assuming equal variances, add the option **unequal**:

```
. ttest drink, by(belong) unequal
Two-sample t test with unequal variances      member: Number of obs =     47
                                            nonmembe: Number of obs =    196

------------------------------------------------------------------------------
Variable |      Mean    Std. Err.        t     P>|t|     [95% Conf. Interval]
---------+--------------------------------------------------------------------
  member |   24.7234    .7124518   34.7019    0.0000     23.28931     26.1575
nonmembe |   17.7602    .4575013     38.82    0.0000     16.85792    18.66249
---------+--------------------------------------------------------------------
    diff |    6.9632    .8466965   8.22396    0.0000     5.280627    8.645773
------------------------------------------------------------------------------
Satterthwaite's degrees of freedom: 88.219993

           Ho: mean(member) - mean(nonmembe) = diff = 0

     Ha: diff < 0            Ha: diff ~= 0             Ha: diff > 0
       t =   8.2240            t =   8.2240              t =   8.2240
   P < t =   1.0000      P > |t| =   0.0000          P > t =   0.0000
```

Adjusting for unequal variances does not alter our basic conclusion that members and nonmembers are significantly different. We can further check this conclusion by trying a nonparametric Mann-Whitney *U* test, also known as a Wilcoxon rank-sum test. Assuming that the rank distributions have similar shape, the rank-sum test here indicates that we can reject the null hypothesis of equal population medians:

```
. ranksum drink, by(belong)

Two-sample Wilcoxon rank-sum (Mann-Whitney) test

  belong |      obs    rank sum    expected
 --------+---------------------------------
  member |       47        8535        5734
nonmembe |      196       21111       23912
 --------+---------------------------------
combined |      243       29646       29646

unadjusted variance    187310.67
adjustment for ties      -472.30
                       ----------
adjusted variance      186838.36

Ho: drink(belong==member) = drink(belong==nonmembe)
              z =   6.480
    Prob > |z| =   0.0000
```

One-Way Analysis of Variance (ANOVA)

Analysis of variance (ANOVA) provides another way, more general than *t* tests, to test for differences among means. The simplest type, one-way ANOVA, tests whether the means of *y* differ across categories of *x*. For example,

```
. oneway drink belong, tabulate

Belong to   |  Summary of 33-point drinking scale
fraternity/ |
sorority    |        Mean   Std. Dev.       Freq.
 -----------+------------------------------------
     member |   24.723404   4.8843233          47
   nonmembe |   17.760204   6.4050179         196
 -----------+------------------------------------
      Total |   19.106996   6.7221166         243

                        Analysis of Variance
    Source              SS         df      MS            F     Prob > F
-----------------------------------------------------------------------
Between groups      1838.08426      1   1838.08426     48.69     0.0000
 Within groups      9097.13385    241   37.7474433
-----------------------------------------------------------------------
    Total           10935.2181    242   45.1868517

Bartlett's test for equal variances:  chi2(1) = 4.8378  Prob>chi2 = 0.028
```

The **tabulate** option produces a table of means and standard deviations in addition to the analysis of variance table itself. One-way ANOVA with a dichotomous *x* variable is mathematically equivalent to a two-sample *t* test, and its *F* statistic equals the corresponding *t* statistic squared. **oneway** offers more options and processes faster, but it lacks an **unequal** option for abandoning the equal-variances assumption.

oneway formally tests the equal-variances assumption, using Bartlett's χ^2. A low Bartlett's probability implies that ANOVA's equal-variance assumption is implausible, in which case we should not trust the ANOVA *F* test results. In the previous **oneway** ***drink belong*** example, Bartlett's $P = .028$ casts doubt on the ANOVA finding of a significant difference between means.

ANOVA's real value lies not in two-sample comparisons but, rather, in more complicated comparisons of three or more means. For example, we could test whether mean drinking behavior varies by year in college. Figure 5.2 shows these four distributions as boxplots. The following command tests whether the four means are equal.

```
. oneway drink year, tabulate scheffe
Year in     |  Summary of 33-point drinking scale
college     |        Mean   Std. Dev.       Freq.
------------+------------------------------------------------------
   Freshman |      18.975   6.9226033          40
   Sophomor |   21.169231   6.5444853          65
     Junior |   19.453333   6.2866081          75
     Senior |   16.650794   6.6409257          63
------------+------------------------------------------------------
      Total |   19.106996   6.7221166         243

                        Analysis of Variance
    Source              SS         df       MS            F     Prob > F
-------------------------------------------------------------------------
Between groups      666.200518      3   222.066839      5.17     0.0018
 Within groups      10269.0176    239   42.9666008
-------------------------------------------------------------------------
    Total           10935.2181    242   45.1868517

Bartlett's test for equal variances:  chi2(3) =  0.5103  Prob>chi2 = 0.917

            Comparison of 33-point drinking scale by Year in college
                                (Scheffe)
Row Mean-|
Col Mean |   Freshman   Sophomor     Junior
---------+---------------------------------
Sophomor |    2.19423
         |      0.429
         |
  Junior |    .478333    -1.7159
         |      0.987      0.498
         |
  Senior |   -2.32421   -4.51844   -2.80254
         |      0.382      0.002      0.103
```

We can reject the hypothesis of equal means ($P = .0018$), but not the hypothesis of equal variances ($P = .917$). The latter (supported by the boxplots of Figure 5.2) is "good news" regarding the ANOVA's validity.

Figure 5.2

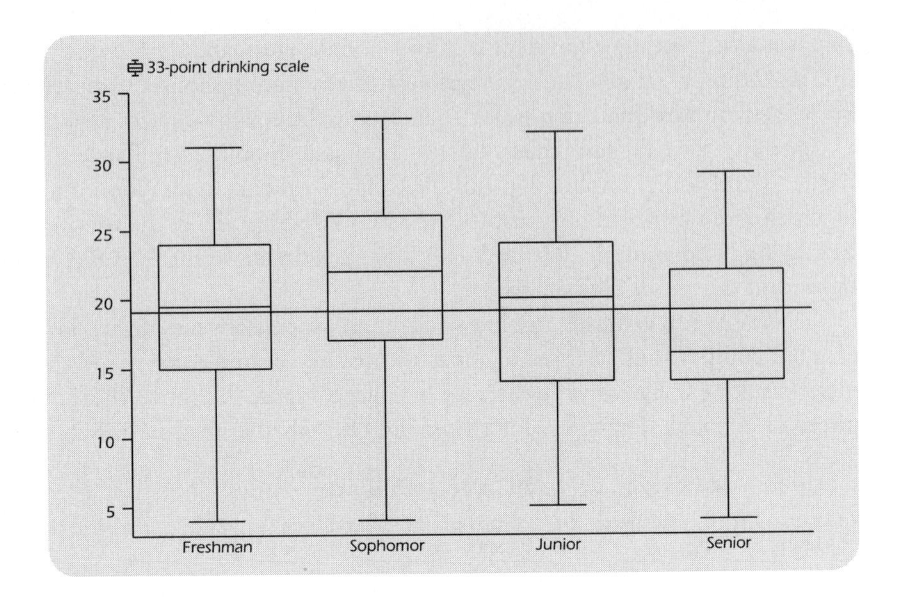

The **scheffe** option (Scheffé multiple-comparison test) produces a table showing the differences between each pair of means. The freshman mean equals 18.975 and the sophomore mean equals 21.16923, so the sophomore-freshman difference is 21.16923 – 18.975 = 2.19423, not statistically distinguishable from zero (P = .429). Of the six contrasts in this table, only the senior-sophomore difference, 16.6508 – 21.1692 = –4.5184, is significant (P = .002). Thus our overall conclusion that these four groups' means are not the same arises mainly from the contrast between seniors (the lightest drinkers) and sophomores (the heaviest).

oneway offers three multiple-comparison options: **scheffe**, **bonferroni** or **sidak** (see Stata reference manuals for definitions). The Scheffé test remains valid under a wider variety of conditions, although it is sometimes less sensitive.

The Kruskal-Wallis test (**kwallis**), a K-sample generalization of the two-sample rank-sum test, provides a nonparametric alternative to one-way ANOVA. It tests the null hypothesis of equal population medians:

```
. kwallis drink, by(year)

Test: Equality of populations (Kruskal-Wallis Test)

    year          _Obs    _RankSum
Freshman            40     4914.00
Sophomor            65     9341.50
  Junior            75     9300.50
  Senior            63     6090.00

chi-squared =     14.453 with 3 d.f.
probability =      0.0023
```

Here, the **kwallis** results (P = .0023) agree with our **oneway** findings of significant differences in *drink* by year in college. Kruskal-Wallis is generally safer than ANOVA if we have reason to doubt ANOVA's equal-variances or normality assumptions, or if we suspect problems caused by outliers. **kwallis**, like **ranksum**, makes the weaker assumption of similar-shaped distributions within each group. In principle, **ranksum** and **kwallis** should produce similar results when applied to two-sample comparisons, but in practice this is true only if the data contain no ties. At present, Stata's version of **ranksum** incorporates a better way to adjust for ties, which has not yet been implemented for **kwallis**. This makes **ranksum** preferable for two-sample problems.

Two- and N-Way Analysis of Variance

One-way ANOVA examines how the means of one measurement variable y vary across categories of a second variable x. N-way ANOVA generalizes this approach to deal with two or more categorical x variables. For example, we might consider how drinking behavior varies not only by fraternity or sorority membership, but also by gender. We start by examining a two-way table of means:

```
. tabulate belong gender, summarize(drink) means
                    Means of 33-point drinking scale

Belong to   | Gender (male)
fraternity/ |
sorority    |    Female        Male |     Total
 -----------+ ----------------------+----------
     member | 22.444444   26.137931 | 24.723404
   nonmembe | 16.517241     19.5625 | 17.760204
 -----------+ ----------------------+----------
      Total | 17.313433   21.311927 | 19.106996
```

It appears that in this sample males drink more than females, and members drink more than nonmembers. The member-nonmember difference is slightly greater among males. Stata's N-way ANOVA command, **anova**, can test for significant differences among these means attributable to belonging to a fraternity or sorority, gender, or the interaction of belonging and gender (written *belong*gender*):

```
. anova drink belong gender belong*gender
                          Number of obs =       243     R-squared     =  0.2221
                          Root MSE      = 5.96592     Adj R-squared =  0.2123
                 Source |  Partial SS    df       MS           F     Prob > F
          --------------+----------------------------------------------------
                  Model |  2428.67237     3  809.557456      22.75     0.0000
                        |
                 belong |   1406.2366     1   1406.2366      39.51     0.0000
                 gender |  408.520097     1  408.520097      11.48     0.0008
          belong*gender |  3.78016612     1  3.78016612       0.11     0.7448
                        |
               Residual |  8506.54574   239  35.5922416
          --------------+----------------------------------------------------
                  Total |  10935.2181   242  45.1868517
```

In this example of "two-way factorial ANOVA", the output shows significant main effects for *belong* ("*P* = .0000") and *gender* (*P* = .0008), but their interaction contributes little to the model (*P* = .7448). This interaction cannot be distinguished from zero, so we might prefer to estimate a simpler model without the interaction term (results not shown):

```
. anova drink belong gender
```

To include any interaction term with **anova**, specify the variable names joined by *. Unless the number of observations with each combination of *x* values is the same (a condition called "balanced data"), it can be hard to interpret the main effects in a model that also includes interactions. This does not mean that the main effects in such models are unimportant, however. Regression analysis can help to make sense of complicated ANOVA results, as illustrated in the following section.

Analysis of Covariance (ANOCOVA)

Analysis of covariance (ANOCOVA) extends *N*-way ANOVA to encompass a mix of categorical and continuous *x* variables. This is accomplished through the **anova** command, if we specify which variables are continuous. For example, when we include *gpa* (college grade point average) among the independent variables we find it too is related to drinking behavior:

```
. anova drink belong gender gpa, continuous(gpa)
                         Number of obs =     218     R-squared     =  0.2970
                         Root MSE      = 5.68939     Adj R-squared =  0.2872

               Source |  Partial SS    df       MS           F     Prob > F
           -----------+----------------------------------------------------
                Model |  2927.03087     3  975.676958      30.14     0.0000
                      |
               belong |  1489.31999     1  1489.31999      46.01     0.0000
               gender |  405.137843     1  405.137843      12.52     0.0005
                  gpa |    407.0089     1    407.0089      12.57     0.0005
                      |
             Residual |  6926.99206   214  32.3691218
           -----------+----------------------------------------------------
                Total |  9854.02294   217  45.4102439
```

From this analysis we know that a significant relation exists between *drink* and *gpa*, when we control for *belong* and *gender*. Beyond their *F* tests for statistical significance, however, ANOVA and ANOCOVA ordinarily do not provide much descriptive information about how variables are related. Regression, with its explicit model and parameter estimates, does a better descriptive job. Because ANOVA and ANOCOVA amount to special cases of regression, we could restate these analyses in regression form. Stata does so automatically if we add the **regress** option to **anova**. For instance, we might want regression to understand results from the following ANOCOVA:

```
. anova drink belong gender belong*gender gpa,
       continuous(gpa) regress
  Source |       SS       df       MS                  Number of obs =     218
---------+------------------------------               F(  4,   213) =   22.57
   Model |  2933.45823     4  733.364558               Prob > F      =  0.0000
Residual |   6920.5647   213  32.4909141               R-squared     =  0.2977
---------+------------------------------               Adj R-squared =  0.2845
   Total |  9854.02294   217  45.4102439               Root MSE      =  5.7001

------------------------------------------------------------------------------
   drink        Coef.   Std. Err.       t     P>|t|      [95% Conf. Interval]
------------------------------------------------------------------------------
_cons        27.47676   2.439962     11.261   0.000       22.6672    32.28633
belong
       1     6.925384   1.286774      5.382   0.000      4.388942    9.461826
       2    (dropped)
gender
       1    -2.629057   .8917152     -2.948   0.004     -4.386774   -.8713407
       2    (dropped)
gpa         -3.054633   .8593498     -3.555   0.000     -4.748552   -1.360713
belong*gender
     1 1    -.8656158   1.946211     -0.445   0.657     -4.701916    2.970685
     1 2    (dropped)
     2 1    (dropped)
     2 2    (dropped)
------------------------------------------------------------------------------
```

With the **regress** option, we get the **anova** output formatted as a regression table. The top part gives the same overall *F* test and R^2 as a standard ANOVA table. The bottom part describes the following regression:

> We construct a separate dummy variable {0,1} representing each category of each *x* variable, except for the highest categories, which are dropped. Interaction terms (if specified in the command's variable list) are constructed from the products of every possible combination of these dummy variables. Regress *y* on all these dummy variables and interactions and also on any continuous variables specified in the command line.

The previous example therefore corresponds to a regression of *drink* on four *x* variables:

1. A dummy coded 1 = fraternity/sorority member, 0 otherwise.
2. A dummy coded 1 = female, 0 otherwise (highest category of *gender*, male, gets dropped).
3. The continuous variable *gpa*.
4. An interaction term coded 1 = sorority female, 0 otherwise.

Interpret the individual dummy variables' regression coefficients as effects on predicted or mean *y*. For example, the coefficient on the first category of *gender* (female) equals –2.6629057. This informs us that the mean drinking scale levels for females are 2.6629057 points lower than those of males with the same grade point average and membership status. And we know that among students of the same gender and membership status, mean drinking scale values decline by 3.054633 with each one-point increase in grades. Note also that we have confidence intervals and individual *t* tests for each coefficient; there is much more information in the **anova, regress** output than in the ANOVA table alone.

Whether or not we use the **regress** option, after any **anova** or **oneway** command we can employ the **predict** command as described in Chapter 6. **predict** obtains predicted *y* values, residuals, and a variety of diagnostic measures for detecting outliers or influential observations. **predict** works with **anova** and **oneway**, just as it does with ordinary least squares regression (**regress**) because the three share a common mathematical framework.

6

Linear Regression Analysis

Stata offers an exceptionally broad range of regression procedures, from elementary to sophisticated. This chapter introduces **`regress`** and related commands that perform simple and multiple ordinary least squares (OLS) regression. A follow-up command, **`predict`**, calculates predicted values, residuals, and diagnostic statistics such as leverage and Cook's *D*. With options, **`regress`** can also accomplish other analyses including weighted least squares and provide tests of user-specified hypotheses. Regression with dummy variables, interaction, or polynomial terms and stepwise variable selection is also covered in this chapter, along with a first look at residual analysis.

Example Commands

```
. regress y x
```

Performs ordinary least squares (OLS) regression of variable *y* on a single predictor variable *x*.

```
. regress y x if ethnic == 3 & income > 50
```

Regresses *y* on *x* using only that subset of the data for which variable *ethnic* equals 3 and *income* is greater than 50.

```
. predict yhat
```

Generates a new variable (here arbitrarily named *yhat*) equal to the predicted values from the most recent regression.

```
. predict e, resid
```

Generates a new variable (arbitrarily named *e*) equal to the residuals from the most recent regression.

```
. graph y yhat x, connect(.s) symbol(Oi)
```

Draws a scatterplot with regression line using the variables *y*, *yhat*, and *x*.

```
. graph e yhat, twoway box yline(0)
```

Draws a residual versus predicted plot using the variables *e* and *yhat*. An alternative, more automatic, way to draw such plots employs the **rvfplot** (residual-versus-fitted) command, discussed in Chapter 7.

```
. regress y x1 x2 x3
```

Performs multiple regression of *y* on three predictor variables, *x1*, *x2*, and *x3*.

```
. regress y x1 x2 x3, robust
```

Calculates robust (Huber/White) estimates of standard errors. See the *User's Guide* for details. The **robust** option works with some other model fitting commands as well.

```
. regress y x1 x2 x3, beta
```

Performs multiple regression and shows standardized regression coefficients (beta weights) in the output table.

```
. correlate x1 x2 x3 y
```

Displays a matrix of Pearson correlations, using only observations with no missing values on all variables specified. Adding the option **covariance** produces a variance-covariance matrix instead of correlations.

```
. pwcorr x1 x2 x3 y, sig
```

Displays a matrix of Pearson correlations, using pairwise deletion of missing values and showing probabilities from *t* tests (of H_0:$\rho = 0$) on each correlation.

```
. graph x1 x2 x3 y, matrix half
```

Draws a scatterplot matrix. Because their variable lists are the same, this example yields a scatterplot matrix having the same organization as the correlation matrix produced by the preceding **pwcorr** command. Listing the dependent (*y*) variable last produces a matrix in which the bottom row is a series of *y*-versus-*x* plots.

```
. test x1 x2
```

Performs an *F* test of the null hypothesis that coefficients on *x1* and *x2* both equal zero, in the most recent regression model.

```
. sw regress y x1 x2 x3, pr(.05)
```

Performs stepwise regression using backward elimination until all remaining predictors are significant at the .05 level. All listed predictors are entered on the first iteration. Thereafter, each iteration drops one predictor with the highest *P* value, until all predictors remaining have probabilities below the "probability to retain," `pr(.05)`. Options permit forward or hierarchical selection. Similar stepwise variants exist for many other model-fitting commands; type `help sw`.

```
. regress y x1 x2 x3 [aweight = w]
```

Performs weighted least squares (WLS) regression of *y* on *x1*, *x2*, and *x3*. Variable *w* holds the analytical weights, which work as if we had multiplied each variable and the constant by $\sqrt{w}$, then performed an ordinary regression. Analytical weights are often employed to correct for heteroscedasticity when the *y* and *x* variables are means, rates, or proportions, and *w* is the number of individuals making up each aggregate observation (for example, city or school) in the data. If the *y* and *x* variables are individual-level, and the weights indicate numbers of replicated observations, then use frequency weights `[fweight = w]` instead.

```
. regress y1 y2 x (x z)
. regress y2 y1 z (x z)
```

Estimates the reciprocal effects of *y1* and *y2*, using instrumental variables *x* and *z*. The first parts of these commands specify the structural equations:

$$y1 = \alpha_0 + \alpha_1 y2 + \alpha_2 x + \varepsilon_1$$

$$y2 = \beta_0 + \beta_1 y2 + \beta_2 w + \varepsilon_2$$

The parentheses in the commands enclose variables that are exogenous to all of the structural equations. `regress` accomplishes two-stage least squares (2SLS) in this example.

The Regression Table

File *states90.dta* contains educational data on the U.S. states and District of Columbia, including these variables:

```
. describe state csat expense percent income high college
      region
```

```
  1. state        str20  %20s                State
 14. csat         int    %9.0g               Mean composite SAT score
 18. expense      int    %9.0g               Per pupil expenditures prim&sec
 17. percent      byte   %9.0g               % HS graduates taking SAT
 19. income       long   %10.0g              Median household income
 20. high         float  %9.0g               % over 25 w/HS diploma
 21. college      float  %9.0g               % over 25 w/bachelor's degree +
  2. region       byte   %9.0g      region   Geographical region
```

Government and political leaders sometimes use mean SAT scores to make pointed comparisons between the educational systems of different U.S. states. For example, one debate has concerned whether SAT scores are higher in states that spend more money on education. We might try to address this question by regressing mean composite SAT score (*csat*) on per-pupil expenditures (*expense*). The appropriate Stata command has the form **regress** ***y x***, where *y* is the predicted or dependent variable, and *x* the predictor or independent variable.

```
. regress csat expense

  Source |       SS       df       MS                  Number of obs =      51
---------+------------------------------               F(  1,    49) =   13.61
   Model |  48708.3001     1  48708.3001               Prob > F      =  0.0006
Residual |   175306.21    49  3577.67775               R-squared     =  0.2174
---------+------------------------------               Adj R-squared =  0.2015

   Total |   224014.51    50   4480.2902               Root MSE      =  59.814
------------------------------------------------------------------------------
    csat |      Coef.   Std. Err.       t     P>|t|       [95% Conf. Interval]
---------+--------------------------------------------------------------------
 expense |  -.0222756    .0060371     -3.690   0.001      -.0344077   -.0101436
   _cons |   1060.732    32.7009      32.437   0.000       995.0175    1126.447
------------------------------------------------------------------------------
```

This regression tells an unexpected story: The more money a state spends on education, the lower its students' mean SAT scores. Any causal interpretation is premature at this point, but the regression table does convey a variety of information about the linear statistical relation between *csat* and *expense*. At the upper right of the regression table is the overall F test, based on the sums of squares at upper left. This F test evaluates the null hypothesis that coefficients on all x variables in the model (here there is only one x variable, *expense*) equal zero. The F statistic, 13.61 with 1 and 49 degrees of freedom, leads easily to rejection of this null hypothesis ($P = .0006$). `Prob > F` means "the probability of a greater F" statistic if we drew samples randomly from a population in which the null hypothesis is true.

At the upper right, we also see the coefficient of determination, $R^2 = .2174$. Per-pupil expenditures explain about 22% of the variance in states' mean composite SAT scores. Adjusted R^2, $R^2_a = .2015$, takes into account the complexity of the model relative to the complexity of the data. This adjusted statistic is often more informative for research.

The lower half of the regression table gives the estimated model itself. We find coefficients (slope and y-intercept) in the first column, here yielding the prediction equation:

predicted *csat* = 1060.732 – .0222756*expense*

The second column lists estimated standard errors of the coefficients. These form the basis for *t* tests (columns 3–4) and confidence intervals (columns 5–6) for each regression coefficient. The *t* statistics (coefficients divided by their standard errors) test null hypotheses that the corresponding population coefficients equal zero. At the α = .05 significance level, we could reject this null hypothesis regarding both the coefficient on *expense* (P = .001) and the *y*-intercept (".000", really meaning P < .0005). Stata's modeling commands calculate 95% confidence intervals routinely, but we can ask for some other level by specifying the **level( )** option, as shown in the following:

```
. regress csat expense, level(99)
```

Because these data do not represent a random sample from some larger population of U.S. states, hypothesis tests and confidence intervals lack their usual meanings. They are discussed in this chapter anyway for illustration purposes.

The term `_cons` stands for the regression constant, usually set at one. Stata automatically includes a constant unless we tell it not to. The **nocons** option causes Stata to suppress the constant, performing regression through the origin. For example,

```
. regress y x, nocons
```

Or,

```
. regress y x1 x2 x3, nocons
```

In certain advanced applications, you might have to specify your own constant. If the "independent variables" include a user-supplied constant (named *c*, for example), employ the **hascons** option instead of **nocons**:

```
. regress y c x, hascons
```

Using **nocons** in this situation would result in a misleading *F* test and R^2. Consult the *Stata Reference Manual* or **help regress** for more about **hascons**.

Multiple Regression

Multiple regression allows us to estimate how *expense* predicts *csat*, while adjusting for a number of other possible predictor variables. We can incorporate other predictors of *csat* simply by listing these variables in the command:

```
. regress csat expense percent income high college

  Source |       SS       df       MS                  Number of obs =      51
---------+------------------------------               F(  5,    45) =   42.23
   Model |  184663.309     5  36932.6617               Prob > F      =  0.0000
Residual |  39351.2012    45  874.471137               R-squared     =  0.8243
---------+------------------------------               Adj R-squared =  0.8048
   Total |   224014.51    50   4480.2902               Root MSE      =  29.571
```

```
------------------------------------------------------------------------------
    csat |      Coef.   Std. Err.       t     P>|t|       [95% Conf. Interval]
---------+--------------------------------------------------------------------
 expense |   .0033528   .0044709      0.750   0.457      -.005652    .0123576
 percent |  -2.618177   .2538491    -10.314   0.000     -3.129455   -2.106898
  income |   .0001056   .0011661      0.091   0.928      -.002243    .0024542
    high |   1.630841    .992247      1.644   0.107      -.367647    3.629329
 college |   2.030894   1.660118      1.223   0.228     -1.312756    5.374544
   _cons |   851.5649   59.29228     14.362   0.000      732.1441    970.9857
------------------------------------------------------------------------------
```

This gives us the multiple regression equation:

predicted *csat* = 851.56 + .00335*expense* – 2.618*percent* + .0001*income* + 1.63*high* + 2.03*college*

Controlling for four other variables weakens the coefficient on *expense* from –.0223 to .00335, which is no longer statistically distinguishable from zero. The unexpectedly negative relation between *expense* and *csat* found in our earlier simple regression has been explained away by the other predictors.

Only the coefficient on *percent* (percentage of high school graduates taking the SAT) attains significance at the .05 level. We could interpret this "fourth-order partial regression coefficient" (so called because its calculation adjusts for four other predictors) as follows:

> $b_2 = -2.618$: Predicted mean SAT scores decline by 2.618 points, with each one-point increase in the percentage of high school graduates taking the SAT—if *expense, income, high,* and *college* do not change.

Taken together, the five *x* variables in this model explain about 80% of the variance in states' mean composite SAT scores ($R^2_a = .8048$). In contrast, our earlier simple regression with *expense* as the only predictor explained about 20% of the variance in *csat.*

To obtain standardized regression coefficients (beta weights) with any regression, add the **beta** option. Standardized coefficients are what we would see in a regression with all variables transformed to standard scores (means = 0, standard deviations = 1).

```
. regress csat expense percent income high college, beta

  Source |       SS       df       MS                  Number of obs =      51
---------+------------------------------               F(  5,    45) =   42.23
   Model |  184663.309     5  36932.6617               Prob > F      =  0.0000
Residual |  39351.2012    45  874.471137               R-squared     =  0.8243
---------+------------------------------               Adj R-squared =  0.8048
   Total |   224014.51    50   4480.2902               Root MSE      =  29.571
```

```
------------------------------------------------------------------------------
    csat |      Coef.   Std. Err.       t     P>|t|                      Beta
---------+--------------------------------------------------------------------
 expense |   .0033528    .0044709     0.750   0.457                   .070185
 percent |  -2.618177    .2538491   -10.314   0.000                 -1.024538
  income |   .0001056    .0011661     0.091   0.928                  .0101321
    high |   1.630841     .992247     1.644   0.107                  .1361672
 college |   2.030894    1.660118     1.223   0.228                  .1263952
   _cons |   851.5649    59.29228    14.362   0.000                         .
------------------------------------------------------------------------------
```

The standardized regression equation is:

$$\text{predicted } csat^* = .07expense^* - 1.0245percent^* + .01income^* + .136high^* + .126college^*$$

where *csat**, *expense**, and so forth denote these variables in standard-score form. We might interpret the standardized coefficient on *percent*, for example, as follows:

> $b_2^* = 1.0245$: Predicted mean SAT scores decline by 1.0245 standard deviations, with each one-standard deviation increase in the percentage of high school graduates taking the SAT—if *expense, income, high,* and *college* do not change.

The F and t tests, R^2, and all other aspects of the regression remain the same.

Predicted Values and Residuals

After any regression, the **predict** command can obtain predicted values, residuals, and other case statistics. Suppose we have just done a regression of state composite SAT scores on their strongest single predictor:

```
. regress csat percent
```

Now to create a new variable called *yhat*, containing predicted *y* values from this regression, type

```
. predict yhat

. label variable yhat "Predicted mean SAT score"
```

Through the **resid** option we can also create another new variable containing the residuals, here named *e:*

```
. predict e, resid

. label variable e "Residual"
```

We might instead have obtained the same predicted *y* and residuals through two **generate** commands:

```
. generate yhat0 = _b[_cons] + _b[percent]*percent

. generate e0 = csat - yhat
```

Stata stores coefficients and other details from the most recent regression. Thus `_b[` *varname* `]` holds the coefficient on independent variable *varname*. `_b[_cons]` holds the coefficient on `_cons`, that is, the *y*-intercept. These stored values are useful in programming and some advanced applications, but for most purposes **predict** saves us the trouble of generating *yhat0* and *e0* "by hand" in this fashion.

Residuals contain information about where the model fits poorly, and so are important for diagnostic or troubleshooting analysis. You can begin such analysis just by sorting and examining the residuals. Negative residuals occur when our model over-predicts the observed values. That is, in these states the mean SAT scores are lower than we might expect based on what percentage of seniors took the test. To see the five lowest residuals,

```
. sort e

. list state percent csat yhat e in 1/5

                state   percent       csat       yhat          e
  1.   South Carolina        58        832   894.3333   -62.3333
  2.    West Virginia        17        926   986.0953  -60.09526
  3.   North Carolina        57        844   896.5714   -52.5714
  4.            Texas        44        874   925.6666  -51.66666
  5.           Nevada        25        919   968.1905  -49.19049
```

The four lowest residuals belong to southern states, suggesting that we might improve our model, or better understand variation in mean SAT scores, by somehow taking region into account.

Positive residuals occur when actual *y* values are higher than predicted. Because the data have been sorted by *e*, to list the five highest residuals we add the qualifier **in 46/l** (meaning 46th through **l**ast observations).

```
. list state percent csat yhat e in 46/l

                state   percent       csat       yhat          e
 46.           Kansas        10       1039   1001.762   37.23806
 47.    Massachusetts        79        896   847.3333   48.66673
 48.      Connecticut        81        897   842.8571   54.14292
 49.     North Dakota         6       1073   1010.714   62.28567
 50.    New Hampshire        75        921   856.2856   64.71434
 51.             Iowa         5       1093   1012.952   80.04758
```

predict also derives other case statistics from the most recently estimated model. Options particularly useful after **regress** are the following:

`. predict new`	Predicted values of *y*

`. predict new, cooksd`	Cook's *D* influence measures
`. predict new, hat`	Diagonal elements of hat matrix (leverage)
`. predict new, resid`	Residuals
`. predict new, rstandard`	Standardized residuals
`. predict new, rstudent`	Studentized residuals
`. predict new, stdf`	Standard errors of predicted individual *y*, sometimes called the standard errors of forecast or the standard errors of prediction
`. predict new, stdp`	Standard errors of predicted mean *y*
`. predict new, stdr`	Standard errors of residuals

When using **predict**, substitute a new variable name of your choosing for the *new* shown. For example, to obtain Cook's *D* influence measures,

```
. predict D, cooksd
```

Or you can use hat matrix diagonals,

```
. predict h, hat
```

The names of variables created by **predict** (such as *yhat, e, D, h*) are arbitrary and invented by the user. As with other elements of Stata commands, we could abbreviate these options to the minimum number of letters it takes to identify them uniquely. For example,

```
. predict e, resid
```

could be shortened to

```
. pre e, re
```

Basic Graphs for Regression

This section introduces some elementary graphs you can use to represent a regression model or examine its fit. Chapter 7 describes more specialized graphs that aid post-regression diagnostic work.

In simple regression such as **regress *csat percent***, predicted values lie on the line defined by the regression equation. By plotting and connecting predicted values, we can make that line visible (Figure 6.1):

Figure 6.1

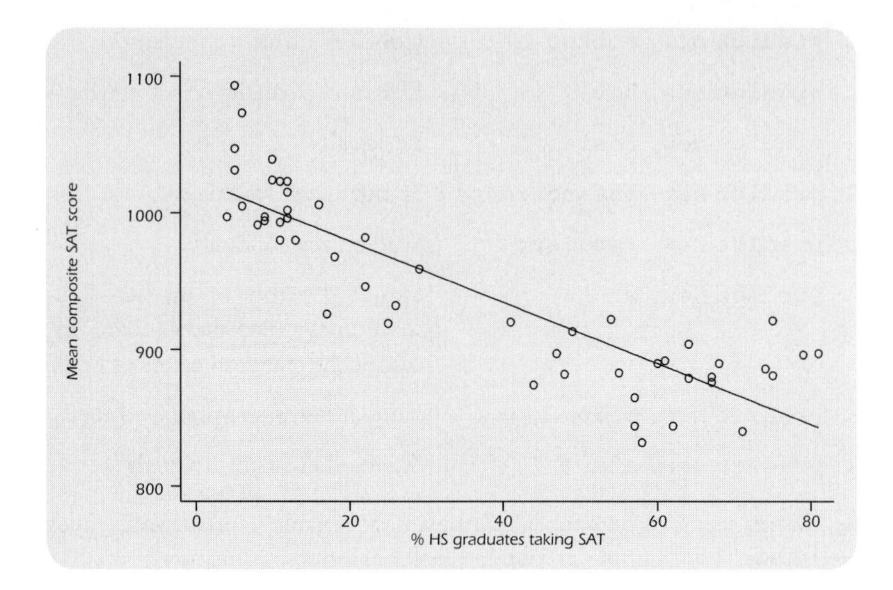


```
. graph csat yhat percent, connect(.s) symbol(Oi) ylabel xlabel
```

connect(.s) tells **graph** not to connect the first-named *y* variable (*csat*), but to connect the second-named (*yhat*) smoothly. **symbol(Oi)** specifies that *csat* should be plotted with large circles, and *yhat* invisibly—so we see only the connecting line. **connect(ss)**, for instance, would have connected both the *csat* and *y* values (producing a messy graph). **symbol(OT)** would have plotted *csat* as large circles and *yhat* as large triangles.

Residual versus predicted values plots are useful diagnostic tools. Figure 6.2 shows an example that includes marginal box plots and a horizontal line at the mean residual, 0.

```
. graph e yhat, twoway box yline(0) ylabel xlabel
```

Figure 6.2 reveals that our present model overlooks an obvious pattern in the data. The residuals or prediction errors appear to be mostly positive (because of too-high predictions) at first, then mostly negative, followed by mostly positive residuals again. Later sections will seek a model that better fits this pattern.

predict can generate two kinds of standard errors for the predicted *y* values, which have two different applications. These applications are sometimes distinguished by the names "confidence intervals" and "prediction intervals":

1. A "confidence interval" in this context expresses our uncertainty in estimating the conditional mean of *y* at a given *x* value (or a given combination of *x* values, in multiple regression). Standard errors for this purpose are obtained through

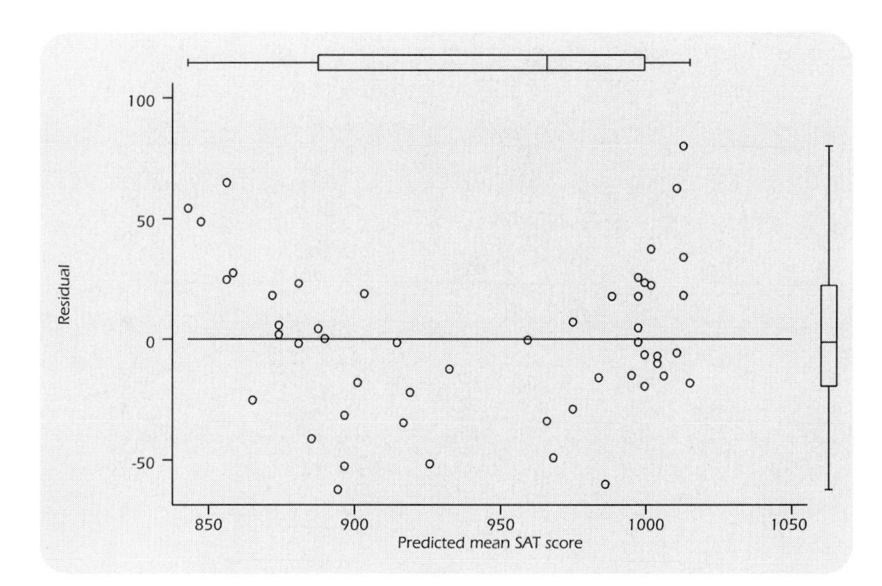

Figure 6.2

```
. predict SE, stdp
```

Select an appropriate *t* value. With 49 degrees of freedom, for 95% confidence we should use $t = 2.01$, found by looking up the *t* distribution in a textbook or simply by asking Stata:

```
. display invt(49,.95)
2.0095752
```

Then the lower confidence limit is approximately

```
. generate low1 = yhat - 2.01*SE
```

And the upper confidence limit is

```
. generate high1 = yhat + 2.01*SE
```

To graph these confidence limits as bands in a simple regression (Figure 6.3), issue a command such as

```
. graph csat yhat low1 high1 percent, connect(.sss)
        symbol(Oiii) ylabel xlabel
```

Confidence bands in simple regression have an hourglass shape, narrowest at the mean of *x*.

Figure 6.3

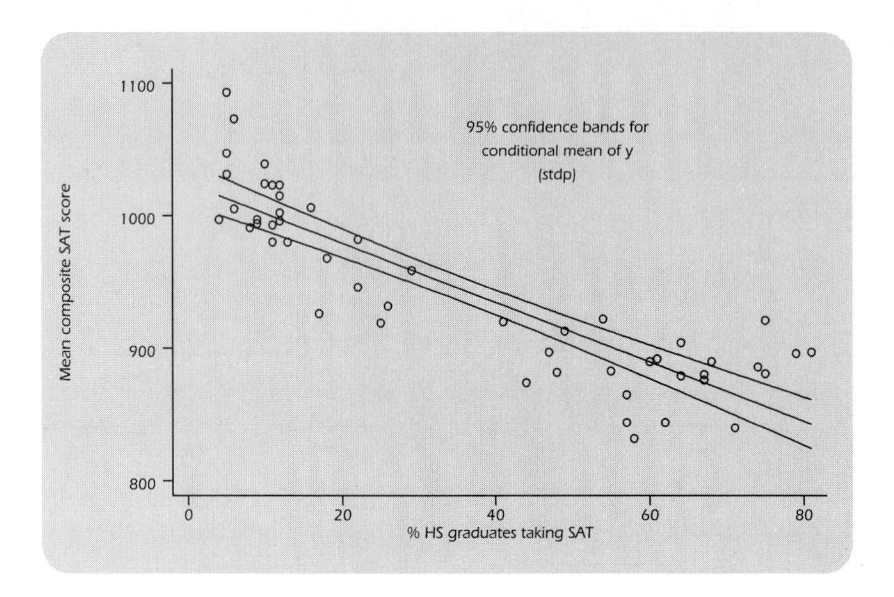

2. A "prediction interval" expresses our uncertainty in estimating the unknown value of *y* for an individual observation with known *x* value(s). Standard errors for this purpose are obtained through a command such as this:

```
. predict SEyhat, stdf
```

To find and graph these prediction bands (Figure 6.4):

```
. generate low2 = yhat - 2.014*SEyhat

. generate high2 = yhat + 2.014*SEyhat

. graph csat yhat low2 high2 percent, connect(.sss)
        symbol(Oiii) ylabel xlabel
```

Predicting the *y* values of individual observations (Figure 6.4) inherently involves greater uncertainty, and hence wider bands, than does predicting the conditional mean of *y* values (Figure 6.3). Prediction intervals, like confidence intervals, will be narrowest at the mean of *x*.

As with other confidence intervals and hypothesis tests in OLS regression, the standard errors and bands just described depend on the assumption of independent and identically distributed errors. Figure 6.2 has cast doubt on this assumption, so the results in Figures 6.3 and 6.4 could be misleading.

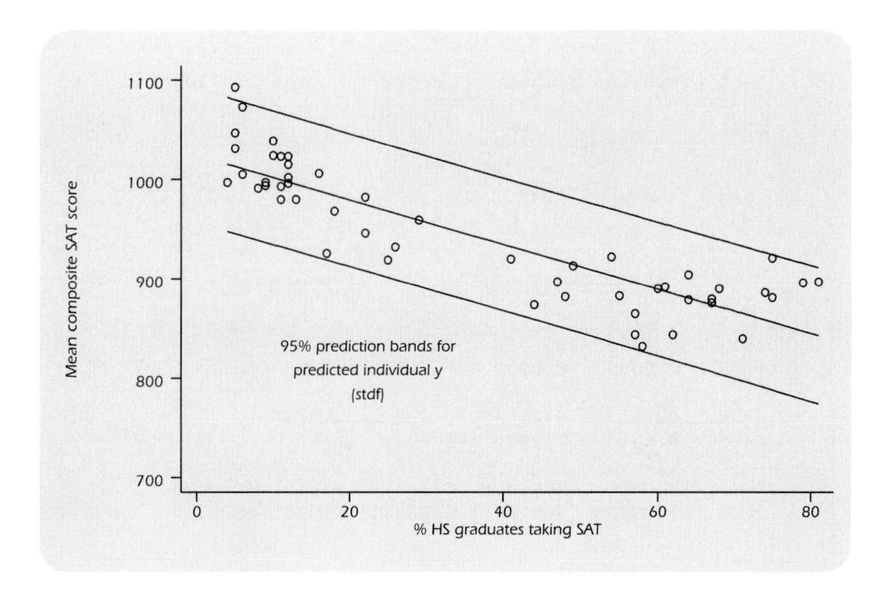

Figure 6.4

Correlations

correlate obtains Pearson product-moment correlations between variables.

```
. correlate csat expense percent income high college
(obs=51)

        |     csat  expense  percent   income     high  college
--------+------------------------------------------------------
    csat|   1.0000
 expense|  -0.4663   1.0000
 percent|  -0.8758   0.6509   1.0000
  income|  -0.4713   0.6784   0.6733   1.0000
    high|   0.0858   0.3133   0.1413   0.5099   1.0000
 college|  -0.3729   0.6400   0.6091   0.7234   0.5319   1.0000
```

correlate uses only a subset of the data that has no missing values on any of the variables listed (with these particular variables, it does not matter because no observations have missing values). Thus **correlate** fits into the context of multiple regression analysis. **regress** with the same variable list will use the same subset of data as **correlate**. Analysts not moving on to regression or other multivariable techniques, however, might prefer to find correlations based on all the data available for each variable pair. The command **pwcorr** (pairwise correlation) accomplishes this and can also furnish *t*-test probabilities for the null hypotheses that each particular correlation equals zero.

```
. pwcorr csat expense percent income high college, sig

             |     csat  expense  percent   income     high  college
-------------+------------------------------------------------------
        csat |   1.0000
             |
             |
     expense |  -0.4663   1.0000
             |   0.0006
             |
     percent |  -0.8758   0.6509   1.0000
             |   0.0000   0.0000
             |
      income |  -0.4713   0.6784   0.6733   1.0000
             |   0.0005   0.0000   0.0000
             |
        high |   0.0858   0.3133   0.1413   0.5099   1.0000
             |   0.5495   0.0252   0.3226   0.0001
             |
     college |  -0.3729   0.6400   0.6091   0.7234   0.5319   1.0000
             |   0.0070   0.0000   0.0000   0.0000   0.0001
             |
```

It is worth mentioning here that if we drew many random samples from a population in which all variables really had 0 correlations, about 5% of the sample correlations would nonetheless test "statistically significant" at the .05 level. Analysts who review many individual hypothesis tests, such as those in a **pwcorr** matrix, to identify the handful that are significant at the .05 level, therefore run a much higher than .05 risk of making a Type I error. This problem is called the "multiple comparison fallacy." **pwcorr** offers two methods, Bonferroni and Šidák, for adjusting significance levels to account for multiple comparisons. Of these, the Šidák method is more precise:

```
. pwcorr csat expense percent income high college, sidak sig

             |     csat  expense  percent   income     high  college
-------------+------------------------------------------------------
        csat |   1.0000
             |
             |
     expense |  -0.4663   1.0000
             |   0.0084
             |
     percent |  -0.8758   0.6509   1.0000
             |   0.0000   0.0000
             |
      income |  -0.4713   0.6784   0.6733   1.0000
             |   0.0072   0.0000   0.0000
             |
        high |   0.0858   0.3133   0.1413   0.5099   1.0000
             |   1.0000   0.3180   0.9971   0.0020
             |
     college |  -0.3729   0.6400   0.6091   0.7234   0.5319   1.0000
             |   0.1004   0.0000   0.0000   0.0000   0.0009
             |
```

Comparing the test probabilities in this table with those of the previous **pwcorr** provides some idea of how much adjustment occurs. In general, the more variables we correlate, the more the adjusted probabilities will exceed their unadjusted counterparts. See the *Stata Reference Manual* discussion of **oneway** for the formulas involved.

correlate itself offers several other important options. Typing the following produces a matrix of variances and covariances instead of correlations:

```
. correlate w x y z, covariance
```

Typing the following after a regression analysis displays the matrix of correlations between estimated coefficients, sometimes used to diagnose multicollinearity (see Chapter 7):

```
. correlate, _coef
```

The following command will display the estimated coefficients' variance-covariance matrix, from which standard errors are derived:

```
. correlate, _coef covariance
```

Pearson correlation coefficients measure how well an OLS regression line fits the data. They consequently share the assumptions and weaknesses of OLS and, like OLS, should generally not be interpreted without first reviewing the corresponding scatterplots. A scatterplot matrix provides a quick way to do this, using the same organization as the correlation matrix. Figure 6.5 shows a scatterplot matrix corresponding to the **pwcorr** correlation matrix given earlier. Only the lower-triangular half of the matrix is drawn, and dots are used as plotting symbols (**symbol(.)**) to make these tiny graphs more readable.

```
. graph csat expense percent income high college, matrix
      half label symbol(.)
```

To obtain a scatterplot matrix corresponding to a **corr** correlation matrix, from which all observations having missing values have been dropped, we need to qualify the command. If all the variables had some missing values, we could type a command such as this:

```
. graph csat expense percent income high college if csat != .
      & expense != . & income != . & high != . & college != . ,
      matrix half label symbol(.)
```

To reduce the likelihood of confusion and mistakes, it makes sense just to create a new dataset containing that subset of the observations that have no missing values:

Figure 6.5

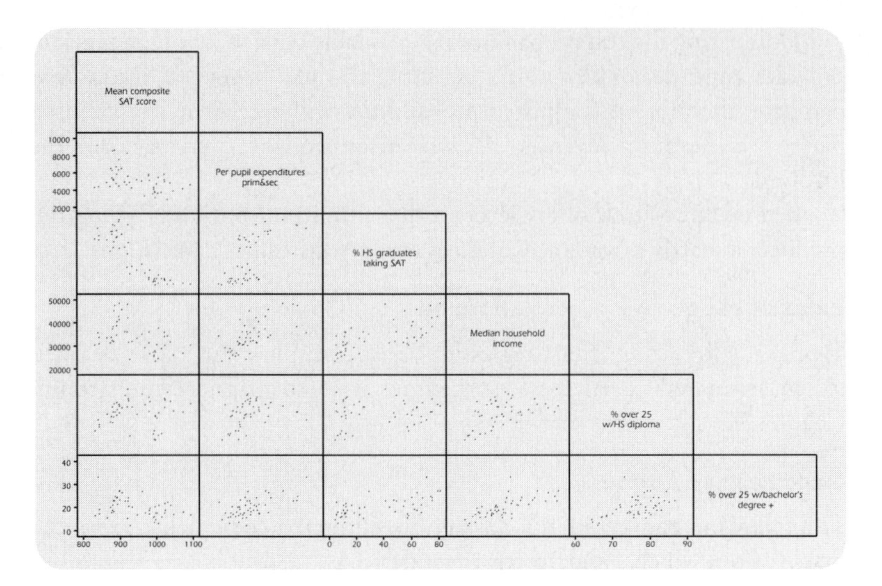


```
. keep if csat !=. & expense != . & income != . & high != .
      & college !=.

. save nmvstate
```

In this example, we immediately saved the reduced dataset with a new name, to avoid inadvertently writing over and losing the information in the old, more complete dataset. An equivalent way to eliminate missing values uses | ("or") instead of & ("and"):

```
. drop if csat ==. | expense == . | income == . | high == .
      | college ==.

. save nmvstate
```

In addition to Pearson correlations, Stata also calculates several rank-based correlations. These can be employed as measures of association between ordinal variables, or as outlier-resistant alternatives to Pearson correlation for measurement variables. To obtain the Spearman rank correlation between *csat* and *expense*, equivalent to the Pearson correlation if these variables were transformed into ranks, use the following:

```
. spearman csat expense

 Number of obs =        51
Spearman's rho =        -0.4282
Test of Ho: csat and expense independent
      Pr > |t| =        0.0017
```

Kendall's τ_a (tau-a) and τ_b (tau-b) rank correlations can be found easily for these data, although with larger datasets their calculation becomes slow:

```
. ktau csat expense

  Number of obs =        51
Kendall's tau-a =        -0.2925
Kendall's tau-b =        -0.2932
Kendall's score =      -373
    SE of score =       123.095    (corrected for ties)
Test of Ho: csat and expense independent
       Pr > |z| =         0.0025   (continuity corrected)
```

For comparison, here is the Pearson correlation with its (unadjusted) P-value:

```
. pwcorr csat expense, sig

          |     csat  expense
----------+------------------
     csat |   1.0000
          |
          |
  expense |  -0.4663   1.0000
          |   0.0006
          |
```

Here both **spearman** (–.4282) and **pwcorr** (–.4663) yield higher correlations than **ktau** (–.2925 or –.2932). All three agree that the null hypothesis of no association can be rejected.

Hypothesis Tests

With every regression, Stata displays two kinds of hypothesis tests. Like most common hypothesis tests, these regression tests begin from the assumption that observations in the sample at hand were drawn randomly and independently from an infinitely large population. The standard tests **regress** provides are the following:

1. Overall F test: The F statistic at the upper right in the regression table evaluates the null hypothesis that in the population, coefficients on all the model's x variables equal zero.

2. Individual t tests: The third and fourth columns of the regression table contain t tests for each individual regression coefficient. These evaluate the null hypotheses that in the population, the coefficients on each particular x variable equal zero.

The displayed t tests are two-sided; for a one-sided test, divide the output's P-value in half.

In addition to these standard *F* and *t* tests, Stata can perform *F* tests of user-specified hypotheses. The **test** command refers back to the most recent model-fitting command such as **anova** or **regress**. For example, individual *t* tests from the following regression report that neither the percent of adults with at least high school diplomas (*high*) nor the percent with college degrees (*college*) has a statistically significant individual effect on composite SAT scores:

```
. regress csat expense percent income high college
```

Conceptually, however, both predictors measure the level of education attained by a state's population, and for some purposes we might want to test the null hypothesis that *both* have zero effect. To do this, we begin by repeating the multiple regression **quietly**, because we do not need to see its full output again. Then use the **test** command:

```
. quietly regress csat expense percent income high college

. test high college
 ( 1)   high = 0.0
 ( 2)   college = 0.0

        F(  2,      45) =      3.32
             Prob > F =      0.0451
```

Unlike the individual null hypotheses, the joint hypothesis that coefficients on *high* and *college* both equal zero can reasonably be rejected ($P = .0451$). Such tests on subsets of coefficients are useful when we have several conceptually related predictors or when individual coefficient estimates appear unreliable because of multicollinearity (Chapter 7).

test could duplicate the overall *F* test:

```
. test expense percent income high college
```

test could also duplicate the individual-coefficient tests:

```
. test expense

. test percent

. test income
```

And so forth. Applications of **test** more useful in advanced work include the following:

1. Test whether a coefficient equals a specified constant. For example, to test the null hypothesis that the coefficient on *income* equals 1 ($H_0: \beta_3 = 1$), instead of the usual null hypothesis that it equals 0 ($H_0: \beta_3 = 0$):

```
. test income = 1
```

2. Test whether two coefficients are equal. For example, the following command evaluates the null hypothesis $H_0: \beta_4 = \beta_5$:

```
. test high = college
```

3. Finally, **test** understands some algebraic expressions, so we could request something like the following, which would test $H_0: \beta_3 = (\beta_4 + \beta_5) / 100$:

```
. test income = (high + college)/100
```

Consult **help test** for more information and examples.

Dummy Variables

Categorical variables can be incorporated as predictors in a regression, if we first re-express their categories as one or more {0,1} dichotomies called "dummy variables." For example, we have reason to suspect that regional differences exist in states' mean SAT scores. The **tabulate** command will generate one dummy variable for each category of the tabulated variable if we add a **gen( )** option. Below, we create four dummy variables from the four-category variable *region*. The dummies are named *reg1*, *reg2*, and so on. *reg1* equals 1 for Western states and 0 for others; *reg2* equals 1 for Northeastern states and 0 for others, and so forth.

```
. tab region, gen(reg)

Geographical|
region      |      Freq.     Percent        Cum.
------------+-----------------------------------
       West |         13       26.00       26.00
    N. East |          9       18.00       44.00
      South |         16       32.00       76.00
    Midwest |         12       24.00      100.00
------------+-----------------------------------
      Total |         50      100.00
```

```
. desc reg1-reg4

  22. reg1         byte   %8.0g                 region==West
  23. reg2         byte   %8.0g                 region==N. East
  24. reg3         byte   %8.0g                 region==South
  25. reg4         byte   %8.0g                 region==Midwest
```

```
. tab reg1

region==West|      Freq.     Percent        Cum.
------------+-----------------------------------
          0 |         37       74.00       74.00
          1 |         13       26.00      100.00
------------+-----------------------------------
      Total |         50      100.00

. tab reg2

region==N.  |
East        |      Freq.     Percent        Cum.
------------+-----------------------------------
          0 |         41       82.00       82.00
          1 |          9       18.00      100.00
------------+-----------------------------------
      Total |         50      100.00
```

Regressing *csat* on one dummy variable, *reg2* (Northeast), is equivalent to performing a two-sample *t* test of whether the mean *csat* is the same across categories of *reg2*. That is, is the mean *csat* the same in the Northeast as in other U.S. states?

```
. regress csat reg2

  Source |       SS       df       MS                  Number of obs =      50
---------+------------------------------               F(  1,    48) =    9.50
   Model |  35191.4017     1  35191.4017               Prob > F      =  0.0034
Residual |  177769.978    48  3703.54121               R-squared     =  0.1652
---------+------------------------------               Adj R-squared =  0.1479
   Total |   212961.38    49  4346.15061               Root MSE      =  60.857

------------------------------------------------------------------------------
    csat |      Coef.   Std. Err.       t     P>|t|       [95% Conf. Interval]
---------+--------------------------------------------------------------------
    reg2 |   -69.0542   22.40167     -3.083   0.003     -114.0958    -24.01262
   _cons |   958.6098   9.504224    100.861   0.000      939.5002     977.7193
------------------------------------------------------------------------------
```

The dummy variable coefficient's *t* statistic ($t = -3.083$, $P = .003$) indicates a significant difference. According to this regression, mean SAT scores are 69.0542 points lower (because $b = -69.0542$) among Northeastern states. We get exactly the same result ($t = 3.083$, $P = .003$) from a simple *t* test, which also shows the means as 889.5556 (Northeast) and 958.6098 (other states), a difference of 69.0542.

```
. ttest csat, by(reg2)

Two-sample t test with equal variances                 0: Number of obs =      41
                                                       1: Number of obs =       9

------------------------------------------------------------------------------
Variable |      Mean    Std. Err.       t     P>|t|       [95% Conf. Interval]
---------+--------------------------------------------------------------------
       0 |  958.6098    10.36563   92.4797   0.0000        937.66     979.5595
       1 |  889.5556    4.652094   191.216   0.0000      878.8278     900.2833
---------+--------------------------------------------------------------------
    diff |   69.0542    22.40167   3.08255   0.0034      24.01262     114.0958
------------------------------------------------------------------------------
Degrees of freedom: 48
```

```
                      Ho: mean(0) - mean(1) = diff = 0
     Ha: diff < 0             Ha: diff ~= 0              Ha: diff > 0
       t =    3.0825            t =    3.0825              t =    3.0825
   P < t =    0.9983       P > |t| =    0.0034          P > t =    0.0017
```

This conclusion dramatically reverses, however, if we control for the percentage of students taking the test. We do so by regressing *csat* on both *reg2* and *percent*:

```
. regress csat reg2 percent

  Source |       SS       df       MS                  Number of obs =      50
---------+------------------------------               F(  2,    47) =  107.18
   Model |  174664.983     2  87332.4916               Prob > F      =  0.0000
Residual |  38296.3969    47  814.816955               R-squared     =  0.8202
---------+------------------------------               Adj R-squared =  0.8125
   Total |   212961.38    49  4346.15061               Root MSE      =  28.545

------------------------------------------------------------------------------
    csat |      Coef.   Std. Err.       t     P>|t|      [95% Conf. Interval]
---------+--------------------------------------------------------------------
    reg2 |   57.52437   14.28326      4.027   0.000      28.79016    86.25858
 percent |  -2.793009   .2134796    -13.083   0.000     -3.222475   -2.363544
   _cons |   1033.749   7.270285    142.188   0.000      1019.123    1048.374
------------------------------------------------------------------------------
```

The Northeastern region variable now has a statistically significant positive coefficient ($b = 57.52437$, $P < .0005$). Here is what happened: Mean SAT scores among Northeastern states really are lower, but they are lower *because higher percentages of students take this test in the Northeast.* A comparatively smaller, more "elite" group of students, often less than 20% of high school seniors, take the SAT in many of the non-Northeast states. In all Northeastern states, however, large majorities (64% to 81%) do so. Once we adjust for differences in the percentages taking the test, SAT scores actually tend to be higher in the Northeast.

To understand dummy variable regression results, it often helps to write out the regression equation, substituting zeroes and ones. For Northeastern states, the equation is approximately:

$$\begin{aligned}\text{predicted } \textit{csat} &= 1033.7 + 57.5\textit{reg2} - 2.8\textit{percent} \\ &= 1033.7 + 57.5 \times 1 - 2.8\textit{percent} \\ &= 1091.2 - 2.8\textit{percent}\end{aligned}$$

For other states, predicted *csat* is 57.5 points lower at any given level of *percent*:

$$\begin{aligned}\text{predicted } \textit{csat} &= 1033.7 + 57.5 \times 0 - 2.8\textit{percent} \\ &= 1033.7 - 2.8\textit{percent}\end{aligned}$$

Dummy variables in models such as this are termed "intercept dummy variables" because they describe a shift in the y-intercept or constant.

From a categorical variable with k categories, we can define k dummy variables, but one of these will be redundant. Once we know a state's values on the West,

Northeast, and Midwest dummy variables, for example, we can already guess its value on the South variable. Furthermore, no more than $k - 1$ of the dummy variables—three, in the case of *region*—can be included in a regression. If we try to include all the possible dummies, Stata will automatically drop one because multicollinearity otherwise makes the calculation impossible:

```
. regress csat reg1 reg2 reg3 reg4 percent

  Source |       SS       df       MS                  Number of obs =      50
---------+------------------------------               F(  4,    45) =   64.61
   Model |  181378.099     4  45344.5247               Prob > F      =  0.0000
Residual |  31583.2811    45  701.850691               R-squared     =  0.8517
---------+------------------------------               Adj R-squared =  0.8385
   Total |   212961.38    49  4346.15061               Root MSE      =  26.492

------------------------------------------------------------------------------
    csat |      Coef.   Std. Err.       t     P>|t|       [95% Conf. Interval]
---------+--------------------------------------------------------------------
    reg1 |    -49.573   14.55516     -3.406   0.001      -78.8886    -20.2574
    reg2 |  (dropped)
    reg3 |  -59.09936   13.85155     -4.267   0.000     -86.99781   -31.20092
    reg4 |  -25.79985   16.96365     -1.521   0.135     -59.96639    8.366693
 percent |  -2.546058   .2140196    -11.896   0.000     -2.977116   -2.115001
   _cons |   1073.438   17.80172     60.300   0.000      1037.583    1109.292
------------------------------------------------------------------------------
```

The model—including R^2, F tests, predictions, and residuals—remains essentially the same regardless of which dummy variable we (or Stata) choose to omit. Interpretation of the coefficients, however, occurs with reference to that omitted category. In this example, the Northeast dummy variable (*reg2*) was omitted. The negative regression coefficients on *reg1*, *reg3*, and *reg4* tell us that, at any given level of *percent*, the predicted mean SAT scores are as follows:

49.573 points lower in the West (*reg1* = 1) than in the Northeast

59.09936 points lower in the South (*reg3* = 1) than in the Northeast

25.79985 points lower in the Midwest (*reg4* = 1) than in the Northeast

The West and South both differ significantly from the Northeast in this respect; the Midwest does not.

An alternative command, **areg**, estimates the same model without going through dummy variable creation. Instead, it "absorbs" the effect of a *k*-category variable such as *region*. The model's fit, F test on the absorbed variable, and other key aspects of the results are the same as those we could obtain through explicit dummy variables. Note that **areg** does not provide estimates of the coefficients on individual dummy variables, however.

```
. areg csat percent, absorb(region)

                                                    Number of obs =        50
                                                    F(  1,     45) =    141.52
                                                    Prob > F      =    0.0000
                                                    R-squared     =    0.8517
                                                    Adj R-squared =    0.8385
                                                    Root MSE      =    26.492

------------------------------------------------------------------------------
    csat |      Coef.   Std. Err.       t     P>|t|     [95% Conf. Interval]
---------+--------------------------------------------------------------------
 percent |  -2.546058   .2140196    -11.896   0.000    -2.977116   -2.115001
   _cons |   1035.445   8.386889    123.460   0.000     1018.553    1052.337
------------------------------------------------------------------------------
  region |          F(3,45) =       9.465   0.000          (4 categories)
```

Although its output is less informative than regression with explicit dummy variables, **areg** does have two advantages. First, its ease speeds exploratory work, where it provides quick feedback about whether a dummy variable approach is worthwhile. Second, when the variable of interest has many values, creating dummies for each of them could lead to too many variables or too large a model for our particular Stata configuration. **areg** thus works around the usual limitations on dataset and matrix size.

Explicit dummy variables have other advantages, however, including ways to model interaction effects. Interaction terms called "slope dummy variables" can be formed by multiplying a dummy times a measurement variable. For example, to model an interaction between Northeast/other region and *percent:*

```
. gen reg2perc = reg2 * percent
(1 missing value generated)
```

The new variable, *reg2perc*, equals *percent* for Northeastern states and zero for all other states. We can include this interaction term among the regression predictors:

```
. regress csat reg2 percent reg2perc

  Source |       SS       df       MS                  Number of obs =        50
---------+------------------------------               F(  3,     46) =     82.27
   Model |   179506.19     3  59835.3968               Prob > F      =    0.0000
Residual |  33455.1897    46  727.286733               R-squared     =    0.8429
---------+------------------------------               Adj R-squared =    0.8327
   Total |   212961.38    49  4346.15061               Root MSE      =    26.968

------------------------------------------------------------------------------
    csat |      Coef.   Std. Err.       t     P>|t|     [95% Conf. Interval]
---------+--------------------------------------------------------------------
    reg2 |  -241.3574   116.6278     -2.069   0.044     -476.117   -6.597821
 percent |  -2.858829   .2032947    -14.062   0.000     -3.26804   -2.449618
reg2perc |   4.179666   1.620009      2.580   0.013     .9187559    7.440576
   _cons |   1035.519   6.902898    150.012   0.000     1021.624    1049.414
------------------------------------------------------------------------------
```

The interaction is statistically significant ($t = 2.58$, $P = .013$). Because this analysis includes both intercept (*reg2*) and slope (*reg2perc*) dummy variables, it is worthwhile to write out the equations. The regression equation for Northeastern states is approximately as follows:

$$\begin{aligned} \text{predicted } csat &= 1035.5 - 241.4 reg2 - 2.9 percent + 4.2 reg2perc \\ &= 1035.5 - 241.4 \times 1 - 2.9 percent + 4.2 \times 1 \times percent \\ &= 794.1 + 1.3 percent \end{aligned}$$

For other states it is this:

$$\begin{aligned} \text{predicted } csat &= 1035.5 - 241.4 \times 0 - 2.9 percent + 4.2 \times 0 \times percent \\ &= 1035.5 - 2.9 percent \end{aligned}$$

An interaction means that the effect of one variable changes, depending on the values of some other variable. From this regression, it appears that *percent* has a relatively weak and positive effect among Northeastern states, whereas its effect is stronger and negative among the rest. To visualize this model, we might draw a scatterplot with separate regression lines for the Northeast and other states and represent those states by different plotting symbols. A quick but crude plot could be obtained by typing

```
. predict yhat1

. graph csat yhat1 percent
```

A nicer version, shown in Figure 6.6, takes more steps. We start by generating predicted values separately from the Northeast and elsewhere regression equations (naming them *yhatNE* and *yhatelse*). Next we generate separate variables holding *csat* values, naming these *csatNE* and *csatelse*. Finally, we put the elements together for Figure 6.6.

```
. generate yhatNE = 794.1 + 1.3*percent if reg2 == 1
(42 missing values generated)

. generate yhatelse = 1035.5 - 2.9*percent if reg2 == 0
(10 missing values generated)

. generate csatNE = csat if reg2 == 1
(42 missing values generated)

. generate csatelse = csat if reg2 == 0
(10 missing values generated)

. label variable csatNE "Northeastern states"
```

Figure 6.6

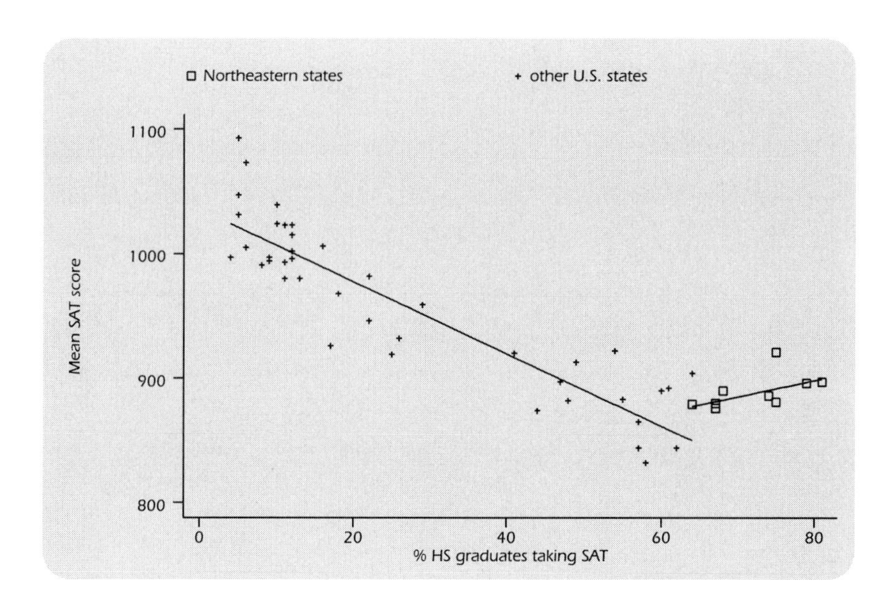


```
. label variable csatelse "other U.S. states"

. graph csatNE csatelse yhatNE yhatelse percent, connect(..ss)
      symbol(Spii) ylabel xlabel l1(Mean SAT score)
```

Figre 6.6 shows the striking difference, captured by our interaction effect, between Northeast and other states. This raises the question of what other regional differences exist. Figure 6.7 takes a step toward exploring this question, by drawing a *csat* – *percent* scatterplot with different symbols for each region. In this plot, the Midwestern states, with one exception (Indiana), seem to have their own steeply negative regional pattern at the left side of the graph. Southern states are the most heterogeneous group.

```
. generate csatW = csat if reg1 == 1
(38 missing values generated)

. label variable csatW "Western states"

. generate csatS = csat if reg3 == 1
(35 missing values generated)

. label variable csatS "Southern states"

. generate csatMW = csat if reg4 == 1
(39 missing values generated)
```

Figure 6.7

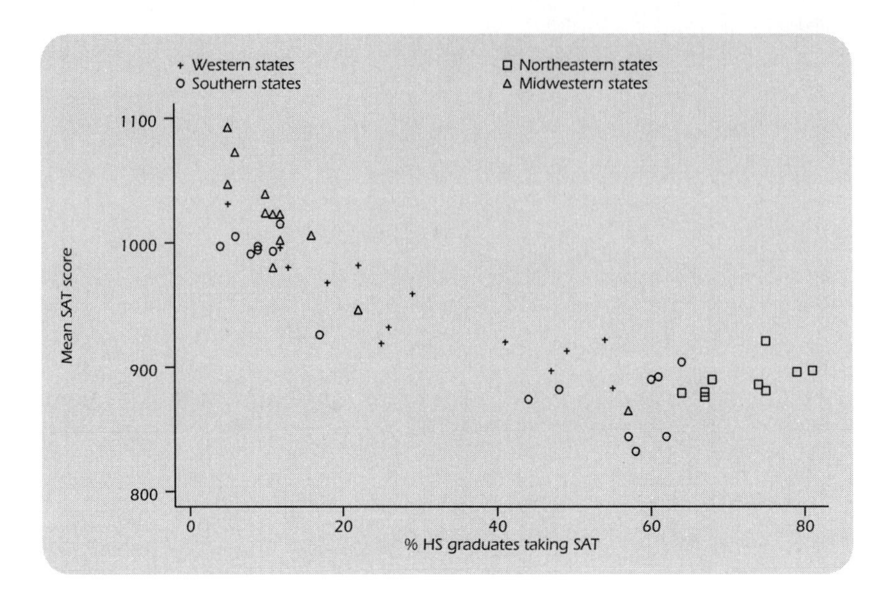


```
. label variable csatMW "Midwestern states"

. graph csatW csatNE csatS csatMW percent, ylabel xlabel
      symbol(pSOT) l1(Mean SAT score)
```

Stepwise Regression

With the regional dummy variable terms added to our data, we now have many possible predictors of *csat.* This results in an overly complicated model, with several coefficients not statistically distinguishable from zero:

```
. regress csat expense percent income college high reg1
      reg2 reg2perc reg3

  Source |       SS       df       MS               Number of obs =      50
---------+------------------------------            F(  9,    40) =   49.51
   Model |  195420.517     9  21713.3908            Prob > F      =  0.0000
Residual |   17540.863    40  438.521576            R-squared     =  0.9176
---------+------------------------------            Adj R-squared =  0.8991
   Total |   212961.38    49  4346.15061            Root MSE      =  20.941
```

```
------------------------------------------------------------------------------
    csat |      Coef.   Std. Err.       t     P>|t|     [95% Conf. Interval]
---------+--------------------------------------------------------------------
 expense |  -.0022508    .0041333     -0.545   0.589    -.0106045     .006103
 percent |   -2.93786    .2302596    -12.759   0.000    -3.403232   -2.472488
  income |  -.0004919    .0010255     -0.480   0.634    -.0025645    .0015806
 college |   3.900087    1.719409      2.268   0.029     .4250318    7.375142
    high |   2.175542    1.171767      1.857   0.071     -.192688    4.543771
    reg1 |  -33.78456    9.302983     -3.632   0.001    -52.58659   -14.98253
    reg2 |  -143.5149    101.1244     -1.419   0.164    -347.8949    60.86509
reg2perc |   2.506616    1.404483      1.785   0.082    -.3319506    5.345183
    reg3 |  -8.799205    12.54658     -0.701   0.487    -34.15679    16.55838
   _cons |   839.2209    76.35942     10.990   0.000     684.8927     993.549
------------------------------------------------------------------------------
```

We might now try to simplify this model, dropping first that predictor with the highest t probability (*income*, $P = .634$), then re-estimating the model and deciding whether to drop something further. Through this process of backward elimination we seek a more parsimonious model, one that is simpler but fits almost equally well. Ideally, this strategy is pursued with attention both to the statistical results and to the substantive or theoretical implications of keeping or discarding certain variables.

For analysts in a hurry, stepwise methods provide ways to automate the process of model selection. They work either by subtracting predictors from a complicated model, or by adding predictors to a simpler one, according to some pre-set statistical criteria. Stepwise methods cannot consider the substantive or theoretical implications of their choices, nor can they do much troubleshooting to evaluate possible weaknesses in the models produced at each step. Despite their drawbacks, stepwise methods meet certain practical needs and have been widely used.

For automatic backward elimination, we issue a **sw regress** command that includes all our possible predictor variables and a maximum P value required to retain them. Setting the P-to-retain criteria as **pr(.05)** ensures that only predictors having coefficients that are significantly different from zero at the .05 level or less are kept in the model.

```
. sw regress csat expense percent income college high reg1
      reg2 reg2perc reg3, pr(.05)
                      begin with full model
p = 0.6341 >= 0.0500  removing income
p = 0.5273 >= 0.0500  removing reg3
p = 0.4215 >= 0.0500  removing expense
p = 0.2107 >= 0.0500  removing reg2
  Source |       SS       df       MS                  Number of obs =      50
---------+------------------------------               F(  5,    44) =   91.01
   Model |  194185.761     5  38837.1521               Prob > F      =  0.0000
Residual |  18775.6194    44  426.718624               R-squared     =  0.9118
---------+------------------------------               Adj R-squared =  0.9018
   Total |   212961.38    49  4346.15061               Root MSE      =  20.657
```

```
------------------------------------------------------------------------------
    csat |      Coef.   Std. Err.       t     P>|t|       [95% Conf. Interval]
---------+--------------------------------------------------------------------
    reg1 |  -30.59218    8.479395     -3.608   0.001     -47.68128    -13.50309
 percent |  -3.119155    .1804553    -17.285   0.000     -3.482839    -2.755471
reg2perc |   .5833272    .1545969      3.773   0.000      .2717577     .8948967
 college |   3.995495    1.359331      2.939   0.005      1.255944     6.735046
    high |   2.231294    .8178968      2.728   0.009      .5829313     3.879657
   _cons |    806.672    49.98744     16.137   0.000      705.9289     907.4151
------------------------------------------------------------------------------
```

sw regress dropped first *income*, then *reg3*, *expense*, and finally *reg2* before settling on the final model. Although it has four fewer coefficients, this final model has almost the same R^2 (.9118 versus .9176) and a higher R^2_a (.9018 versus .8991) compared with the earlier version.

If, instead of *P*-to-retain, **pr(.05)**, we specify a *P*-to-enter value such as **pe(.05)**, then **sw regress** performs forward inclusion (starting with an "empty" or constant-only model) instead of backward elimination. Other stepwise options include hierarchical selection and locking certain predictors into the model. For example, the following command specifies that the first term (***x1***) should be locked into the model and not subject to possible removal:

```
. sw regress y x1 x2 x3, pr(.05) lockterm1
```

Typing the following command calls for forward inclusion of any predictors found significant at the .10 level, but with variables *x4*, *x5*, and *x6* treated as one unit—either entered or left out together.

```
. sw regress y x1 x2 x3 (x4 x5 x6), pe(.10)
```

The following command invokes backward elimination with a $P = .20$ criterion:

```
. sw regress y x1 x2 x3 (x4 x5 x6) x7, pr(.20) hier
```

The hierarchical (**hier**) option specifies that the terms are ordered: Consider dropping the last term (*x7*) first, and stop if it is not dropped. If *x7* is dropped, next consider the second-to-last term, (*x4 x5 x6*) , and so forth.

Many other Stata commands besides **regress** also have stepwise versions that work in a similar manner. The list of available stepwise procedures includes the following:

sw cnreg	Censored normal regression
sw cox	Cox proportional hazard model regression
sw ereg	Exponential regression
sw fit	OLS regression with diagnostic statistics
sw glm	General linear models

`sw logistic`	Logistic regression
`sw ologit`	Ordered logistic regression
`sw oprobit`	Ordered probit regression
`sw poisson`	Poisson regression
`sw probit`	Probit regression
`sw qreg`	Quantile regression
`sw regress`	OLS regression
`sw tobit`	Tobit regression
`sw weibull`	Weibull regression

Type **`help sw`** for more details about stepwise options and logic.

Polynomial Regression

Earlier in this chapter, Figures 6.1 and 6.2 revealed what appears to be a distinctly curvilinear relation between mean composite SAT score (*csat*) and the percentage of high school seniors taking the test (*percent*). Figure 6.6 illustrates one way to model the upturn in SAT scores at high *percent* values: as a phenomenon peculiar to the Northeastern states. The interaction model fit reasonably well (R^2_a= .8327). But Figure 6.8, a residuals versus predicted values plot for the regression shown in Figure 6.6, still exhibits signs of trouble.

Chapter 8 presents a variety of techniques for curvilinear and nonlinear regression. "Curvilinear regression" here refers to OLS (for example, **`regress`**) employing nonlinearly-transformed *y* or *x* variables. Although curvilinear regression fits a curved model to the data, this model remains linear in the transformed variables. (Nonlinear regression, also discussed in Chapter 8, applies non-OLS methods to fit models that cannot be linearized through transformation.)

One simple type of curvilinear regression, called polynomial regression, often succeeds in fitting U or inverted-U shaped curves. It includes as predictors both the independent variable and its square (and possibly higher powers if necessary). Because the *csat* – *percent* relation appears somewhat U-shaped, we generate a new variable equal to *percent* squared, then include *percent* and $percent^2$ as predictors of *csat.* Figure 6.9 graphs the resulting curve.

Figure 6.8

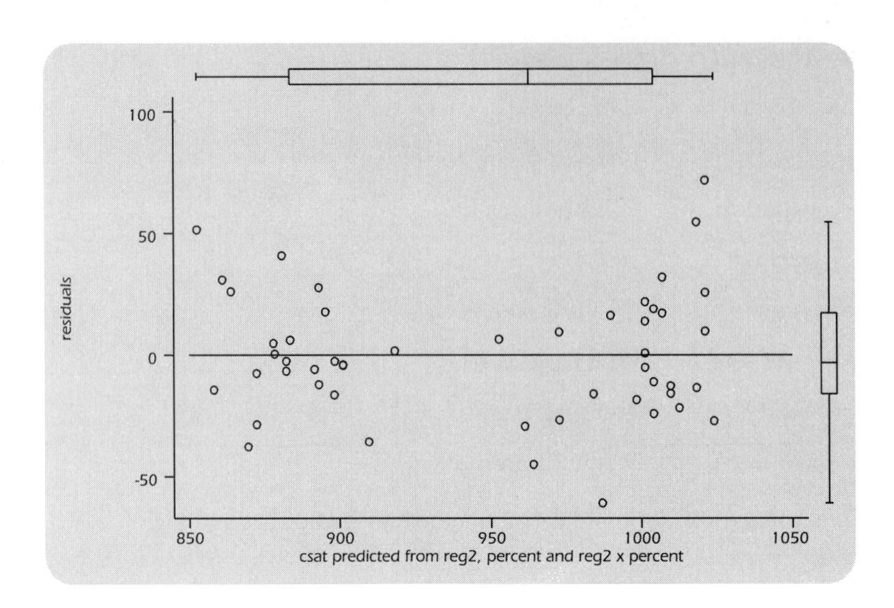


```
. generate percent2 = percent^2

. regress csat percent percent2
  Source |       SS       df       MS                  Number of obs =      51
---------+------------------------------               F(  2,    48) =  153.48
   Model |  193721.829     2  96860.9146               Prob > F      =  0.0000
Residual |  30292.6806    48  631.097513               R-squared     =  0.8648
---------+------------------------------               Adj R-squared =  0.8591
   Total |   224014.51    50   4480.2902               Root MSE      =  25.122

------------------------------------------------------------------------------
    csat |      Coef.   Std. Err.       t     P>|t|       [95% Conf. Interval]
---------+--------------------------------------------------------------------
 percent |  -6.111993   .6715406     -9.101   0.000      -7.462216    -4.76177
percent2 |   .0495819   .0084179      5.890   0.000       .0326566    .0665072
   _cons |   1065.921   9.285379    114.796   0.000       1047.252    1084.591
------------------------------------------------------------------------------

. predict yhat2

. graph csat yhat2 percent, connect(.s) symbol(Oi)
      ylabel xlabel
```

The polynomial model matches the data slightly better than our interaction model (R^2_a = .8591 versus .8327). Because the curvilinear pattern is now mostly gone from a residual versus predicted values plot (Figure 6.10), the usual assumption of independent, identically distributed errors also appears more plausible relative to this polynomial model.

Figure 6.9

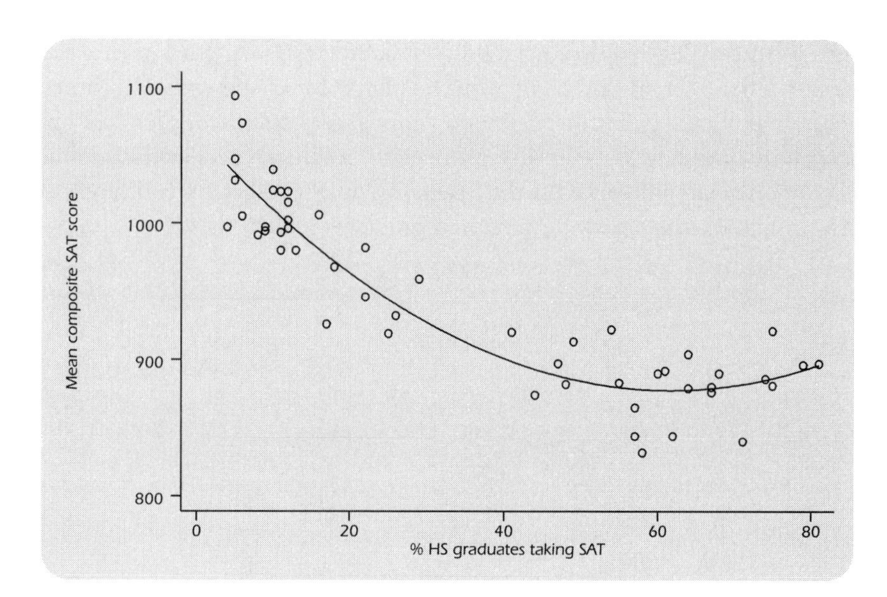

In Figures 6.6 and 6.9, we have two alternative models for the observed upturn in SAT scores at high levels of student participation. Statistical evidence seems to lean toward the polynomial model at this point. For serious research, however, we ought to choose between similar-fitting alternative models on substantive as well as statistical grounds. Which model seems more useful, or makes more sense? Which, if either, regression model suggests or corresponds to a good real-world explanation for the upturn in test scores at high levels of student participation?

Figure 6.10

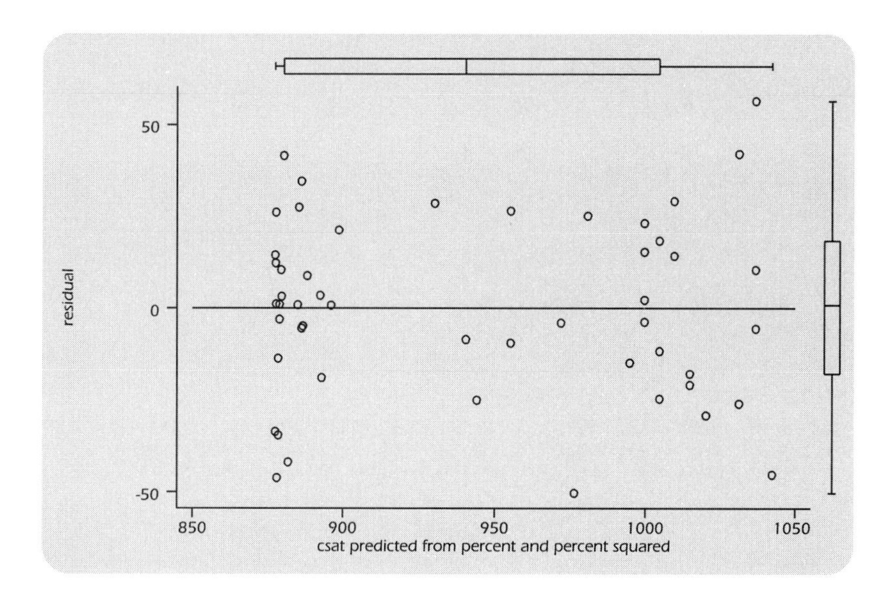

Although it can closely fit sample data, polynomial regression also has several statistical weakness. The different powers of x can be highly correlated with each other, giving rise to multicollinearity. Furthermore, polynomial regression tends to track observations that have unusually large positive or negative x values, so a few data points can exert disproportionate influence on the results. Chapter 7 takes a second look at this example, using tools that check for potential problems such as these.

7

Regression Diagnostics

Do the data give us any reason to distrust our regression results? Can we find better ways to specify the model, or to estimate its parameters? Careful diagnostic work, checking for potential problems and evaluating the plausibility of key assumptions, forms a crucial step in modern data analysis. We fit an initial model, but then look closely at our results for signs of trouble or ways in which the model needs improvement. Many of the general methods introduced in earlier chapters, such as boxplots, normality tests, or just sorting and listing the data, prove useful for troubleshooting. Stata also provides a toolkit of more specialized methods designed for this purpose.

Example Commands

`. regdw y x1 x2 x3, t(time)`

Performs OLS regression of *y* on *x1*, *x2*, and *x3*, including a Durbin–Watson test for first-order autocorrelation of errors. `t( )` specifies a variable indicating the time at which each observation was recorded; this variable could also be one of the predictors. We generally assume that the observations are equally spaced in time.

`. corc y x1 x2 x3, t(time)`

Regresses *y* on *x1*, *x2*, and *x3*, using the Cochrane–Orcutt correction for first-order autocorrelation, which assumes equal spacing in time.

. fit ***y x1 x2 x3***

Performs OLS regression of *y* on *x1*, *x2*, and *x3*. **fit** accomplishes the same thing as **regress**, except that after **fit** we can obtain a wider variety of diagnostic statistics and plots—including many of those described by Belsley, Kuh, and Welsch (1980) and Cook and Weisberg (1982). *All the following examples assume that we first have run* **fit**.

. ovtest, rhs

Performs the Ramsey regression specification error test (*RESET*) for omitted variables. The option **rhs** calls for using powers of the right-hand-side variables (from the recent **fit**), instead of powers of predicted *y* (default).

. hettest

Performs Cook and Weisberg's test for heteroscedasticity. If we have reason to suspect that heteroscedasticity is a function of a particular predictor *x1*, we could focus on that by typing **hettest** ***x1***.

. rvfplot

Graphs the residuals versus the fitted (predicted) values of *y*.

. rvpplot ***x1***

Graphs the residuals against values of predictor *x1*.

. avplot ***x1***

Constructs an added-variable plot (also called a partial-regression or leverage plot) showing the relation between *y* and *x1*, both adjusted for other *x* variables in the recent **fit** command. Such plots help visually identify outliers and influence points.

. avplots

Draws and combines in one image all of the added-variable plots from the recent **fit**.

. cprplot ***x1***

Constructs a component-plus-residual plot (also known as a partial residual plot) showing the adjusted relation between *y* and predictor *x1*. Such plots help detect nonlinearities in the data.

. acprplot ***x1*****, connect(m) bands(7)**

Constructs an augmented component-plus-residual plot (also known as an augmented partial residual plot), often better than **cprplot** in screening for nonlinearities. The options **connect(m) bands(7)** call for connecting the cross-medians of seven vertical bands with line segments. Alternatively, we might ask for a lowess-smoothed curve with bandwidth 0.5, by specifying the options **`connect(k) bwidth(.5)`**.

`. lvr2plot`

Constructs a leverage-versus-squared-residual plot (also known as an L-R plot).

`. fpredict` ***`new`*****`, covratio`**

Generates a new variable equal to Belsley, Kuh, and Welsch's *COVRATIO* statistic. *COVRATIO* measures the *i*th case's influence on the variance-covariance matrix of the estimated coefficients.

`. fpredict` ***`new`*****`, dfits`**

Generates a new variable equal to Welsch and Kuh's influence statistic *DFFITS*. Other available statistics that summarize each case's overall influence on the fitted model include Cook's distance *D* (**`fpredict`** ***`new`*****`, cooksd`**) and Welsch's distance *W* (**`fpredict`** ***`new`*****`, welsch`**).

`. dfbeta`

Automatically generates *DFBETAS* for each predictor variable in the previous `fit`. *DFBETAS* measure how much each case affects each coefficient.

Autocorrelation

Autocorrelation refers to correlation between sequential values of a variable. It often exists when the data comprise a time or spatial series. After a regression analysis, autocorrelation can persist among the residuals, indicating that we have omitted some variable that affects adjacent y values in similar ways. Autocorrelated errors lead to inefficient coefficient estimates, biased standard error estimates, and invalid t or F tests.

"First-order autocorrelation" is the simplest type: Correlation occurs between successive errors, e_t and e_{t+1}. The Durbin–Watson statistic provides a well-known test for positive first-order autocorrelation. To obtain this statistic, perform regression with **`regdw`** substituted for the usual **`regress`** command. The syntax and output of **`regdw`** resemble those of **`regress`**. An error message results if the data do not match the test's basic assumptions:

1. A Durbin–Watson test examines the data sequentially, thus only makes sense if the observations have some inherent order, as does a time series. To obtain the test, we need to specify what variable denotes "time" (or the observation's place in a sequence). We assume that the observations are equally spaced in this sequence, for example, at equal time intervals.

2. **regdw** cannot proceed if any variables in the model have missing values.

A small time-series dataset (adapted from Brown, Kane, and Roodman 1994) relevant to the global warming hypothesis illustrates the test for autocorrelation:

```
. use co2, clear
(Global warming (Brown 1994))

. describe
Contains data from c:\data\co2.dta
  obs:            29                        Global warming (Brown 1994)
 vars:             3                        26 Dec 1996 16:43
 size:           406 (96.5% of memory free)
-------------------------------------------------------------------------------
   1. year        int     %8.0g             Year
   2. co2         float   %9.0g             Atmospheric CO2, ppm
   3. temp        float   %9.0g             Global mean temp, C
-------------------------------------------------------------------------------
Sorted by:  year
```

These data, graphed in Figure 7.1, show the steady buildup of carbon dioxide (CO_2) in the earth's atmosphere due to burning of fossil fuels. Accompanying this buildup has been an erratic upward trend in the mean global temperature, which many scientists believe reflects "global warming" caused by the CO_2 buildup. Note the use of the **rescale** option with the following **graph** command, to draw separately scaled left- and right-hand axes for the CO_2 and temperature time series.

```
. graph temp co2 year, connect(ll) ylabel rlabel rescale
      xlabel(1965,1970,1975,1980,1985,1990)
```

co2 rises almost linearly with *year*; indeed the two have a near-perfect .9972 correlation:

```
. correlate
(obs=29)

       |     year      co2     temp
-------+---------------------------
   year|   1.0000
    co2|   0.9972   1.0000
   temp|   0.7519   0.7620   1.0000
```

Both *year* and *co2* correlate with *temp*, suggesting that we could model *temp* almost equally well as a function of carbon dioxide buildup, or simply as a linear upward trend caused by some other factor that increases with time.

Because these are time series data, it seems prudent to check for autocorrelation when we regress *temp* on *co2*. **regdw** accomplishes both, revealing that carbon dioxide concentration explains about 58% of the sample variance in mean global temperature. The **t( )** option, which is actually not optional, identifies the "time" variable.

Figure 7.1

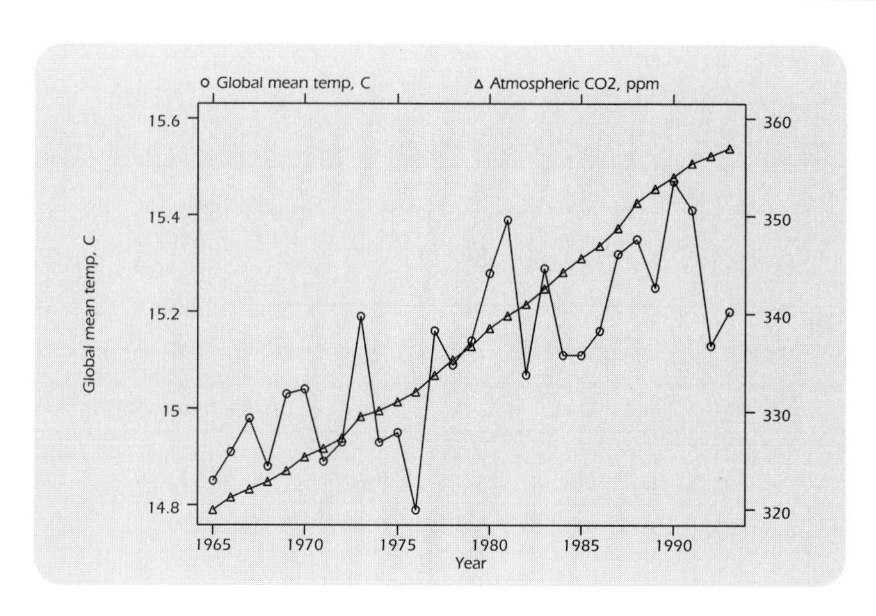


```
. regdw temp co2, t(year)
  Source |       SS       df       MS                  Number of obs =      29
---------+------------------------------               F(  1,    27) =   37.38
   Model |  .541649057     1  .541649057               Prob > F      =  0.0000
Residual |  .391233554    27  .014490132               R-squared     =  0.5806
---------+------------------------------               Adj R-squared =  0.5651
   Total |  .932882611    28  .033317236               Root MSE      =  .12037

------------------------------------------------------------------------------
    temp |      Coef.   Std. Err.       t     P>|t|       [95% Conf. Interval]
---------+--------------------------------------------------------------------
     co2 |   .0117341   .0019192      6.114   0.000       .0077962    .0156721
   _cons |   11.15195   .6483844     17.200   0.000       9.821576    12.48233
------------------------------------------------------------------------------
Durbin-Watson Statistic =   1.879804
```

Most regression texts include tables of critical values for the Durbin–Watson statistic. Consulting such a table tells us that, with 29 observations and one *x* variable, the critical values are $d_L = 1.12$ and $d_U = 1.25$. In this example we cannot reject the null hypothesis that there is no first-order autocorrelation among the errors because our calculated Durbin–Watson statistic exceeds d_U $(1.8798 > 1.25)$. If the Durbin–Watson statistic had been less than 1.12, we would instead have rejected this null hypothesis and concluded that significant positive autocorrelation exists.

Suppose that we had, in fact, found significant autocorrelation. One way to proceed then involves regression with a Cochrane–Orcutt correction. This can be accomplished through the **corc** command:

```
. corc temp co2, t(year)
Iteration 0:   rho = 0.0000
Iteration 1:   rho = 0.0324

(Cochrane-Orcutt regression)

  Source |       SS       df       MS                  Number of obs =      28
---------+------------------------------               F(  1,    26) =   29.61
   Model |  .441111207     1  .441111207               Prob > F      =  0.0000
Residual |  .387322448    26  .014897017               R-squared     =  0.5325
---------+------------------------------               Adj R-squared =  0.5145
   Total |  .828433655    27  .030682728               Root MSE      =  .12205

------------------------------------------------------------------------------
    temp |      Coef.   Std. Err.       t     P>|t|       [95% Conf. Interval]
---------+--------------------------------------------------------------------
     co2 |   .0114231   .0020992      5.442   0.000       .0071081    .0157381
  _inter |   11.25907   .7105931     15.845   0.000        9.79842    12.71971
---------+--------------------------------------------------------------------
     rho |     0.0332     0.1961      0.169   0.867        -0.3692      0.4356
------------------------------------------------------------------------------

Durbin-Watson statistic (original)     1.879804
Durbin-Watson statistic (transformed) 1.953272
```

The estimated autocorrelation coefficient appears at the bottom of the regression table. This model incorporates a relation between successive errors:

$$temp_t = 11.36 + .01\,co2 + u_t \qquad\qquad u_t = .0332u_{t-1} + e_t$$

Again indicating the absence of autocorrelation here, the estimated coefficient (`rho` = .0332) is not significantly different from zero. Its 95% confidence interval is

$$-.3692 < \rho < .4356$$

Both the Durbin–Watson and Cochrane–Orcutt procedures consider only first-order autocorrelation. If the data contain more complicated patterns such as cycles or longer lags, then we might need the full toolkit of time series regression methods to proceed. Simple graphs of residuals versus time or observation number help alert analysts to potential problems, however. Figure 7.2 shows a graph of the residuals from our previous **regdw** model against year. No obvious patterns are visible, supporting our earlier conclusion that first-order autocorrelation poses no threat to this regression.

```
. quietly regdw temp co2, t(year)

. predict e, resid

. graph e year, ylabel yline(0) border connect(l)
      xlabel(1965,1970,1975,1980,1985,1990)
```

Figure 7.2

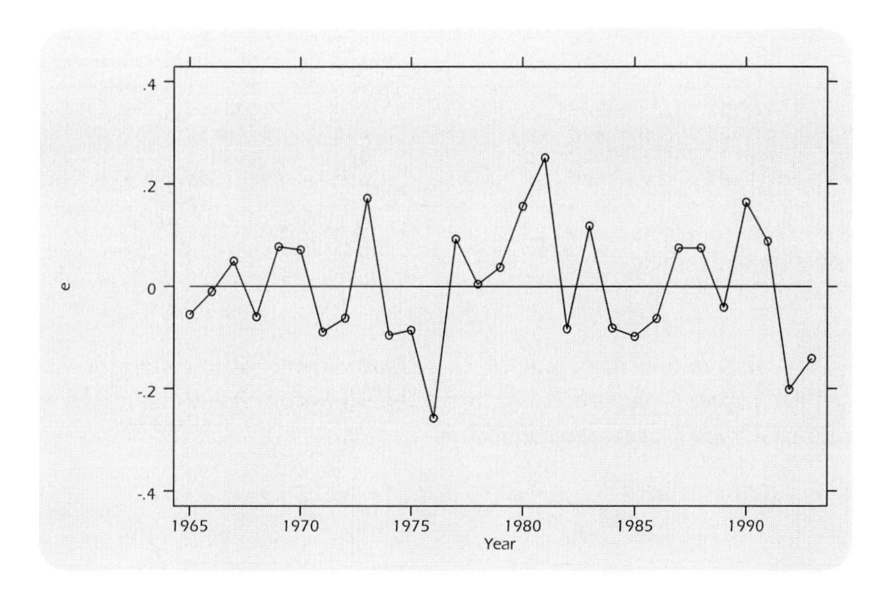

SAT Score Regression, Revisited

Chapter 6 examined regression models for mean state Scholastic Aptitude Test scores. Several possible predictors, including interaction and polynomial terms, were considered over the course of Chapter 6. The remainder of this chapter continues with this example, focusing on a relatively simple model that explains about 92% of the variance in mean state SAT scores. It contains three predictors: *percent* (percentage of high school graduates taking the test), *percent2* (*percent* squared), and *high* (percentage of adults with a high school diploma).

```
. use states90, clear
(U.S. states data 1990-91)

. gen percent2 = percent^2

. regress csat percent percent2 high
  Source |       SS       df       MS                  Number of obs =      51
---------+------------------------------               F(  3,    47) =  193.37
   Model |  207225.103     3  69075.0343               Prob > F      =  0.0000
Residual |  16789.4069    47  357.221424               R-squared     =  0.9251
---------+------------------------------               Adj R-squared =  0.9203
   Total |   224014.51    50   4480.2902               Root MSE      =   18.90
```

```
------------------------------------------------------------------------------
    csat |      Coef.   Std. Err.       t     P>|t|       [95% Conf. Interval]
---------+--------------------------------------------------------------------
 percent |  -6.520312    .5095805    -12.795   0.000     -7.545455   -5.495168
percent2 |   .0536555    .0063678      8.426   0.000      .0408452    .0664658
    high |   2.986509    .4857502      6.148   0.000      2.009305    3.963712
   _cons |   844.8207    36.63387     23.061   0.000      771.1228    918.5185
------------------------------------------------------------------------------
```

The regression equation is

$$\text{predicted } csat = 844.82 - 6.52 percent + .05 percent2 + 2.99 high$$

Figure 7.3 depicts interrelations among these four variables as a scatterplot matrix. As noted in Chapter 6, the squared term *percent2* allows our regression model to fit the obviously curvilinear *csat – percent* scatter.

```
. graph percent percent2 high csat, matrix half symbol(o)
```

An alternative to **regress**, the **fit** command also accomplishes OLS regression. While performing the regression, however, **fit** unobtrusively calculates and stores certain results that will make subsequent diagnostic work easier.

```
. fit csat percent percent2 high

  Source |       SS       df       MS                  Number of obs =      51
---------+------------------------------               F(  3,    47) =  193.37
   Model |  207225.103     3  69075.0343               Prob > F      =  0.0000
Residual |  16789.4069    47  357.221424               R-squared     =  0.9251
---------+------------------------------               Adj R-squared =  0.9203
   Total |   224014.51    50   4480.2902               Root MSE      =   18.90

------------------------------------------------------------------------------
    csat |      Coef.   Std. Err.       t     P>|t|       [95% Conf. Interval]
---------+--------------------------------------------------------------------
 percent |  -6.520312    .5095805    -12.795   0.000     -7.545455   -5.495168
percent2 |   .0536555    .0063678      8.426   0.000      .0408452    .0664658
    high |   2.986509    .4857502      6.148   0.000      2.009305    3.963712
   _cons |   844.8207    36.63387     23.061   0.000      771.1228    918.5185
------------------------------------------------------------------------------
```

Two hypothesis tests, available after **fit**, perform checks on the model specification. The omitted-variables test **ovtest** essentially regresses y on the x variables and also the second, third, and fourth powers of predicted y (after standardizing $\hat{y}$ to have mean 0 and variance 1). It then performs an F test of the null hypothesis that all three coefficients on those powers of $\hat{y}$ equal zero. If we reject this null hypothesis, further polynomial terms would improve the model. With the *csat* regression, we need not reject the null hypothesis:

```
. ovtest

Ramsey RESET test using powers of the fitted values of csat
       Ho:  model has no omitted variables
                 F(3, 44) =      1.48
                 Prob > F =      0.2319
```

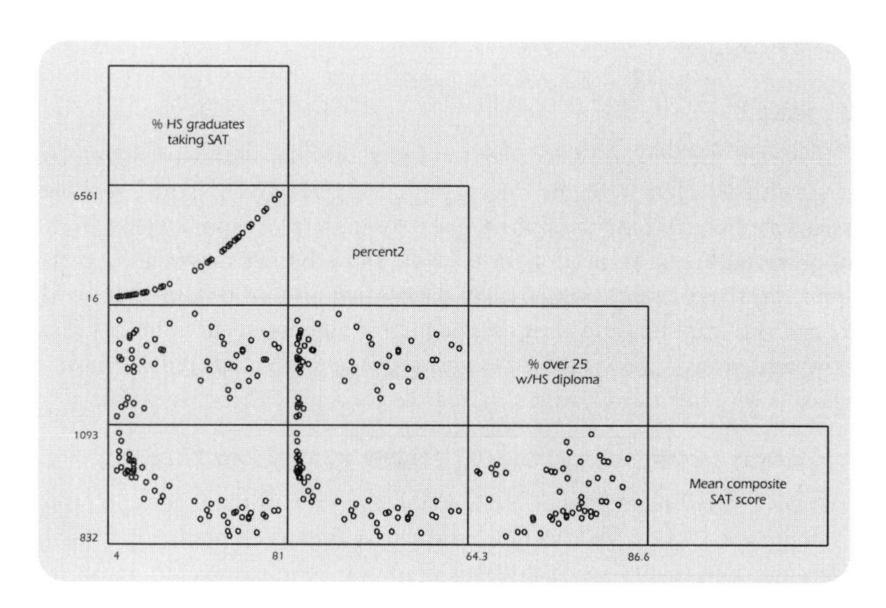

Figure 7.3

A heteroscedasticity test, **hettest**, tests the assumption of constant error variance by examining whether squared standardized residuals are linearly related to $\hat{y}$ (see Cook and Weisberg, 1994, for discussion and example). Results from the *csat* regression suggest that in this instance we should reject the null hypothesis of constant variance.

```
. hettest
Cook-Weisberg test for heteroscedasticity using fitted values of csat
     Ho: Constant variance
         chi2(1)      =     4.86
         Prob > chi2  =     0.0274
```

"Significant" heteroscedasticity means that our standard errors and hypothesis tests might be invalid. In Figure 7.4 we will see why this result occurs.

Diagnostic Plots

Chapter 6 demonstrated how **predict** can generate new variables such as predicted values and residuals after a **regress** command. **fpredict** serves this purpose after **fit**. Scatterplots of residuals versus predicted values give us a simple diagnostic display (Figures 6.2, 6.8 and 6.10). We could obtain a residual-versus-predicted graphs for the recent **fit** regression by these three commands:

```
. fpredict yhat3
```

```
. fpredict e3, resid

. graph e3 yhat3
```

After **fit**, however, there is an easier way to obtain residual-versus-predicted (or "residual-versus-fitted") plots: type the single command **rvfplot**. Figure 7.4 shows a residual versus predicted plot with optional one-way scatterplots and boxplots in the margins, a horizontal line at 0, neatly labeled axes, and a border drawn around the plot. This plot shows residuals symmetrically distributed around 0 (consistent with the normal-errors assumption) and with no evidence of outliers or curvilinearity. The largest positive and negative residuals occur with above-average predicted values of *y*, however, which is why **hettest** earlier rejected the constant-variance hypothesis.

```
. rvfplot, oneway twoway box yline(0) ylabel xlabel box border
```

Residual-versus-fitted plots provide a one-graph overview of the regression residuals. For more detailed study, we can plot residuals against each predictor variable separately through a series of "residual-versus-predictor" commands. To graph the residuals against predictor *high* (not shown):

```
. rvpplot high
```

The one-variable graphs described in Chapter 3 can also be employed for residual analysis. For example, we could use boxplots to check the residuals for outliers or skew, or quantile-normal plots to evaluate the assumption of normal errors.

Added-variable plots are valuable diagnostic tools, known by different names including partial-regression leverage plots, adjusted partial residual plots, or adjusted

Figure 7.4

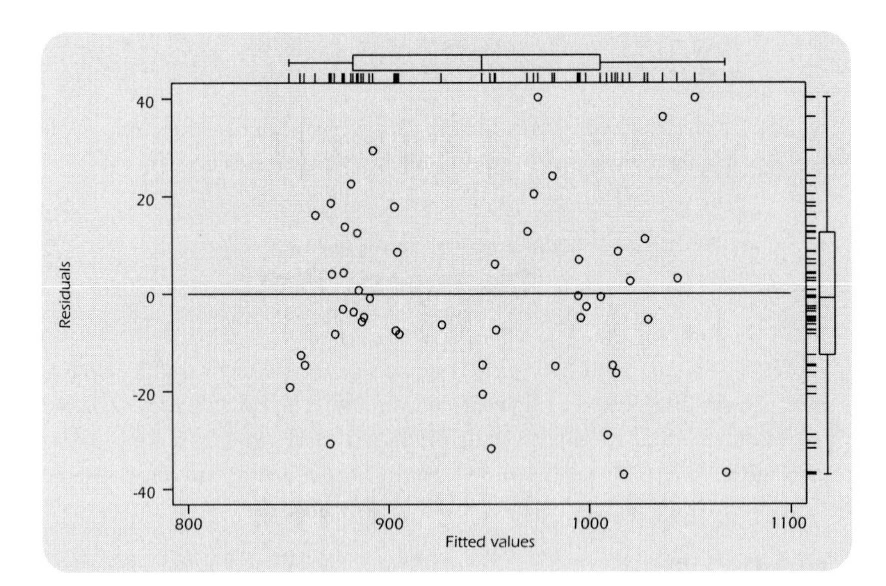

variable plots. They depict the relation between *y* and one *x* variable, adjusting for the effects of other *x* variables. If we regressed *y* on *x2* and *x3*, and likewise regressed *x1* on *x2* and *x3*, then took the residuals from each regression and graphed these residuals in a scatterplot, we would thereby obtain an added-variable plot for the relation between *y* and *x1*, adjusted for *x2* and *x3*. An **avplot** command performs the necessary calculations automatically, so we can obtain the adjusted-variable plot for predictor *high*, for example, just by typing

```
. avplot high
```

To speed the process further, we could type **avplots** to obtain a complete set of tiny added-variable plots with each of the predictor variables in the preceding regression. Figure 7.5 shows the results from the regression of *csat* on *percent*, *percent2*, and *high*. The lines drawn in added-variable plots have slopes equal to the corresponding partial regression coefficients. For example, the slope of the line at lower left in Figure 7.5 equals 2.99, which is the coefficient on *high*.

```
. avplots, ylabel xlabel border
```

Added-variable plots help uncover observations exerting a disproportionate influence on the regression model. In simple regression with one *x* variable, ordinary scatterplots suffice for this purpose. In multiple regression, however, the signs of influence become more subtle. An observation with an unusual combination of values on several *x* variables might have high leverage, or potential to influence the regression, even though none of its individual *x* values is unusual by itself. High-leverage observations

Figure 7.5

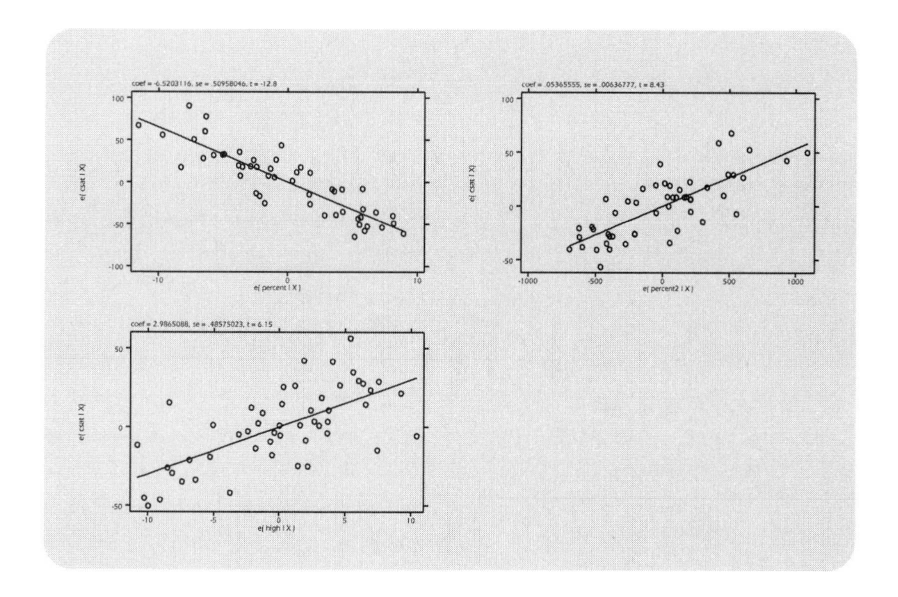

show up in added-variable plots as points horizontally distant from the rest of the data. We saw no such problems in the Figure 7.5, however.

With added-variable plots, as with any scatterplot, we can use observation identifiers as the plotting symbols. Suppose that we did notice an outlier in an added-variable plot and wanted to learn which observation this was. In the states' SAT scores example, we might do so by using state names as the plotting symbols. For readability these names are plotted at 110% of normal size in Figure 7.6.

```
. avplot high, symbol([state]i) psize(110)
```

Component-plus-residual plots, produced by commands of the form **cprplot x1**, take a different approach to graphing multiple regression. The component-plus residual plot for variable *x1* graphs each observation's residual plus its component predicted from *x1* against values of *x1*.

$$e_i + b_1 \, x1_i$$

Such plots can help diagnose nonlinearities and suggest alternative functional forms. An augmented component-plus-residual plot (Mallows, 1986) works somewhat better, although both types often seem inconclusive. Figure 7.7 shows an augmented component-plus-residual plot from the regression of *csat* on *percent*, *percent2*, and *high*.

```
. acprplot high, connect(m) bands(5) ylabel xlabel border
```

The straight line in Figure 7.7 corresponds to the regression model. The segmented curve connects cross-medians (**connect(m)**) within five vertical bands

Figure 7.6

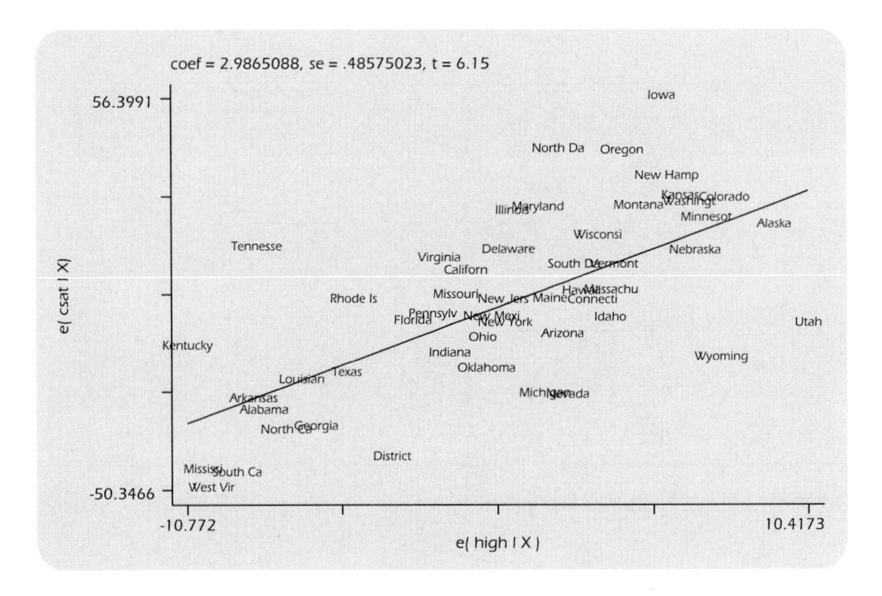

Figure 7.7

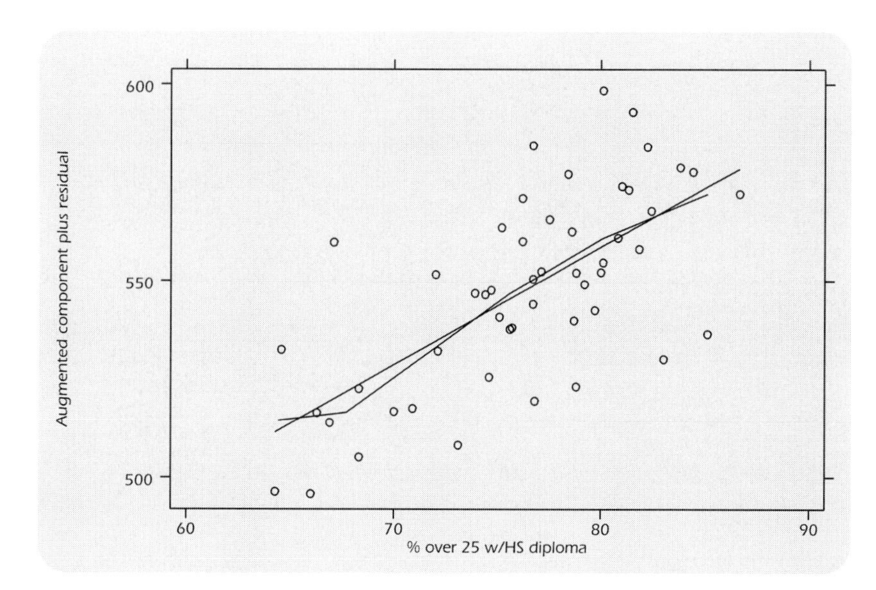

(**bands(5)**). If these cross-medians showed some curved pattern unlike the linear regression model, we would have reason to doubt the model's adequacy. In Figure 7.7, however, the component-plus-residuals median closely resembles the regression model itself. This plot therefore reinforces the conclusion we drew earlier from Figure 7.4: The present regression model adequately accounts for all nonlinearity visible in the raw data (Figure 7.3), so that none remains apparent in its residuals.

As its name implies, a leverage-versus-squared-residuals plot graphs leverage (hat matrix diagonals) against the residuals squared. Figure 7.8 shows such a plot for the *csat* regression. To identify individual observations we make values of *state*, at 110% of their default size, the plotting symbols:

```
. lvr2plot, ylabel xlabel border symbol([state]) psize(110)
```

Lines in a leverage-versus-squared-residuals plot mark the means of leverage (horizontal line) and squared residuals (vertical line). Leverage tells us how much potential an observation has to influence the regression, based on its particular combination of *x* values—extreme *x* values or unusual combinations give observations high leverage. Large squared residuals indicate observations with *y* values much different than those predicted by the regression model. Connecticut, Massachusetts, and Mississippi have the greatest potential leverage, but the model fits them relatively well. (This is not necessarily good. Sometimes, although not here, high-leverage observations exert so much influence that they control the regression, and it *must* fit them well.) Iowa and Tennessee are poorly fit, but have less potential influence. Utah stands out as one observation that is both ill fit and potentially influential.

Figure 7.8

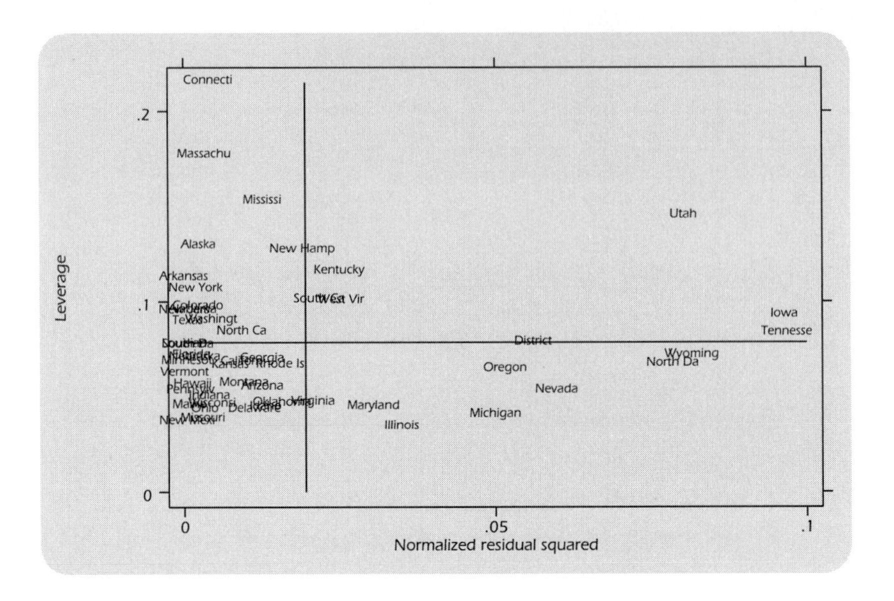


```
. list csat yhat3 percent high e3 if state == "Utah"

         csat       yhat3    percent       high          e3
 45.     1031    1067.712          5       85.1   -36.71239
```

Only 5% of Utah students took the SAT, and 85.1% of the state's adults graduated from high school. This unusual combination of near-extreme values on both *x* variables is the source of the state's leverage and leads our model to predict mean SAT scores 36.7 points higher than what Utah students actually achieved. To see exactly how much difference this one observation makes, we could repeat the regression using Stata's "not equal to" qualifier `!=` to set Utah aside. Because *state* is a string variable, we enclose its value "Utah" in double quotes:

```
. regress csat percent percent2 high if state != "Utah"

  Source |       SS       df       MS                Number of obs =      50
---------+------------------------------             F(  3,    46) =  202.67
   Model |  201097.423     3  67032.4744             Prob > F      =  0.0000
Residual |  15214.0968    46  330.741235             R-squared     =  0.9297
---------+------------------------------             Adj R-squared =  0.9251
   Total |   216311.52    49  4414.52082             Root MSE      =  18.186

------------------------------------------------------------------------------
    csat |      Coef.   Std. Err.       t     P>|t|       [95% Conf. Interval]
---------+--------------------------------------------------------------------
 percent |  -6.778706   .5044217    -13.439   0.000     -7.794054   -5.763357
percent2 |   .0563562   .0062509      9.016   0.000      .0437738    .0689387
    high |   3.281765   .4865854      6.744   0.000      2.302319     4.26121
   _cons |   827.1159   36.17138     22.867   0.000      754.3067    899.9252
------------------------------------------------------------------------------
```

In the $n = 50$ (instead of $n = 51$) regression, all three coefficients have strengthened a bit since we deleted an ill-fit observation. The general conclusions remain unchanged, however.

Chambers et al. (1983) provide more detailed examples and explanations of diagnostic plots and other graphical methods for data analysis.

Diagnostic Case Statistics

After using **fit**, we can obtain a variety of diagnostic statistics through the **fpredict** command, which has the following options:

`. fpredict new`	Predicted values of *y*.
`. fpredict new, cooksd`	Cook's *D* influence measures.
`. fpredict new, covratio`	*COVRATIO* influence measures.
`. fpredict new, dfbeta(x1)`	*DFBETAS* measuring each case's influence on the regression coefficient on predictor *x1*.
`. fpredict new, dfits`	*DFFITS* influence measures.
`. fpredict new, hat`	Diagonal elements of hat matrix (leverage).
`. fpredict new, leverage`	Hat matrix diagonals (same as **hat**).
`. fpredict new, residuals`	Residuals.
`. fpredict new, rstandard`	Standardized residuals.
`. fpredict new, rstudent`	Studentized residuals.
`. fpredict new, stdf`	Standard errors of predicted individual *y*, sometimes called the standard error of forecast or the standard error of prediction.
`. fpredict new, stdp`	Standard errors of predicted mean *y*.
`. fpredict new, stdr`	Standard errors of residuals.
`. fpredict new, welsch`	Welsch's distance influence measures.

When using **fpredict**, substitute new variable names of your choosing for *new* in the previous examples. The variables created by **predict** or **fpredict** are all case statistics, meaning that they have values for each observation in the data. Diagnostic work usually begins by calculating the predicted values and residuals.

There is some overlap in purpose among other **fpredict** statistics. Many attempt to measure how much each observation influences regression results. "Influencing regression results," however, could refer to several different things—effects on the *y*-intercept, on a particular slope coefficient, on all the slope coefficients, or on the estimated standard errors, for example. Consequently, we have a variety of alternative case statistics designed to measure influence.

Standardized and studentized residuals (**rstandard** and **rstudent**) help to identify outliers among the residuals—observations that particularly contradict the regression model. Studentized residuals have the most straightforward interpretation. They correspond to the *t* statistic we would obtain by including in the regression a dummy predictor coded 1 for that observation and 0 for all others. Thus, they test whether a particular observation significantly shifts the *y* intercept.

Hat matrix diagonals (**hat**) measure leverage, meaning the potential to influence regression coefficients. Observations possess high leverage when their *x* values (or their combination of *x* values) are unusual. Several other statistics measure actual influence on coefficients. *DFBETAS* (obtained either through **fpredict *new*, dfbeta(*x1*)**, or more conveniently just by typing **dfbeta**) indicate by how many standard errors the coefficient on *x1* would change, if observation *i* were dropped from the regression. Cook's *D* (**cooksd**), Welsch's distance (**welsch**), and *DFFITS* (**dfits**), on the other hand, all try to summarize how much observation *i* influences the regression model as a whole—or equivalently, how much observation *i* influences the set of predicted values. *COVRATIO* measures the influence of the *i*th observation on the estimated standard errors. Below we generate a full set of diagnostic statistics including *DFBETAS* for all three predictors. Note that **fpredict** supplies labels automatically for the variables it creates.

```
. fpredict standard, rstandard

. fpredict student, rstudent

. fpredict h, hat

. fpredict D, cooksd

. fpredict DFFITS, dfits

. fpredict W, welsch

. fpredict COVRATIO, covratio
```

```
. dfbeta

DFpercen:   DFbeta(percent)
DF1:        DFbeta(percent2)
DFhigh:     DFbeta(high)

. describe standard-DFhigh

  25. standard     float   %9.0g                Standardized residuals
  26. student      float   %9.0g                Studentized residuals
  27. h            float   %9.0g                Leverage
  28. D            float   %9.0g                Cook's D
  29. DFFITS       float   %9.0g                Dfits
  30. W            float   %9.0g                Welsch distance
  31. COVRATIO     float   %9.0g                Covratio
  32. DFpercen     float   %9.0g                DFbeta(percent)
  33. DF1          float   %9.0g                DFbeta(percent2)
  34. DFhigh       float   %9.0g                DFbeta(high)

. summarize standard-DF1

Variable |     Obs        Mean   Std. Dev.        Min         Max
---------+-------------------------------------------------------------
standard |      51   -.0031359   1.010579  -2.099976    2.233379
 student |      51     -.00162   1.032723  -2.182423    2.336977
       h |      51    .0784314   .0373011   .0336437    .2151227
       D |      51    .0219941   .0364003   .0000135    .1860992
  DFFITS |      51   -.0107348   .3064762   -.896658    .7444486
       W |      51    -.089723   2.278704  -6.854601     5.52468
COVRATIO |      51    1.092452   .1316834   .7607449    1.360136
DFpercen |      51     .000938   .1498813  -.5067295    .5269799
     DF1 |      51   -.0010659   .1370372   -.440771    .4253958
  DFhigh |      51   -.0012204   .1747835  -.6316988    .3414851
```

summarize shows us the minimum and maximum values of each statistic, so we can quickly check whether any are large enough to cause concern. For example, special tables could be used to determine whether the observation with the largest absolute studentized residual (*student*) constitutes a significant outlier. Alternatively, we could apply the Bonferroni inequality and t distribution table: max|*student*| is significant at level α if $|t|$ is significant at α/n. In this example, we have max|*student*| = 2.337 (Iowa) and $n = 51$. For Iowa to be a significant outlier (cause a significant shift in intercept) at $\alpha = .05$, $t = 2.337$ must be significant at .05 / 51:

```
. display .05/51
.00098039
```

Stata's **tprob( )** function can approximate the probability of $|t| > 2.337$, given $df = n-K-1 = 51-3-1 = 47$:

```
. display tprob(47, 2.337)
.02375138
```

The obtained P-value ($P = .0238$) is not below $\alpha/n = .00098$, so Iowa is not a significant outlier at $\alpha = .05$.

Studentized residuals measure the *i*th case's influence on the *y*-intercept. Cook's *D*, *DFFITS*, and Welsch's distance all measure the *i*th case's influence on all coefficients in the model (or, equivalently, on all *n* predicted *y* values). To list the five most influential observations as measured by Cook's *D*:

```
. sort D

. list state yhat3 D DFFITS W in -5/1

              state      yhat3          D      DFFITS          W
 47.   North Dakota   1036.696   .0705921    .5493086   4.020527
 48.        Wyoming   1017.005   .0789454   -.5820746  -4.270465
 49.      Tennessee   974.6981    .111718    .6992343   5.162398
 50.           Iowa    1052.78   .1265392    .7444486    5.52468
 51.           Utah   1067.712   .1860992    -.896658  -6.854601
```

The **in -5/1** qualifier tells Stata to list only the fifth-from-last (-5) through last (lowercase letter "l") observations. Figure 7.9 shows one way to display influence graphically: Symbols in a residual-versus-predicted plot are given sizes ("importance weights") proportional to values of Cook's *D*. Five influential observations stand out, with large positive or negative residuals and high predicted *csat* values.

```
. graph e3 yhat3 [iweight = D], ylabel xlabel yline(0)
```

Although they have different statistical rationales, Cook's *D*, Welsch's distance, and *DFFITS* are closely related, and in practice they tend to flag the same observations as influential. Figure 7.10 shows their similarity in the current example.

Figure 7.9

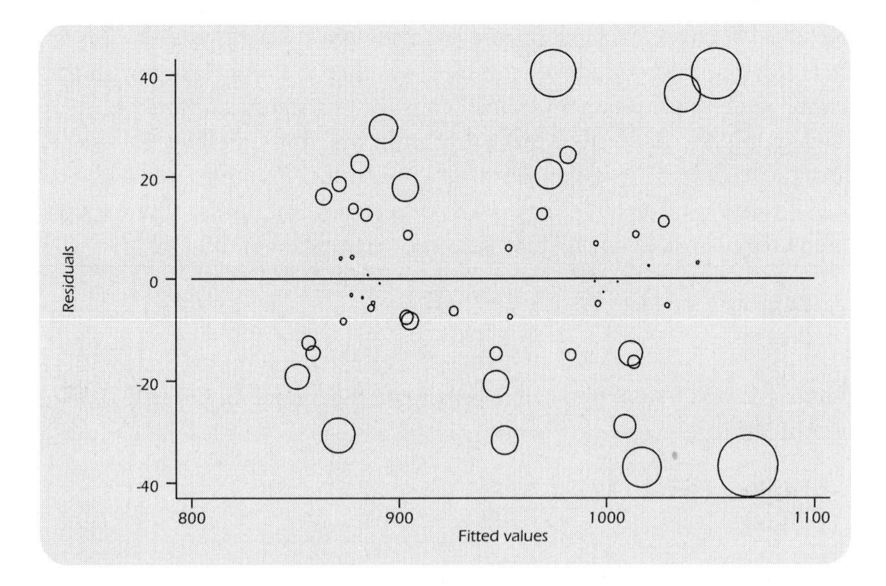

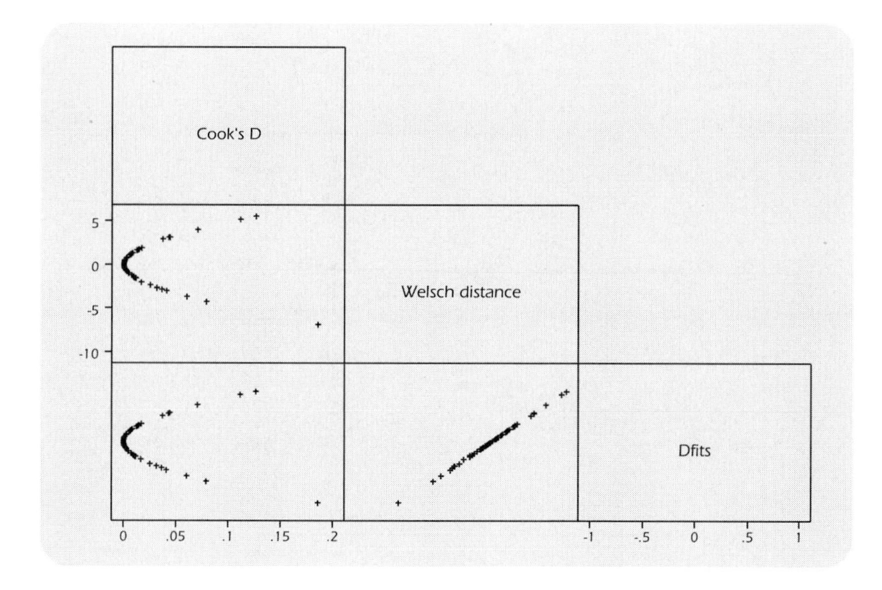

Figure 7.10

```
. graph D W DFFITS, matrix label half symbol(p)
```

DFBETAS indicate how much each observation influences each regression coefficient. Typing **dfbeta** after a **fit** regression automatically generates *DFBETAS* for each predictor. In this example, they were given the names *DFhigh* (*DFBETAS* for predictor *high*), *DFpercen*, and *DF1* (*DFBETAS* for *percent2*). Figure 7.11 shows their distributions as boxplots.

```
. graph DFpercen DF1 DFhigh, box ylabel(-.5,0,.5)
      symbol([state]) psize(110)
```

From left to right, Figure 7.11 plots the distributions of *DFBETAS* for *percent, percent2,* and *high*. (We could more easily distinguish them in color.) The extreme values in each plot belong to Iowa and Utah, which also have the two highest Cook's *D* values. For example, Utah's *DFhigh* = –.63. This tells us that Utah causes the coefficient on *high* to be .63 standard errors lower than it would be, if Utah were set aside. Similarly, *DFpercen* = .53 indicates that with Utah present, the coefficient on *percent* is .53 standard errors higher (because the *percent* regression coefficient is negative, "higher" here means closer to 0) than it otherwise would be. In other words, Utah weakens the apparent effects of both *high* (percent adults with high school diplomas) and *percent* (percentage of seniors taking the SAT).

The most direct way to learn how particular observations affect a regression is to repeat that regression with the observations set aside. For example, we could set aside all states that move one or more coefficients by half a standard error (that is, have absolute *DFBETAS* greater than .5):

Figure 7.11

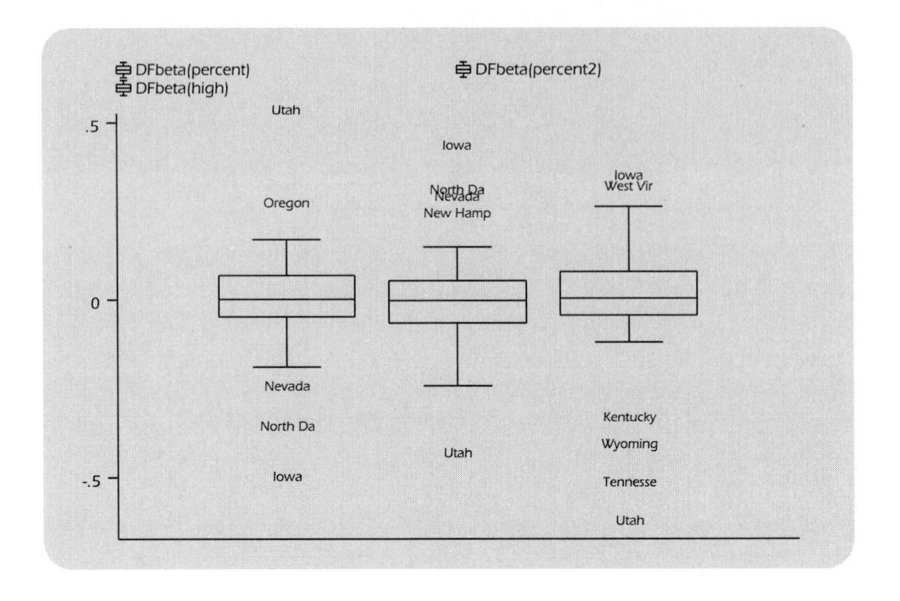


```
. fit csat percent percent2 high if abs(DFpercen) < .5
      & abs(DF1) < .5 & abs(DFhigh) < .5

  Source |       SS       df       MS                  Number of obs =      48
---------+------------------------------               F(  3,    44) =  215.47
   Model |  175366.782     3  58455.5939               Prob > F      =  0.0000
Residual |  11937.1351    44  271.298525               R-squared     =  0.9363
---------+------------------------------               Adj R-squared =  0.9319
   Total |  187303.917    47  3985.18972               Root MSE      =  16.471

------------------------------------------------------------------------------
    csat |      Coef.   Std. Err.       t     P>|t|       [95% Conf. Interval]
---------+--------------------------------------------------------------------
 percent |  -6.510868   .4700719    -13.851   0.000     -7.458235     -5.5635
percent2 |   .0538131    .005779      9.312   0.000      .0421664    .0654599
    high |    3.35664   .4577103      7.334   0.000      2.434186    4.279095
   _cons |   815.0279   33.93199     24.019   0.000      746.6424    883.4133
------------------------------------------------------------------------------
```

Careful inspection will reveal the details in which this regression table (based on $n = 48$) differs from its $n = 51$ or $n = 50$ counterparts seen earlier. Our central conclusion—that mean state SAT scores are well predicted by the percentage of adults with high school diplomas and, curvilinearly, by the percentage of students taking the test—remains unchanged, however.

Although diagnostic statistics draw attention to influential observations and allow us to see how great their influence is, they do not answer the question of whether we should set those observations aside. That requires a substantive decision, based on careful evaluation of the data and research context. In this example, we have no substantive reason to discard any states, and even the most influential of them do not fundamentally change our conclusions.

Using any fixed definition of what constitutes an "outlier," we are liable to see more of them in larger-size samples. For this reason, sample-size-adjusted cutoffs are sometimes recommended for identifying unusual observations. After fitting a regression model with K coefficients (including the constant) based on n observations, we might look more closely at those observations for which any of the following are true:

leverage $h > 2K/n$

Cook's $D > 4/n$

$|DFFITS| > 2\sqrt{K/n}$

$|Welsch's\ W| > 3\sqrt{K}$

$|DFBETAS| > 2/\sqrt{n}$

$|COVRATIO - 1| \geq 3K/n$

The reasoning behind these cutoffs, and the diagnostic statistics more generally, can be found in Cook and Weisberg (1982, 1994) or Belsley, Kuh, and Welsch (1980). For an introduction, see Rawlings (1988).

Multicollinearity

If perfect multicollinearity (linear relation) exists among the predictors, regression equations become unsolvable. Stata handles this by warning the user, then automatically dropping one of the offending predictors. High but not perfect multicollinearity causes more subtle problems. When we add a new x variable that is strongly related to x variables already in the model, signs of possible trouble include the following:

1. Substantially higher standard errors, with correspondingly lower t statistics
2. Unexpected changes in coefficient magnitudes or signs
3. Nonsignificant coefficients despite a high R^2

The matrix of correlations between estimated regression coefficients provides one diagnostic for multicollinearity. This matrix can be displayed after **regress**, **fit**, **anova** or other estimation procedures by typing

```
. correlate, _coef

        |  percent percent2     high    _cons
--------+------------------------------------
 percent|   1.0000
percent2|  -0.9798   1.0000
    high|  -0.2122   0.1798   1.0000
   _cons|   0.0433  -0.0277  -0.9807   1.0000
```

This command refers to the most recently estimated model. For this example, that would be the **fit** regression with $n = 48$ observations.

High correlations between any pair of coefficients indicate potential multicollinearity problems. In this example, the correlations between the constant and the coefficient on *high*, and between coefficients on *percent* and *percent2* (both –.98) give cause for concern. In polynomial regression, or regression with interaction terms, some *x* variables are calculated directly from other *x* variables. Although strictly speaking their relation is nonlinear, it might be close enough to linear to cause statistical problems. Regression attempts to estimate the independent effects of each *x* variable, but cannot do so reliably if the *x* variables vary too closely together.

The problem comes into sharper focus if we regress each *x* on the other *x* variables. Then calculate $1 - R^2$ from this regression to see what fraction of the first *x* variable's variance is independent of the other *x* variables. For example, we find that about 97% of the *high* variance is independent of *percent* and *percent2*:

```
. quietly regress high percent percent2

. display 1-_result(7)
.96942331
```

After regression, `_result(7)` holds the resulting R^2. Similar commands reveal that only 4% of the *percent* variance is independent, however:

```
. quietly regress percent high percent2

. display 1-_result(7)
.04010307
```

Because *percent* and *percent2* are closely related, we cannot estimate their separate effects with nearly as much precision as when we regress *csat* on *percent* alone. That is, the standard error for the coefficient on *percent* increases almost threefold (from .18 to .51) when we go from a regression simple of *csat* on *percent*, to a multiple regression of *csat* on *percent*, *percent* squared, and *high*. Despite this loss of precision, we can still distinguish all the multiple regression coefficients from zero. Moreover, the multiple regression equation constitutes a much better prediction model.

Some authors recommend "centering" as a strategy for reducing multicollinearity in polynomial regression. Centering involves subtracting the mean from *x* variable values before creating the squared term. This creates a new *x* variable centered on zero and much less correlated with its own squared values. The resulting regression fits the same as an uncentered version. By reducing multicollinearity, centering often (*but not always*) yields more precise coefficient estimates with lower standard errors. The following commands perform these steps with the SAT scores example:

```
. summ percent

Variable |     Obs        Mean    Std. Dev.        Min        Max
---------+-----------------------------------------------------
 percent |      51    35.76471    26.19281          4         81

. gen Cperc = percent - _result(3)

. gen Cperc2 = Cperc^2

. summ percent Cperc Cperc2

Variable |     Obs        Mean    Std. Dev.        Min        Max
---------+-----------------------------------------------------
 percent |      51    35.76471    26.19281          4         81
   Cperc |      51    2.24e-07    26.19281  -31.76471   45.23529
  Cperc2 |      51    672.6113    456.0967   27.40831   2046.232

. correlate Cperc Cperc2 high csat
(obs=51)

        |   Cperc   Cperc2     high     csat
--------+------------------------------------
  Cperc|  1.0000
 Cperc2|  0.3791   1.0000
   high|  0.1413  -0.0417   1.0000
   csat| -0.8758  -0.0428   0.0858   1.0000

. fit csat Cperc Cperc2 high

  Source |       SS       df       MS              Number of obs =      51
---------+------------------------------           F(  3,    47) =  193.37
   Model |  207225.103     3  69075.0343           Prob > F      =  0.0000
Residual |   16789.407    47  357.221426           R-squared     =  0.9251
---------+------------------------------           Adj R-squared =  0.9203
   Total |   224014.51    50  4480.2902            Root MSE      =   18.90

------------------------------------------------------------------------------
    csat |      Coef.   Std. Err.       t     P>|t|      [95% Conf. Interval]
---------+--------------------------------------------------------------------
   Cperc |  -2.682362   .1119085    -23.969   0.000     -2.907493   -2.457231
  Cperc2 |   .0536555   .0063678      8.426   0.000      .0408452    .0664659
    high |   2.986509   .4857502      6.148   0.000      2.009305    3.963712
   _cons |   680.2552   37.82329     17.985   0.000      604.1646    756.3458
------------------------------------------------------------------------------
```

```
. correlate, _coef

       |    Cperc   Cperc2      high    _cons
-------+------------------------------------
  Cperc|   1.0000
 Cperc2|  -0.3893   1.0000
   high|  -0.1700   0.1040   1.0000
  _cons|   0.2105  -0.2151  -0.9912   1.0000
```

The overall fit, predictions, and *F* test of a centered model should be identical to those of its uncentered form. In this example the coefficients on *high* and *Cperc2*, and their standard errors and *t* tests, remain likewise unchanged. The coefficient on the centered variable *Cperc*, however, appears different and has a much smaller standard error compared with our previous uncentered regression. In this instance, centering did successfully improve that coefficient estimate's precision.

8

Fitting Curves

This chapter describes three broad approaches to fitting curves:

1. Nonparametric methods, including band regression and lowess smoothing
2. Linear regression with transformed variables ("curvilinear regression")
3. Nonlinear regression

Nonparametric regression often serves as an exploratory tool because it can summarize nonlinear data patterns visually without requiring the analyst to specify a particular model in advance. Transformed variables extend the usefulness of linear parametric methods, such as OLS, to encompass many types of curvilinear relations. Nonlinear regression, on the other hand, requires a different class of methods to estimate parameters of intrinsically nonlinear models.

Example Commands

`. graph` *y x*`, bands(10) connect(m)`

Produces a *y* versus *x* scatterplot with line segments connecting the (median *x*, median *y*) points within 10 equal-width vertical bands (a form of "band regression"). Typing **`connect(s)`** instead of **`connect(m)`** results in the cross-medians being connected by a smooth cubic spline curve instead of by line segments.

```
. graph x1 x2 x3 y, bands(4) connect(s) matrix
```

Produces a scatterplot matrix, with band regression curves (based on four bands) in each plot.

```
. ksm y x, lowess bwidth(.4) gen(newvar)
```

Draws a lowess-smoothed curve on a scatterplot of *y* versus *x*. Lowess calculations use a bandwidth of .4 (40% of the data). Smoothed values are kept and named *newvar*.

```
. regress lny x1 sqrtx2 invx3
```

Transformed-variables regression generates new variables through commands such as the following, then performs regression using these new variables:

```
. generate lny = ln(y)
. generate sqrtx2 = sqrt(x2)
. generate invx3 = 1/x3
```

When, as in this example, the *y* variable is transformed, the predicted values generated by **predict *yhat***, or residuals generated by **predict *e*, resid**, will be likewise in transformed units. For graphing or other purposes we might want to return predicted values or residuals to raw-data units, using inverse transformations such as

```
. replace yhat = exp(yhat)
```

```
. boxcox y x1 x2 x3
```

Finds the maximum-likelihood Box–Cox transformation λ for *y*, assuming that $y^{(\lambda)}$ (see page 101) is a linear function of *x1* – *x3* plus Gaussian, constant-variance errors. To view the regression model estimated by this command, subsequently type **regress**

```
. nl exp2 y x
```

Performs nonlinear least squares to fit an exponential growth model (**exp2**) of the following form:

$$\text{predicted } y = b_1 \, b_2^{\,x}$$

```
. nlpred yhat
```

Calculates predicted values from the most recent **nl** model.

```
. nlpred e, resid
```

Calculates residuals from the most recent **nl** model.

```
. nl log4 y x, init(B0=5, B1=25, B2=.1, B3=50)
```

Performs nonlinear least squares to fit a logistic growth model (**log4**) of this form: predicted $y = b_0 + b_1 /(1 + \exp(-b_2 (x - b_3)))$. Set the initial parameter values for the iterative estimation process at $b_0 = 5$, $b_1 = 25$, and so forth

Band Regression

Nonparametric regression methods generally do not yield an explicit regression equation. They are primarily graphic tools for displaying the relation, possibly nonlinear, between *y* and *x*. Stata can draw a simple kind of nonparametric regression, band regression, onto any scatterplot or scatterplot matrix. For illustration, consider these sobering Cold War data (*missile2.dta*) from MacKenzie (1990). The observations are 48 types of long-range nuclear missiles, deployed by the United States and Soviet Union during their arms race, 1958 to 1990:

```
. use missile2, clear
(Missiles (MacKenzie 1990))
```

```
. describe
Contains data from c:\data\missile2.dta
  obs:            48                          Missiles (MacKenzie 1990)
 vars:             6                          22 Dec 1996 19:44
 size:         1,392 (95.8% of memory free)
-----------------------------------------------------------------------------
   1. missile      str15  %15s                Missile
   2. country      byte   %8.0g     soviet    US or Soviet missile?
   3. year         int    %8.0g               Year of first deployment
   4. type         byte   %8.0g     type      ICBM or submarine-launched?
   5. range        int    %8.0g               Range in nautical miles
   6. CEP          float  %9.0g               Circular Error Probable (miles)
-----------------------------------------------------------------------------
Sorted by:  country  year
```

Variables in the *missile2.dta* include an accuracy measure called the "Circular Error Probable" (*CEP*). *CEP* represents the radius of a bull's eye within which 50% of the missile's warheads should land. Year by year, scientists on both sides worked to improve accuracy (Figure 8.1):

```
. graph CEP year, bands(8) connect(m) ylabel xlabel
```

Figure 8.1 shows *CEP* declining (accuracy increasing) over time. The options **bands(8) connect(m)** instruct **graph** to divide the scatterplot into 8 equal-width vertical bands and draw line segments connecting the points (median *x*, median *y*) within each band. This curve traces how the median of *CEP* changes with *year*.

Figure 8.1

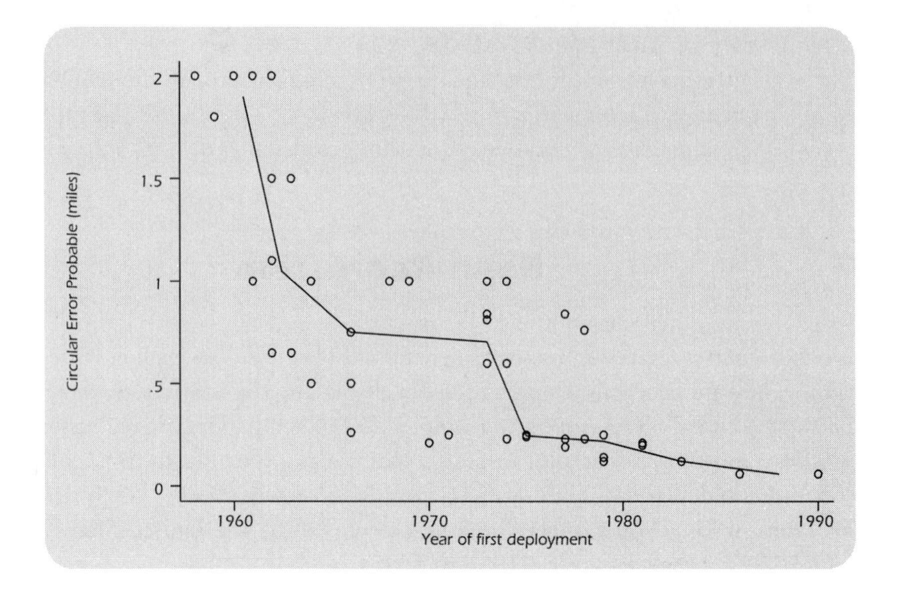

Nonparametric regression does not require the analyst to specify a relationship's functional form in advance. Instead, it helps us explore the data with an "open mind." This often uncovers interesting results, as when we view trends in U.S. and U.S.S.R. missile accuracy separately (Figure 8.2):

```
. graph CEP year, connect(m) bands(8) ylabel xlabel by(country)
```

The shapes of the two curves in Figure 8.2 differ substantially. U.S. missiles became much more accurate in the 1960s, permitting a shift to smaller warheads. Three or more small warheads would fit on the same size of missile that formerly carried one large warhead. The accuracy of Soviet missiles improved more slowly, apparently stalling during the late 1960s to early 1970s, and remained a decade or so behind their American counterparts. To make up for this accuracy disadvantage, Soviet strategy emphasized larger rockets carrying high-yield warheads. Nonparametric regression can assist with "qualitative" description of this sort or serve as a preliminary to fitting parametric models such as those described later.

We can add band regression curves to any scatterplot or scatterplot matrix by specifying the **connect(m) bands( )** options. Band regression's simplicity makes it a convenient exploratory tool, but it possesses one notable disadvantage—the bands have the same width across the range of *x* values, although some of these bands contain few or no observations. With normally distributed variables, for example, data density decreases toward the extremes. Consequently, the left and right endpoints of the band regression curve (which tend to dominate its appearance) often reflect just a

Figure 8.2

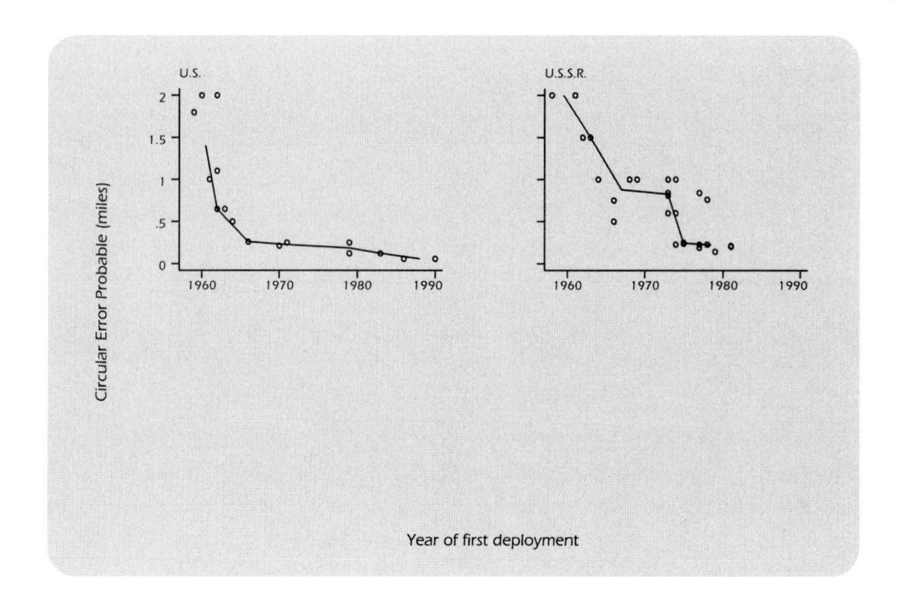

few data points. The next section describes a more sophisticated, computation-intensive approach.

Lowess Smoothing

Lowess stands for **lo**cally **we**ighted **s**catterplot **s**moothing. To obtain a lowess-smoothed graph of *CEP* against *year*, for U.S. missiles only, type

```
. ksm CEP year if country == 0, lowess
```

ksm permits all the usual Stata **graph** options and, with a **gen( )** option, also generates smoothed predicted values as a new variable (Figure 8.3):

```
. ksm CEP year if country == 0, lowess bwidth(.4) ylabel xlabel
      symbol(Oi) gen(lsCEP)
```

Like Figure 8.2, Figure 8.3 shows U.S. missile accuracy improving rapidly during the 1960s and progressing at a more gradual rate in the 1970s and 1980s. Lowess-smoothed values of *CEP* are generated here with the name *lsCEP*. The **bwidth(.4)** option specifies the lowess bandwidth: the fraction of the sample used in smoothing each point. The default is **bwidth(.8)**. The closer `bwidth` is to 1, the greater the degree of smoothing.

Figure 8.3

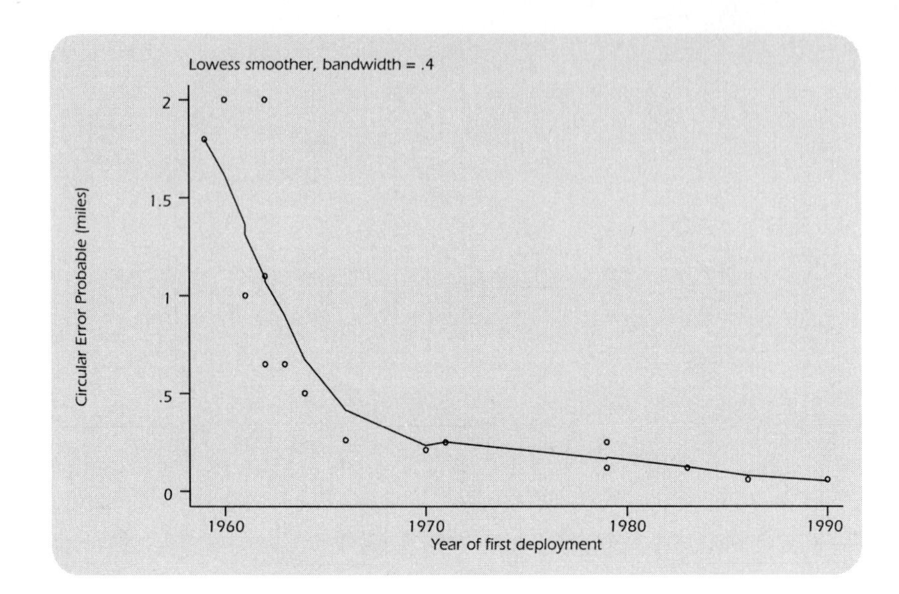

Lowess predicted (smoothed) y values for n observations result from n weighted regressions. Let k represent the half-bandwidth, truncated to an integer. For each y_i, a smoothed value y_i^s is obtained by weighted regression involving only those observations within the interval from $i = \max(1, i - k)$ through $i = \min(i + k, n)$. The jth observation within this interval receives weight w_j according to a tricube function:

$$w_j = (1 - |u_j|^3)^3$$

where

$$u_j = (x_i - x_j) / \Delta$$

Δ stands for the distance between x_i and its furthest neighbor within the interval. Weights equal 1 for $x_i = x_j$, but fall off to zero at the interval's boundaries. Chambers et al. (1983) and Cleveland (1985) provide further details and examples.

ksm is a flexible command. In addition to the usual **graph** options, it offers these special options:

line For running-line least squares smoothing; default is running mean.

weight For Cleveland's tricube weighting function; default is unweighted.

lowess Equivalent to specifying both options **line weight**.

bwidth( ) Specifies the bandwidth. Centered subsets of approximately bwidth × n observations are used for smoothing, except toward the endpoints where smaller, uncentered bands are used. The default is **bwidth(.8)**.

`logit`	Transforms smoothed values to logits.
`adjust`	Adjusts the mean of smoothed values to equal the mean of the original *y* variable; like **`logit`**, useful with dichotomous *y*.
`gen( )`	Creates *newvar* containing smoothed values of *y*.
`nograph`	Suppresses displaying the graph.

Because it requires *n* weighted regressions, lowess smoothing proceeds slowly with large samples.

In addition to smoothing *y–x* scatterplots, **ksm** can be used for exploratory time series smoothing. The file *ice.dta* contains results from the Greenland Ice Sheet Project 2 described in Mayewski, Holdsworth and colleagues (1993) and Mayewski, Meeker and colleagues (1994). Researchers extracted and chemically analyzed an ice core representing more than 100,000 years of climate history. *ice.dta* includes a small fraction of this information: measured non-sea salt sulfate concentration and an index of "Polar Circulation Intensity" since AD 1500.

```
. use ice, clear
(Greenland ice (Mayewski 1994))

. describe
Contains data from c:\data\ice.dta
  obs:           272                        Greenland ice (Mayewski 1993)
 vars:             3                        23 Dec 1996 13:25
 size:         5,984 (96.3% of memory free)
---------------------------------------------------------------------------
   1. year        int    %10.0g             Year
   2. sulfate     double %10.0g             SO4 ion concentration, ppb
   3. PCI         double %6.0g              Polar Circulation Intensity
---------------------------------------------------------------------------
Sorted by:  year
```

To retain more detail from this 272-point time series, we smooth with a relatively narrow bandwidth, only 5% of the sample:

```
. ksm sulfate year, lowess bwidth(.05) gen(smooth) nograph
```

Options in the previous command suppressed the default **ksm** graph but generated smoothed values as a new variable (*smooth*) so that we can construct a better-looking graph (Figure 8.4) directly:

```
. graph sulfate smooth year, connect(ll) symbol(ii) ylabel
      xlabel(1500,1600,1700,1800,1900,1985)
      l1(Sulfate concentration)
```

Figure 8.4

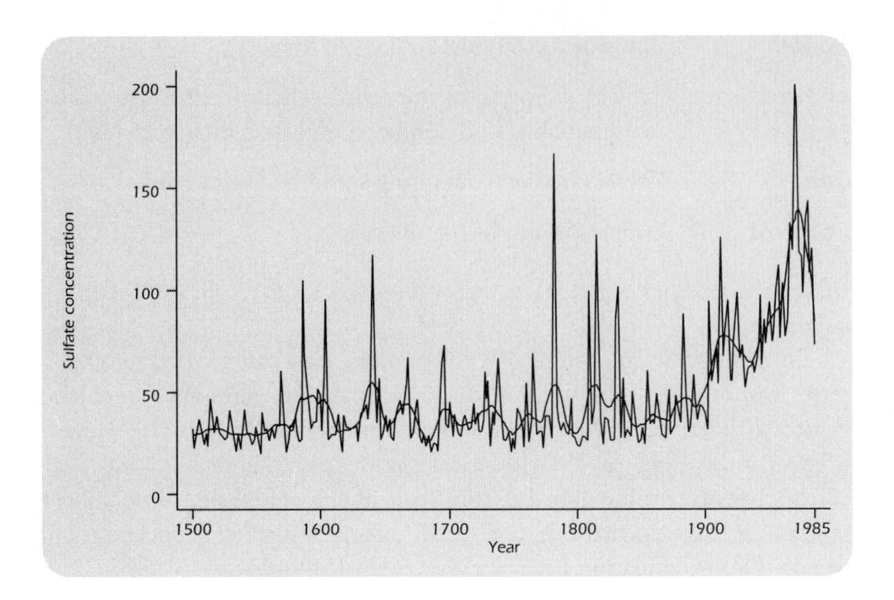

Non-sea salt sulfate (SO_4) reached the Greenland ice after being injected into the atmosphere, chiefly by volcanoes or the burning of fossil fuels such as coal and oil. Both the smoothed and raw curves in Figure 8.4 convey information. The smoothed curve shows oscillations around a slightly rising mean from 1500 through the early 1800s. After 1900, fossil fuels drive the smoothed curve upward, with temporary setbacks after 1929 (the Great Depression) and the early 1970s (combined effects of the U.S. Clean Air Act, 1970; the Arab oil embargo, 1973; and subsequent oil price

Figure 8.5

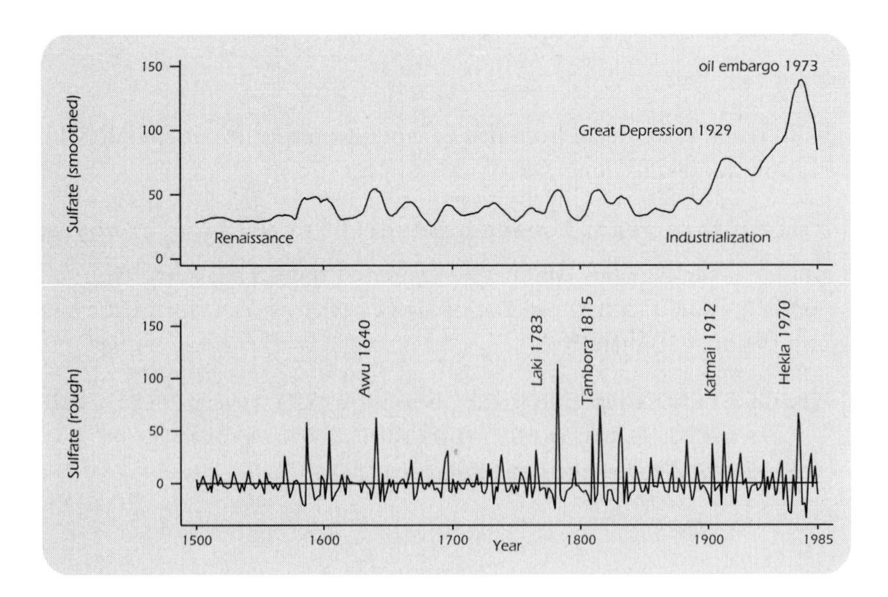

hikes). Most sharp peaks of the raw data have been identified with known volcanic eruptions such as Iceland's Hekla (1970) or Alaska's Katmai (1912).

After smoothing time series data, it is often useful to study the smooth and rough (residual) series separately. Figure 8.5 presents both in a graph enhanced using Stage, the Stata Graphics Editor.

Regression with Transformed Variables (Curvilinear Regression)

By subjecting one or more variables to nonlinear transformation, then including the transformed variable(s) in a linear regression, we implicitly fit a curvilinear model to the underlying data. Chapters 6 and 7 gave one example of this approach, polynomial regression, that incorporates second (and perhaps higher) powers of at least one x variable among the predictors. Other common transformations, applied to x or y variables, include logarithms, other powers, and Box–Cox transformations.

Dataset *tornado.dta* provides a simple illustration involving U.S. tornadoes over the years 1916 to 1986 (from the Council on Environmental Quality, 1988).

```
Contains data from c:\data\tornado.dta
  obs:           71                            Tornados (Env.Qual. 87-88)
 vars:            4                            24 May 1996 12:43
 size:          994 (95.8% of memory free)
-------------------------------------------------------------------------------
   1. year        int    %8.0g                 Year
   2. tornado     int    %8.0g                 Number of tornados
   3. lives       int    %8.0g                 Number of lives lost
   4. avlost      float  %9.0g                 Average lives lost/tornado
-------------------------------------------------------------------------------
Sorted by:  year
```

The number of fatalities decreased over this period, even as the number of recognized tornadoes increased—because of improvements in warnings and our ability to detect more tornadoes now, even those that do little damage. Consequently, the average lives lost per tornado (*avlost*) declined with time, but a linear regression (Figure 8.6) does not describe this trend well. The scatter descends more steeply than the regression line at first, then levels off in the mid-1950s. The regression line actually predicts negative numbers of deaths in later years. Furthermore, average tornado deaths exhibit more variation in early years than later—evidence of heteroscedasticity.

The relation becomes linear, and heteroscedasticity vanishes, if we work instead with the log of average lives lost (Figure 8.7):

Figure 8.6

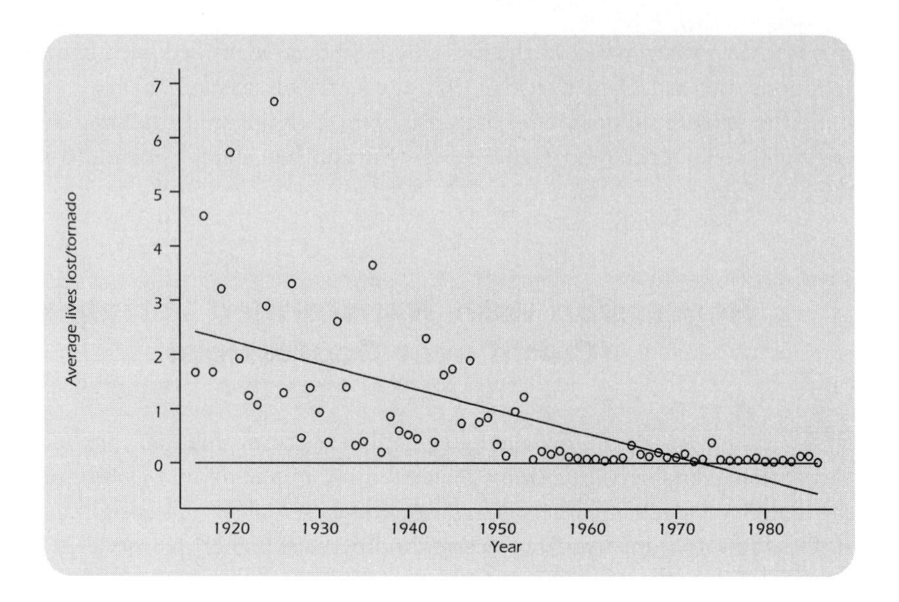


```
. generate lavlost = ln(avlost)

. label variable lavlost "Log of average lives lost"

. quietly regress lavlost year

. predict yhat

. graph lavlost yhat year, connect(.s) symbol(Oi) ylabel
    xlabel(1920,1930,1940,1950,1960,1970,1980)
```

Figure 8.7

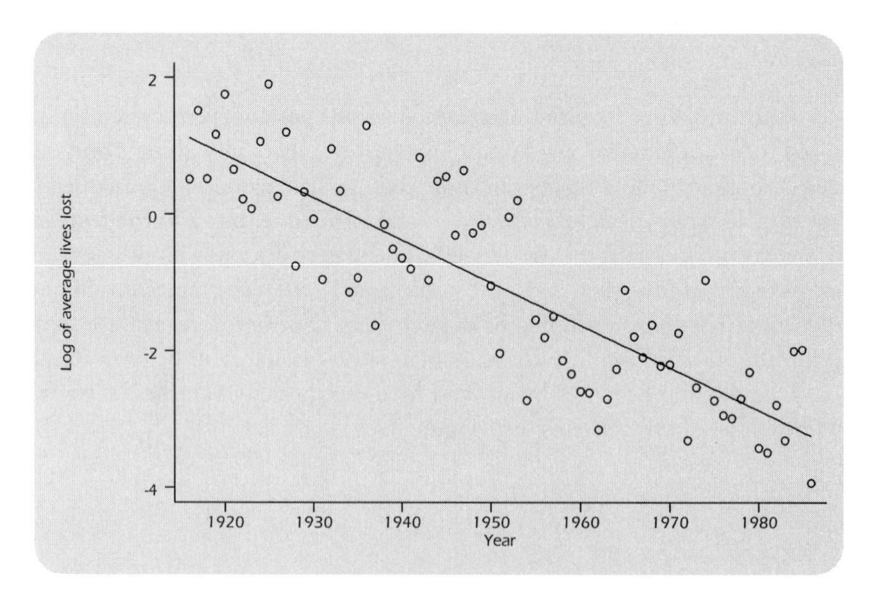

Because we regressed logarithms of lives lost on *year*, the predicted values (*yhat*) are also measured in logarithmic units. Return these predicted values to their natural units (lives lost) by inverse transformation, in this case exponentiating (*e* to power) *yhat:*

```
. replace yhat = exp(yhat)
(71 real changes made)
```

Graphing these inverse-transformed predicted values reveals the regression curve (Figure 8.8):

```
. graph avlost yhat year, yline(0) connect(.s) symbol(Oi)
      ylabel(0,1,2,3,4,5,6,7)
      xlabel(1920,1930,1940,1950,1960,1970,1980)
```

Contrast Figures 8.7 and 8.8 with Figure 8.6 to see how transformation made the analysis both simpler and more realistic.

For a multiple-regression example, we turn to the 109-country data in *nations.dta:*

```
Contains data from c:\data\nations.dta
  obs:           109                           Data on 109 countries
 vars:            15                           22 Dec 1996 20:12
 size:         4,033 (95.7% of memory free)

-----------------------------------------------------------------------
  1. country        str8    %9s                Country
  2. pop            float   %9.0g              1985 population in millions
  3. birth          byte    %8.0g              Crude birth rate/1000 people
```

Figure 8.8

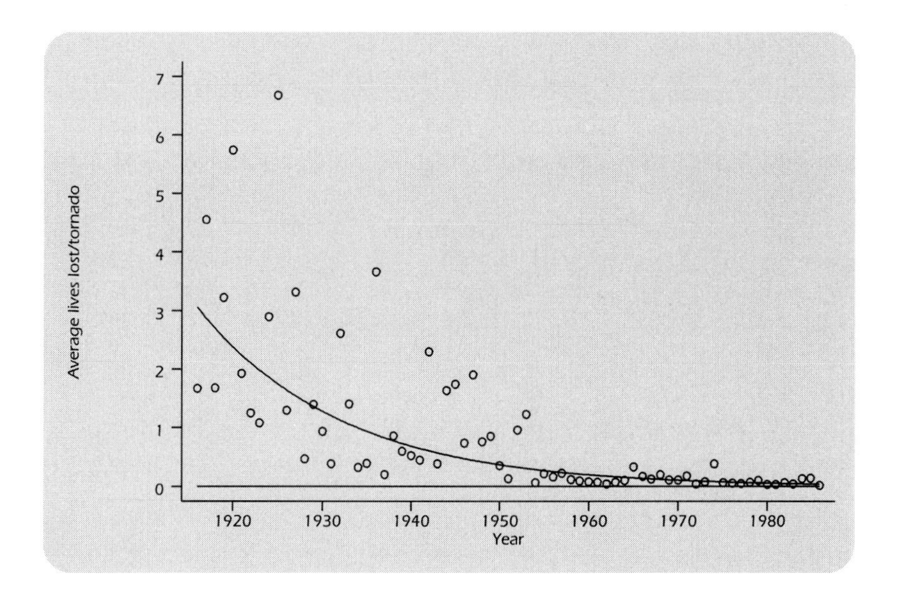

```
  4. death        byte   %8.0g              Crude death rate/1000 people
  5. chldmort     byte   %8.0g              Child (1-4 yr) mortality 1985
  6. infmort      int    %8.0g              Infant (<1 yr) mortality 1985
  7. life         byte   %8.0g              Life expectancy at birth 1985
  8. food         int    %8.0g              Per capita daily calories 1985
  9. energy       int    %8.0g              Per cap energy consumed, kg oil
 10. gnpcap       int    %8.0g              Per capita GNP 1985
 11. gnpgro       float  %9.0g              Annual GNP growth % 65-85
 12. urban        byte   %8.0g              % population urban 1985
 13. school1      int    %8.0g              Primary enrollment % age-group
 14. school2      byte   %8.0g              Secondary enroll % age-group
 15. school3      byte   %8.0g              Higher ed. enroll % age-group
-------------------------------------------------------------------------------
Sorted by:
```

Relations among birth rate, per capita gross national product (GNP), and child mortality are clearly not linear, as can be seen in the scatterplot matrix of Figure 8.9. The skewed *gnpcap* and *chldmort* distributions also present potential leverage and influence problems.

Experimenting with power transformations reveals that the log of *gnpcap* and the square root of *chldmort* have distributions more symmetrical, with fewer outliers or potential leverage points, than the raw variables. More important, these transformations eliminate the nonlinearities, as can be seen in Figure 8.10.

```
. generate loggnp = log10(gnpcap)

. label variable loggnp "Log-10 of per cap GNP"
```

Figure 8.9

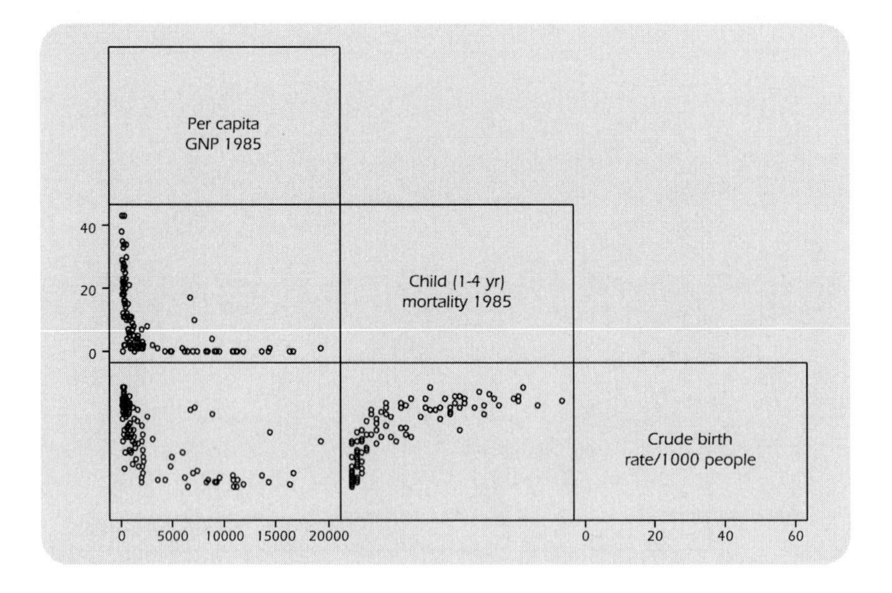

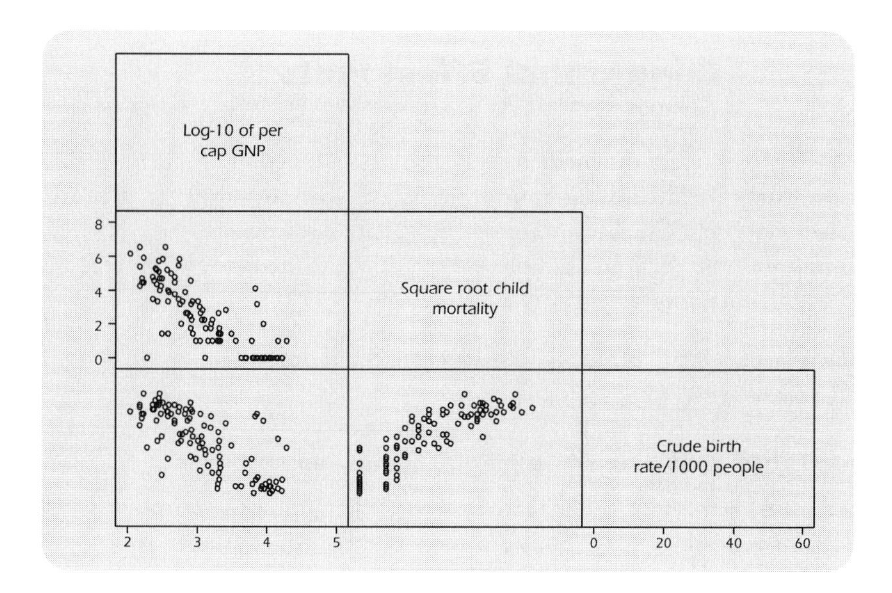

Figure 8.10

```
. generate srmort = sqrt(chldmort)

. label variable srmort "Square root child mortality"

. graph loggnp srmort birth, matrix half label
```

We can now apply linear regression using the transformed variables:

```
. regress birth loggnp srmort

  Source |       SS       df       MS                  Number of obs =     109
---------+------------------------------               F(  2,   106) =  198.06
   Model |  15837.9603     2  7918.98016               Prob > F      =  0.0000
Residual |  4238.18646   106  39.9828911               R-squared     =  0.7889
---------+------------------------------               Adj R-squared =  0.7849
   Total |  20076.1468   108  185.890248               Root MSE      =  6.3232

------------------------------------------------------------------------------
   birth |      Coef.   Std. Err.       t     P>|t|      [95% Conf. Interval]
---------+--------------------------------------------------------------------
  loggnp |  -2.353738   1.686255     -1.396   0.166     -5.696903    .9894258
  srmort |   5.577359    .533567     10.453   0.000       4.51951    6.635207
   _cons |   26.19488   6.362687      4.117   0.000      13.58024    38.80953
------------------------------------------------------------------------------
```

Unlike the raw-data regression (not shown), this transformed-variables version finds that per capita gross national product does not significantly affect birth rate, once we control for child mortality. The transformed-variables regression fits slightly better: $R^2_a = .7849$ instead of .6715. (We can compare R^2_a here because the *y* variable, *birth*, was not transformed.) Leverage plots would confirm that transformations have much reduced the curvilinearity of the raw-data regression.

Conditional Effect Plots

Conditional effect plots trace the predicted values of *y* as a function of one *x* variable, with other *x* variables held constant at arbitrary values such as means, medians, or quartiles. Such plots help visualize a transformed-variables regression.

Continuing with the previous example, we can calculate predicted birth rates as a function of *loggnp*, with *srmort* held at its mean (2.49):

```
. generate yhat1 = _b[_cons] + _b[loggnp]*loggnp +
      _b[srmort]*2.49

. label variable yhat1 "cond. effect loggnp, srmort=mean"
```

The `_b[varname]` terms refer to the regression coefficient on *varname* from this session's most recent regression. `_b[_cons]` is the *y*-intercept or constant.

For a conditional effect plot, graph *yhat1* (after inverse transformation if needed, although it is not needed here) against the untransformed *x* variable (Figure 8.11):

```
. graph yhat1 gnpcap, connect(s) symbol(i) ylabel xlabel
      l1(Birth rate)
```

Figure 8.11

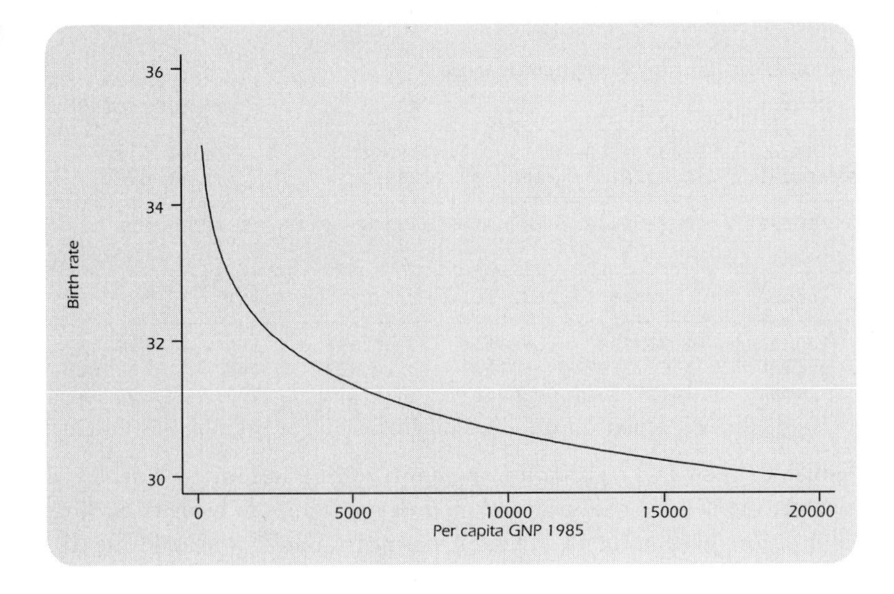

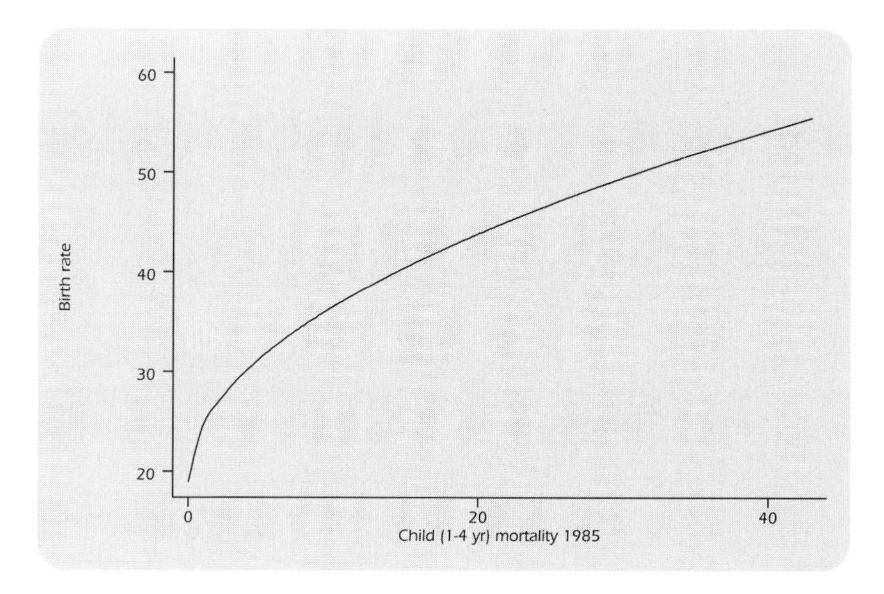

Figure 8.12

Similarly, Figure 8.12 depicts predicted birth rates as a function of *srmort*, with *loggnp* held at its mean (3.089199):

```
. generate yhat2 = _b[_cons] + _b[loggnp]*3.09 +
      _b[srmort]*srmort

. label variable yhat2 "cond. effect srmort, loggnp=mean"

. graph yhat2 chldmort, connect(s) symbol(i) ylabel xlabel
      l1(Birth rate)
```

How can we compare the strength of different *x* variables' effects? Standardized regression coefficients (beta weights) are sometimes used for this purpose, but they imply a specialized definition of "strength" and can easily be misleading. A more substantively meaningful comparison might come from looking at conditional effect plots drawn with identical *y* scales, and with vertical lines marking the 10th and 90th percentiles of the *x*-variable distributions (found by **summarize, detail**). The vertical distance traveled by the regression curve over the middle 80% of *x* values provides a visual comparison of effect magnitude. For example, we might create these two separate plots using **yscale** to control vertical scales and **xline** to draw 10th and 90th-percentile lines:

```
. graph yhat2 chldmort, connect(l) sort symbol(i) ylabel xlabel
      l1(Birth rate) yscale(20,60) xline(0,27)

. graph yhat1 gnpcap, connect(l) sort symbol(i) ylabel xlabel
      l1(Birth rate) yscale(20,60) xline(230,10890)
```

Figure 8.13

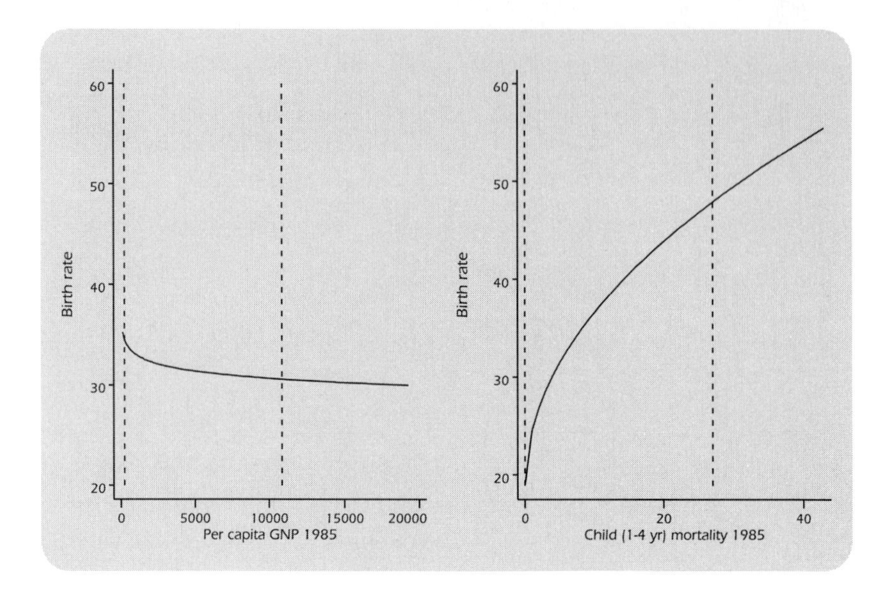

Figure 8.13 combines both plots into a single image (enhanced by Stage), thereby emphasizing the much stronger effect of mortality rate—as separate plots (Figures 8.11 and 8.12) did not.

Nonlinear Regression Example 1

Variable transformations allow fitting some curvilinear relations using the familiar techniques of intrinsically linear models. Intrinsically nonlinear models, in contrast, require a different class of fitting techniques. The **nl** command performs nonlinear regression by iterative least squares. This section introduces it using a dataset of simple examples, *nonlin2.dta*:

```
Contains data from c:\data\nonlin2.dta
  obs:           100                         Artificial data—nonlinear
 vars:             5                         26 Dec 1996 17:36
 size:         2,400 (95.7% of memory free)
-------------------------------------------------------------------------------
   1. x             float  %9.0g             Independent variable
   2. y1            float  %9.0g             y1 = 10 * 1.02^x
   3. y2            float  %9.0g             y2 = 10 * (1 - .95^x)
   4. y3            float  %9.0g             y3 = 5 + 25/(1+exp(-.1*(x-50)))
   5. y4            float  %9.0g             y4 = 5 + 25*exp(-exp(-.1*(x-50)
-------------------------------------------------------------------------------
Sorted by:  x
```

The *nonlin2.dta* data are manufactured, with *y* variables defined as various nonlinear functions of *x*, plus random Gaussian errors. *y1*, for example, represents the exponential growth process $y1 = 10 \times 1.02^x$. Estimating these parameters from the data, **nl** obtains $y1 = 10.615 \times 1.019^x$, reasonably close to the true model:

```
. nl exp2 y1 x
(obs = 100)

Iteration 0:  residual SS =  1890.374
Iteration 1:  residual SS =  1766.356
Iteration 2:  residual SS =  1763.732

  Source |       SS       df       MS                  Number of obs =       100
---------+------------------------------               F(  2,    98) =   3757.81
   Model |  135260.689     2  67630.3447               Prob > F      =    0.0000
Residual |  1763.73154    98  17.9972606               R-squared     =    0.9871
---------+------------------------------               Adj R-squared =    0.9869
   Total |  137024.421   100  1370.24421               Root MSE      =  4.242318
                                                       Res. dev.     =  570.7894
2-param. exp. growth curve, y1=b1*b2^x
------------------------------------------------------------------------------
      y1 |      Coef.   Std. Err.       t     P>|t|       [95% Conf. Interval]
---------+--------------------------------------------------------------------
      b1 |   10.61532   .4557343     23.293   0.000       9.710932    11.51971
      b2 |   1.019341    .000548   1860.200   0.000       1.018253    1.020428
------------------------------------------------------------------------------
 (SE's, P values, CI's, and correlations are asymptotic approximations)
```

To obtain predicted values and residuals after **nl**, use the **nlpred** command. Figure 8.14 graphs the predicted values to show the fit ($R^2 = .987$) between this model and data.

```
. nlpred yhat1

. graph y1 yhat1 x, connect(.s) symbol(Oi) ylabel xlabel
```

The **exp2** part of our **nl exp2 *y1 x*** command specified a particular exponential growth function by calling a brief program named *nlexp2.ado*. Stata includes several such programs, defining the following functions:

exp3	Exponential, $y = b_0 + b_1 b_2^x$
exp2	Exponential, $y = b_1 b_2^x$
exp2a	Negative exponential, $y = b_1 (1 - b_2^x)$
log4	Logistic, $y = b_0 + b_1 / (1 + \exp(-b_2 (x - b_3)))$
log3	Logistic, $y = b_1 / (1 + \exp(-b_2 (x - b_3)))$
gom4	Gompertz, $y = b_0 + b_1 \exp(-\exp(-b_2 (x - b_3)))$
gom3	Gompertz, $y = b_1 \exp(-\exp(-b_2 (x - b_3)))$

Figure 8.14

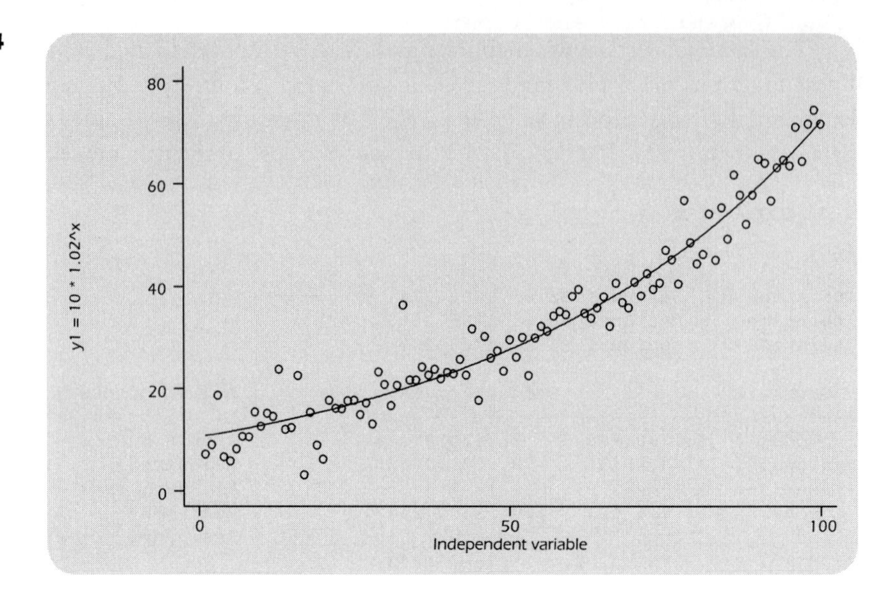

nonlin2.dta contains examples corresponding to **exp2** (*y1*), **exp2a** (*y2*), **log4** (*y3*) and **gom4** (*y4*) functions. Figure 8.15 shows curves fit by **nl** to *y2, y3,* and *y4.*

Users can write further `nl`*`function`* programs of their own. Here, in its entirety, is the `nlexp2.ado` program defining an exponential growth model:

```
*! version 1.1.1  07/22/93
program define nlexp2
   version 3.1
   if "`1'"=="?" {
      global S_2 "2-param. exp. growth curve, $S_E_depv=b1*b2^`2'"
      global S_1 "b1 b2"
* Approximate initial values by regression of log Y on X.
      local exp "[$S_E_wgt $S_E_exp]"
      tempvar Y
      quietly {
         gen `Y' = log($S_E_depv) $S_E_if $S_E_in
         reg `Y' `2' `exp'
      }
      global b1 = exp(_b[_cons])
      global b2 = exp(_b[`2'])
      exit
   }
   replace `1'=$b1*$b2^`2'
end
```

Figure 8.15

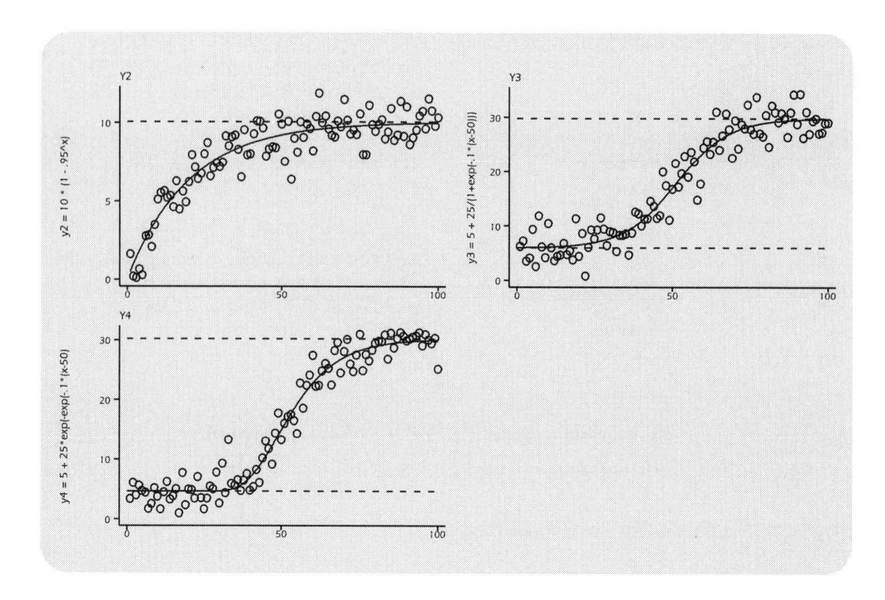

This program finds some approximate initial values of the parameters to be estimated, storing these as "global macros" named *b1* and *b2*. It then calculates an initial set of predicted values, as a "local macro" named *1*, employing the initial parameter estimates and the model equation:

```
replace `1' = $b1 * $b2^`2'
```

Subsequent iterations of **`nl`** will return to this line, calculating new predicted values (replacing the contents of macro *1*) as they refine the parameter estimates *b1* and *b2*. In Stata programs the notation `$b1` means "the contents of global macro *b1*." Similarly, the notation `` `1' `` means "the contents of local macro *1*."

Before attempting to write your own nonlinear function, examine `nllog4.ado`, `nlgom4.ado`, and so on for other examples and consult the manual or **`help nl`** for explanations. Chapter 13 gives further examples of macros and other aspects of Stata programming.

Nonlinear Regression Example 2

Our second example involves real data and illustrates some steps that can help in research. Dataset *lichen.dta* contains measurements of lichen growth observed on the Arctic island of Spitsbergen (from Werner, 1990). These slow-growing symbionts are often used to date rock monuments and other deposits, so their growth rates interest scientists in several fields.

```
Contains data from c:\data\lichen.dta
  obs:            11                           Lichen growth (Werner 1990)
 vars:             8                           22 Dec 1996 20:29
 size:           572 (95.8% of memory free)
-----------------------------------------------------------------------------
  1. locale       str31   %31s                 Locality and feature
  2. point        str1    %9s                  Control point
  3. date         int     %8.0g                Date
  4. age          int     %8.0g                Age in years
  5. rshort       float   %9.0g                Rhizocarpon short axis mm
  6. rlong        float   %9.0g                Rhizocarpon long axis mm
  7. pshort       int     %8.0g                P.minuscula short axis mm
  8. plong        int     %8.0g                P.minuscula long axis mm
-----------------------------------------------------------------------------
Sorted by:
```

Lichens characteristically exhibit a period of relatively fast early growth, gradually slowing, as suggested by the lowess-smoothed curve in Figure 8.16:

. ksm *rlong age*, lowess ylabel xlabel

Lichenometricians seek to summarize and compare such patterns by drawing growth curves. Their growth curves might not employ an explicit mathematical model, but we can fit one here to illustrate the process of nonlinear regression. Gompertz curves are asymmetrical S-curves, widely used to model biological growth:

$$y = b_1 \exp(-\exp(-b_2 (x - b_3)))$$

They might provide a reasonable model for lichen growth.

Figure 8.16

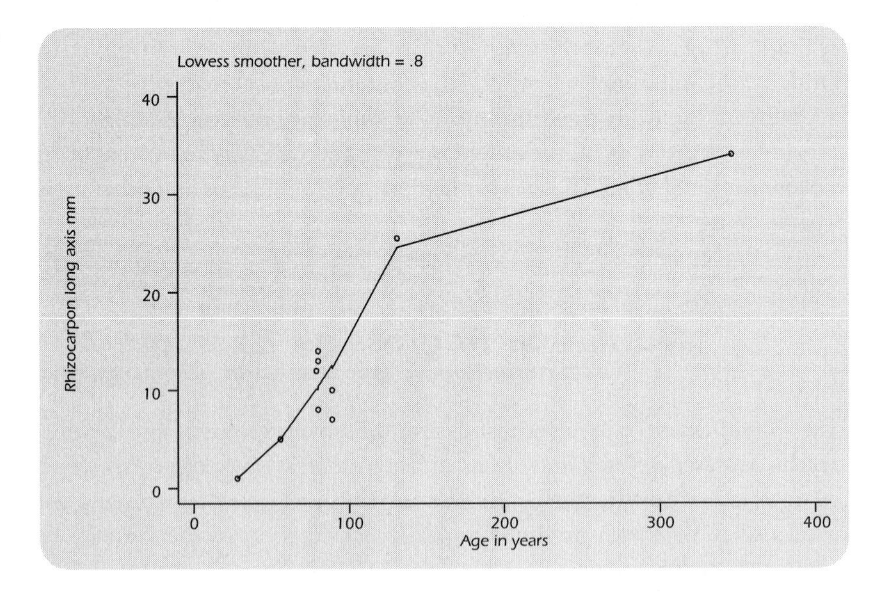

If we intend to graph a nonlinear model, the data should contain a good range of closely spaced *x* values. Actual ages of the 11 lichen samples in *lichen.dta* range from 28 to 346 years. We can create 89 additional artificial observations, with "ages" from 0 to 352 in 4-year increments, by the following commands:

```
. range newage 0 396 100
obs was 11, now 100

. replace age = newage[_n-11] if age == .
(89 real changes made)
```

The first command created a new variable, *newage*, with 100 values ranging from 0 to 396 in 4-year increments. In so doing, we also created 89 new artificial observations, with missing values on all variables except *newage*. The **replace** command substitutes the missing artificial-case *age* values with *newage* values, starting at 0. The first 15 observations in our data now look like this:

```
. list rlong age newage in 1/15
          rlong        age     newage
  1.          1         28          0
  2.          5         56          4
  3.         12         79          8
  4.          8         80         12
  5.         13         80         16
  6.         14         80         20
  7.         10         89         24
  8.          7         89         28
  9.       25.5        131         32
 10.         34        346         36
 11.         34        346         40
 12.          .          0         44
 13.          .          4         48
 14.          .          8         52
 15.          .         12         56
```

```
. summarize rlong age newage
Variable |     Obs        Mean   Std. Dev.       Min        Max
---------+-----------------------------------------------------
   rlong |      11    14.86364   11.31391          1         34
     age |     100      170.68   104.7042          0        352
  newage |     100         198    116.046          0        396
```

We now could **drop *newage***. Only the original 11 observations have nonmissing *rlong* values, so only they will enter into model estimation. Stata can calculate predicted values for any observation with nonmissing *x* values, however, so we can obtain such predictions for both the 11 real observations and the 89 artificial ones. This will allow us to accurately graph the regression curve.

Lichen growth starts with a size close to zero, so we chose the **gom3** Gompertz function rather than **gom4** (which incorporates a nonzero starting level in its parameter b_0). Figure 8.16 suggests an asymptotic upper limit somewhere near 34, so we could include this as our initial estimate of the parameter b_1. Estimation of this model is accomplished by

```
. nl gom3 rlong age, init(B1=34) nolog
(obs = 11)

  Source |       SS        df       MS                  Number of obs =        11
---------+------------------------------                F(  3,      8) =    125.68
   Model |  3633.16112     3  1211.05371                Prob > F      =    0.0000
Residual |  77.0888815     8  9.63611018                R-squared     =    0.9792
---------+------------------------------                Adj R-squared =    0.9714
   Total |     3710.25    11  337.295455                Root MSE      =  3.104208
                                                        Res. dev.     =  52.63435
3-parameter Gompertz function, rlong=b1*exp(-exp(-b2*(age-b3)))
------------------------------------------------------------------------------
   rlong |      Coef.   Std. Err.       t     P>|t|       [95% Conf. Interval]
---------+--------------------------------------------------------------------
      b1 |   34.36637   2.267165     15.158   0.000      29.13828    39.59447
      b2 |   .0217685   .0060807      3.580   0.007      .0077463    .0357907
      b3 |   88.79703   5.632568     15.765   0.000       75.8083    101.7858
------------------------------------------------------------------------------
 (SE's, P values, CI's, and correlations are asymptotic approximations)
```

A **nolog** option suppresses displaying a log of iterations with the output. The regression table's 95% confidence intervals indicate that all three parameter estimates differ significantly from 1.

We obtain predicted values using **nlpred**, and graph these to see the regression curve. The **yline** option allows us to draw in the lower and estimated upper limits (0 and 34.366) of this curve (Figure 8.17).

```
. nlpred yhat

. graph rlong yhat age, connect(.s) symbol(Oi) yline(0,34.366)
      ylabel(0,10,20,30) xlabel(0,100,200,300)
```

Especially when you are working with sparse data or a relatively complex model, nonlinear regression programs can be quite sensitive to their initial parameter estimates. The **init** option with **nl** permits researchers to suggest their own initial values, if the default values supplied by an `nl` *`function`* program do not seem to work. Previous experience with similar data, or publications by other researchers, could help supply suitable initial values. Alternatively, we could estimate through trial and error by employing **generate** to calculate predicted values based on arbitrarily chosen sets of parameter values, and **graph** to compare the resulting predictions with the data.

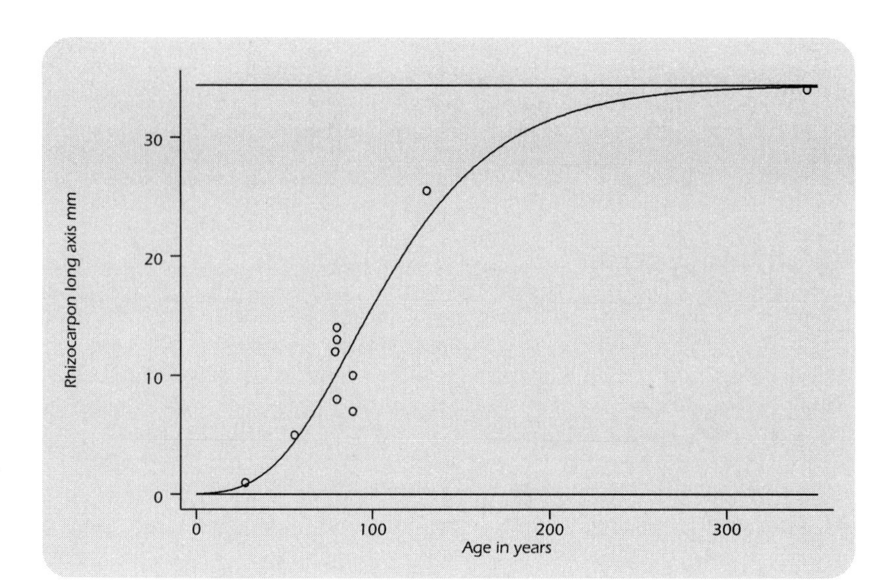

Figure 8.17

9

Robust Regression

Stata's basic **regress**, **fit**, **oneway**, and **anova** commands, among others, perform ordinary least squares (OLS) regression. The popularity of OLS derives partly from its theoretical advantages given "ideal" data. If errors are normally, independently, and identically distributed (normal *i.i.d.*), then OLS is more efficient than any other unbiased estimator. The flip side of this statement often gets overlooked: if errors are not normal, or not *i.i.d.*, then other unbiased estimators might outperform OLS. In fact, the efficiency of OLS degrades quickly in the face of heavy-tailed (outlier-prone) error distributions. Yet such distributions are common in many fields.

OLS tends to track outliers, fitting them at the expense of the rest of the sample. Over the long run, this leads to greater sample-to-sample variation or inefficiency when samples often contain outliers. Robust regression methods aim to achieve almost the efficiency of OLS with ideal data and substantially better-than-OLS efficiency in non-ideal (for example, nonnormal errors) situations. "Robust regression" encompasses a variety of different techniques, each with advantages and drawbacks for dealing with problematic data. This chapter introduces two varieties of robust regression, **rreg** and **qreg**, and briefly compares their results with those of OLS (**regress**).

For clarity, this chapter focuses mostly on two-variable examples, but robust multiple regression or *n*-way ANOVA are straightforward using the same commands. Chapter 13 returns to the topic of robustness, showing how Monte Carlo experiments can be used to evaluate competing statistical techniques.

Example Commands

. rreg ***y x1 x2 x3***

Performs robust regression of *y* on three predictors, using iteratively reweighted least squares with Huber and biweight functions tuned for 95% Gaussian efficiency. Given appropriately configured data, **rreg** can also obtain robust means, confidence intervals, difference of means tests, and ANOVA/ANOCOVA.

. rreg ***y x1 x2 x3*, nolog tune(6) genwt(*rweight*) iterate(10)**

Performs robust regression of *y* on three predictors. The options shown tell Stata not to print the iteration log, to use a tuning constant of 6 (which downweights outliers more steeply than the default 7), to generate a new variable (arbitrarily named *rweight*) holding the final-iteration robust weights for each observation, and to limit the maximum number of iterations.

. qreg ***y x1 x2 x3***

Performs quantile regression, also known as least absolute value (LAV) or minimum *L1*-norm regression, of *y* on three predictors. By default, **qreg** models the conditional .5 quantile (median) of *y* as a linear function of the predictor variables and thus provides "median regression."

. qreg ***y x1 x2 x3*, quantile(.25)**

Performs quantile regression modeling the conditional .25 quantile (first quartile) of *y* as a linear function of *x1*, *x2*, and *x3*.

. bsqreg ***y x1 x2 x3*, rep(100)**

Performs quantile regression, with standard errors estimated by bootstrap data resampling with 100 repetitions (default is **rep(20)**).

. predict *e*, resid

Calculates residual values (arbitrarily named *e*) after any **regress**, **rreg**, **qreg**, or **bsqreg** command. Similarly, **predict *yhat*** calculates the predicted values of *y*. Other **predict** options apply, with some restrictions.

Regression with Ideal Data

To clarify the issue of robustness, we will explore the contrived dataset *robust.dta*:

```
Contains data from c:\data\robust.dta
  obs:          20                        Artificial data-robustness
 vars:          10                        23 Dec 1996 16:11
 size:          880 (96.1% of memory free)
-------------------------------------------------------------------------------
   1. x           float  %4.2f            Normal X
   2. e1          float  %4.2f            Normal errors
   3. y1          float  %4.2f            y1 = 10 + 2*x + e1
   4. e2          float  %4.2f            Normal errors with 1 outlier
   5. y2          float  %4.2f            y2 = 10 + 2*x + e2
   6. x3          float  %4.2f            Normal X with 1 leverage obs.
   7. e3          float  %4.2f            Normal errors with 1 extreme
   8. y3          float  %4.2f            y3 = 10 + 2*x3 + e3
   9. e4          float  %4.2f            Skewed errors
  10. y4          float  %4.2f            y4 = 10 + 2*x + e4
-------------------------------------------------------------------------------
Sorted by:
```

The variables *x* and *e1* each contain 20 random values from independent standard normal distributions. *y1* contains 20 values produced by the regression model

$$y1 = 10 + 2x + e1$$

The commands that manufactured these first three variables are:

```
. clear

. set obs 20

. generate x = invnorm(uniform())

. generate e1 = invnorm(uniform())

. generate y1 = 10 + 2*x + e1
```

With real data, coding mistakes and other aberrations sometimes create wild errors. To simulate this, we might shift the second observation's error from –0.89 to 19.89:

```
. generate e2 = e1

. replace e2 = 19.89 in 2

. generate y2 = 10 + 2*x + e2
```

Similar manipulations produce the other variables in *robust.dta*.

y1 and *x* present an ideal regression problem: The expected value of *y1* really is a linear function of *x*, and errors come from normal, independent, and identical distributions—because we defined them that way. OLS does a good job of estimating the true intercept (10) and slope (2), obtaining the line shown in Figure 9.1:

Figure 9.1

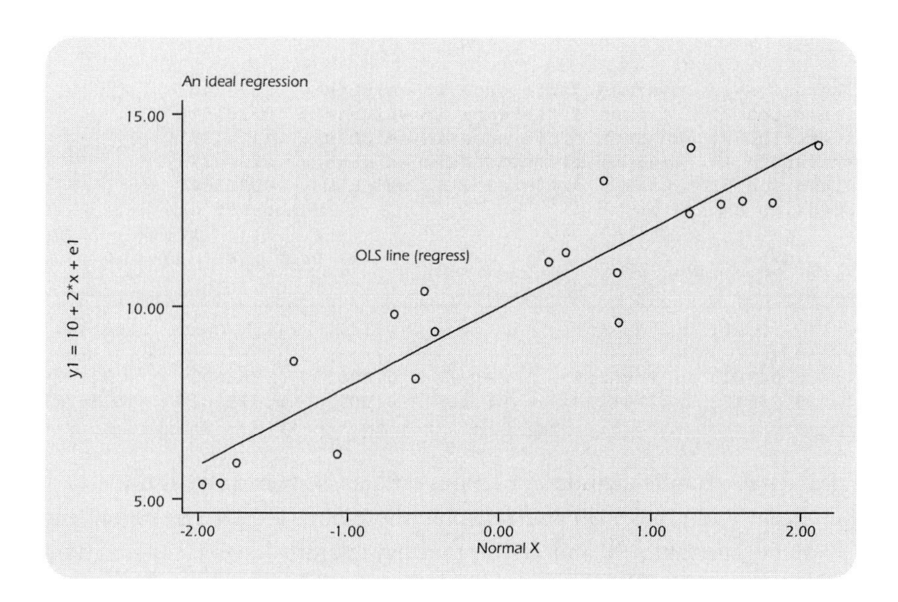


```
. regress y1 x

  Source |       SS       df       MS              Number of obs =      20
---------+------------------------------           F(  1,    18) =  108.25
   Model |  134.059351     1  134.059351           Prob > F      =  0.0000
Residual |    22.29157    18  1.23842055           R-square      =  0.8574
---------+------------------------------           Adj R-square  =  0.8495
   Total |  156.350921    19  8.22899586           Root MSE      =  1.1128

------------------------------------------------------------------------------
      y1 |      Coef.   Std. Err.       t     P>|t|      [95% Conf. Interval]
--------+---------------------------------------------------------------------
       x |   2.048057   .1968465     10.404   0.000      1.634498    2.461616
   _cons |   9.963161   .2499861     39.855   0.000       9.43796    10.48836
------------------------------------------------------------------------------
```

An iteratively reweighted least squares (IRLS) procedure, **rreg**, obtains robust regression estimates. The first **rreg** iteration begins with OLS. Any observations so influential as to have Cook's *D* values greater than 1 are automatically set aside after this first step. Next, weights are calculated for each observation using a Huber function, which downweights observations that have larger residuals, and weighted least squares is performed. After several WLS iterations the weight function shifts to a Tukey biweight (as suggested by Li, 1985), tuned for 95% Gaussian efficiency (see *Regression with Graphics* for details). **rreg** estimates standard errors and tests hypotheses using a pseudovalues method (Street, Carrol, and Ruppert, 1988) that does not assume normality.

```
. rreg y1 x

   Huber iteration 1:  maximum difference in weights = .35774407
   Huber iteration 2:  maximum difference in weights = .02181578
Biweight iteration 3:  maximum difference in weights = .14421371
Biweight iteration 4:  maximum difference in weights = .01320276
Biweight iteration 5:  maximum difference in weights = .00265408
Robust regression estimates                          Number of obs =      20
                                                     F(  1,    18) =   79.96
                                                     Prob > F      =  0.0000

------------------------------------------------------------------------------
      y1 |      Coef.   Std. Err.       t     P>|t|     [95% Conf. Interval]
---------+--------------------------------------------------------------------
       x |   2.047813   .2290049      8.942   0.000      1.566692    2.528935
   _cons |   9.936163   .2908259     34.165   0.000      9.325161    10.54717
------------------------------------------------------------------------------
```

This "ideal data" example includes no serious outliers, so here **rreg** is unneeded. The **rreg** intercept and slope estimates resemble those obtained by **regress** (and are not far from the true values 10 and 2), but they have slightly larger estimated standard errors. Given normal *i.i.d.* errors, as in this example, **rreg** theoretically possesses about 95% of the efficiency of OLS.

rreg and **regress** both belong to the family of *M*-estimators (for maximum-likelihood). An alternative order-statistic strategy called *L*-estimation fits quantiles of *y*, rather than its expectation or mean. For example, we could model how the median (.5 quantile) of *y* changes with *x*. **qreg**, an *L1*-type estimator, accomplishes such quantile regression and provides another method with good resistance to outliers:

```
. qreg y1 x

Iteration  1:  WLS sum of weighted deviations =  17.711531

Iteration  1: sum of abs. weighted deviations =  17.130001
Iteration  2: sum of abs. weighted deviations =  16.858602

Median Regression                                   Number of obs =       20
  Raw sum of deviations     46.84 (about 10.4)
  Min sum of deviations   16.8586                   Pseudo R2     =   0.6401

------------------------------------------------------------------------------
      y1 |      Coef.   Std. Err.       t     P>|t|     [95% Conf. Interval]
---------+--------------------------------------------------------------------
       x |   2.139896   .2590447      8.261   0.000      1.595664    2.684129
   _cons |    9.65342   .3564108     27.085   0.000      8.904628    10.40221
------------------------------------------------------------------------------
```

Although **qreg** obtains reasonable parameter estimates, its standard errors here exceed those of **regress** (OLS) and **rreg**. Given ideal data, **qreg** is the least efficient of these three estimators. The following sections view their performance with less ideal data.

Y Outliers

The variable *y2* is identical to *y1*, but with one outlier caused by the "wild" error of observation #2. OLS has little resistance to outliers, so this shift in observation #2 (at upper left in Figure 9.2) substantially changes the **regress** results:

```
. regress y2 x

  Source |       SS       df       MS                  Number of obs =      20
---------+------------------------------               F(  1,    18) =    0.97
   Model |   18.764271     1   18.764271               Prob > F      =  0.3378
Residual |  348.233471    18  19.3463039               R-square      =  0.0511
---------+------------------------------               Adj R-square  = -0.0016
   Total |  366.997742    19  19.3156706               Root MSE      =  4.3984

------------------------------------------------------------------------------
      y2 |      Coef.   Std. Err.       t     P>|t|       [95% Conf. Interval]
--------+---------------------------------------------------------------------
       x |   .7662304   .7780232     0.985    0.338     -.8683356    2.400796
   _cons |    11.1579   .9880542    11.293    0.000      9.082078    13.23373
------------------------------------------------------------------------------
```

The outlier raises the OLS intercept (from 9.936 to 11.1579) and lessens the slope (from 2.0478 to 0.766). Standard errors quadruple, and the OLS slope (solid line in Figure 9.2) no longer significantly differs from zero.

The outlier has little impact on **rreg**, however, as shown by the dashed line in Figure 9.2. The robust coefficients barely change and remain close to the true parameters 10 and 2. Nor do the robust standard errors increase much.

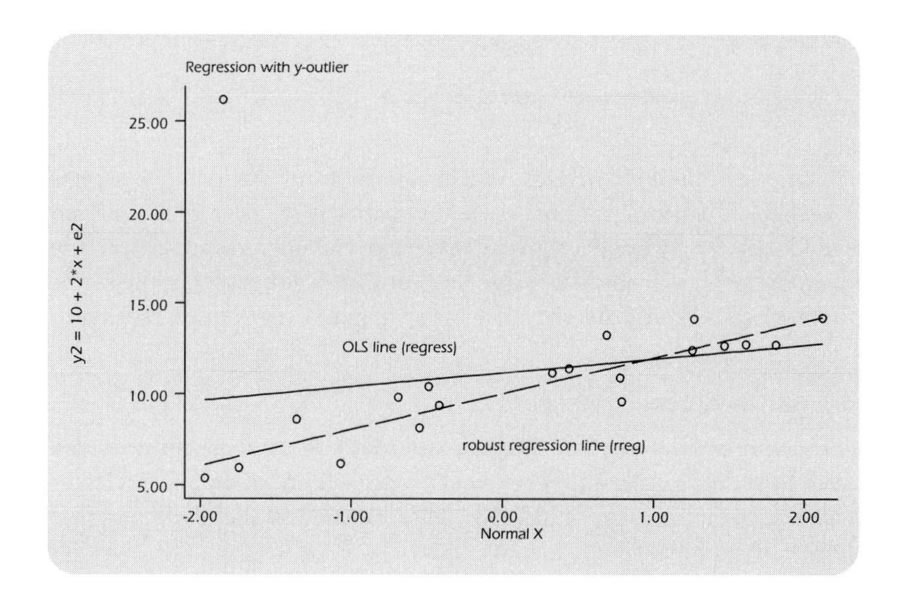

Figure 9.2

```
. rreg y2 x, nolog genwt(rweight)
Robust regression estimates                           Number of obs =      19
                                                      F(  1,     17) =   63.01
                                                      Prob > F      =  0.0000

------------------------------------------------------------------------------
      y2 |      Coef.   Std. Err.       t     P>|t|       [95% Conf. Interval]
---------+--------------------------------------------------------------------
       x |   1.979015   .2493146      7.938   0.000       1.453007    2.505023
   _cons |   10.00897   .3071265     32.589   0.000       9.360986    10.65695
------------------------------------------------------------------------------
```

The **nolog** option caused Stata not to print the iteration log. The **genwt(*rweight*)** option saved robust weights as a variable named *rweight*:

```
. predict resid, resid

. list y2 x resid rweight
          y2        x        resid     rweight
  1.    5.37    -1.97    -.7403071   .94644465
  2.   26.19    -1.85     19.84221           .
  3.    5.93    -1.74    -.6354806   .96037073
  4.    8.58    -1.36     1.262494    .8493384
  5.    6.16    -1.07    -1.731421    .7257631
  6.    9.80    -0.69     1.156554   .87273631
  7.    8.12    -0.55    -.8005085   .93758391
  8.   10.40    -0.49      1.36075   .82606386
  9.    9.35    -0.42       .17222   .99712388
 10.   11.16     0.33     .4979582   .97581674
 11.   11.40     0.44     .5202664   .97360863
 12.   13.26     0.69     1.885513   .68048066
 13.   10.88     0.78    -.6725982   .95572833
 14.    9.58     0.79    -1.992389   .64644918
 15.   12.41     1.26    -.0925257   .99913568
 16.   14.14     1.27     1.617685   .75887073
 17.   12.66     1.47    -.2581189   .99338589
 18.   12.74     1.61    -.4551811   .97957817
 19.   12.70     1.81    -.8909839   .92307041
 20.   14.19     2.12    -.0144787   .99997651
```

Residuals near zero produce weights near one; farther-out residuals get progressively lower weights. Observation #2 has been automatically set aside as too influential because of Cook's $D > 1$; **rreg** assigns its *rweight* as "missing," so this observation has no effect on the final estimates. The same final estimates, although not the correct standard errors or tests, could be obtained by a weighted regression (results not shown):

```
. regress y2 x [aweight = rweight]
```

Applied to the regression of *y2* on *x*, **qreg** also resists the outlier's influence and performs much better than **regress**—but not as well as **rreg**. **qreg** appears less efficient than **rreg**, and in this sample its coefficient estimates are slightly farther from the true values of 10 and 2:

```
. qreg y2 x
Iteration  1:  WLS sum of weighted deviations =  45.211528

Iteration  1: sum of abs. weighted deviations =  41.892111
Iteration  2: sum of abs. weighted deviations =  38.130976
Iteration  3: sum of abs. weighted deviations =  36.563548
Iteration  4: sum of abs. weighted deviations =  36.200356

Median Regression                                   Number of obs =         20
  Raw sum of deviations     56.68 (about 10.88)
  Min sum of deviations 36.20036                    Pseudo R2     =     0.3613

------------------------------------------------------------------------------
      y2 |      Coef.   Std. Err.       t     P>|t|     [95% Conf. Interval]
---------+--------------------------------------------------------------------
       x |   1.821428   .4105944      4.436   0.000      .9588014    2.684055
   _cons |     10.115   .5088526     19.878   0.000      9.045941    11.18406
------------------------------------------------------------------------------
```

Monte Carlo researchers have also noticed that the standard errors calculated by **qreg** sometimes underestimate the true sample-to-sample variation, particularly with smaller samples. As an alternative, Stata provides the command **bsqreg**, which performs the same median or quantile regression as **qreg**, but employs bootstrapping (data resampling) to estimate the standard errors empirically. The option **rep( )** controls the number of bootstrap repetitions. Its default is **rep(20)**, enough for exploratory work. Before reaching "final" conclusions, we might take the time to draw 200 or more bootstrap samples, as done below. Both **qreg** and **bsqreg** fit identical models, but the bootstrapped standard error estimates are somewhat larger.

```
. bsqreg y2 x, rep(200)
(estimating base model)
(bootstrapping
...............................................................
> ..............................................................
> ..............................................................
> .........)

Median Regression, bootstrap(200) SEs               Number of obs =         20
  Raw sum of deviations     56.68 (about 10.88)
  Min sum of deviations 36.20036                    Pseudo R2     =     0.3613

------------------------------------------------------------------------------
      y2 |      Coef.   Std. Err.       t     P>|t|     [95% Conf. Interval]
---------+--------------------------------------------------------------------
       x |   1.821428   .4178074      4.359   0.000      .9436476    2.699209
   _cons |     10.115   .5197453     19.461   0.000      9.023056    11.20694
------------------------------------------------------------------------------
```

X-Outliers (Leverage)

rreg, **qreg**, and **bsqreg** deal comfortably with *y*-outliers, unless the observations with unusual *y* values have unusual *x* values (leverage) too. The *y3* and *x3* variables in *robust.dta* present an extreme example of leverage. Apart from the leverage observation (#2), these variables equal *y1* and *x*.

The high leverage of observation #2, combined with its exceptional *y3* value, make it influential: **regress** and **qreg** both track this outlier, reporting that the "best-fitting" line has a negative slope (Figure 9.3).

```
. regress y3 x3

  Source |       SS       df       MS                  Number of obs =      20
---------+------------------------------               F(  1,    18) =   11.01
   Model |  139.306724     1  139.306724               Prob > F      =  0.0038
Residual |  227.691018    18   12.649501               R-square      =  0.3796
---------+------------------------------               Adj R-square  =  0.3451
   Total |  366.997742    19  19.3156706               Root MSE      =  3.5566

------------------------------------------------------------------------------
      y3 |      Coef.   Std. Err.       t     P>|t|       [95% Conf. Interval]
---------+--------------------------------------------------------------------
      x3 |  -.6212248    .1871973     -3.319   0.004     -1.014512    -.227938
   _cons |   10.80931    .8063436     13.405   0.000      9.115244    12.50337
------------------------------------------------------------------------------

. qreg y3 x3
Iteration  1:  WLS sum of weighted deviations =  57.305936

Iteration  1: sum of abs. weighted deviations =  56.194665

Median Regression                                     Number of obs =      20
  Raw sum of deviations     56.68 (about 10.88)
  Min sum of deviations 56.19466                      Pseudo R2     =  0.0086

------------------------------------------------------------------------------
      y3 |      Coef.   Std. Err.       t     P>|t|       [95% Conf. Interval]
---------+--------------------------------------------------------------------
      x3 |  -.6222217     .347103     -1.793   0.090     -1.351458    .1070146
   _cons |   11.36533    1.419214      8.008   0.000      8.383676    14.34699
------------------------------------------------------------------------------
```

Figure 9.3 illustrates that **regress** and **qreg** are not robust against leverage (*x*-outliers). The **rreg** program, however, not only downweights large-residual observations (which by itself gives little protection against leverage), but also automatically sets aside observations with Cook's *D* (influence) statistics greater than 1. This happens when we regress *y3* on *x3*; **rreg** ignores the one influential observation and produces a more reasonable regression line based on the remaining 19 observations:

Figure 9.3

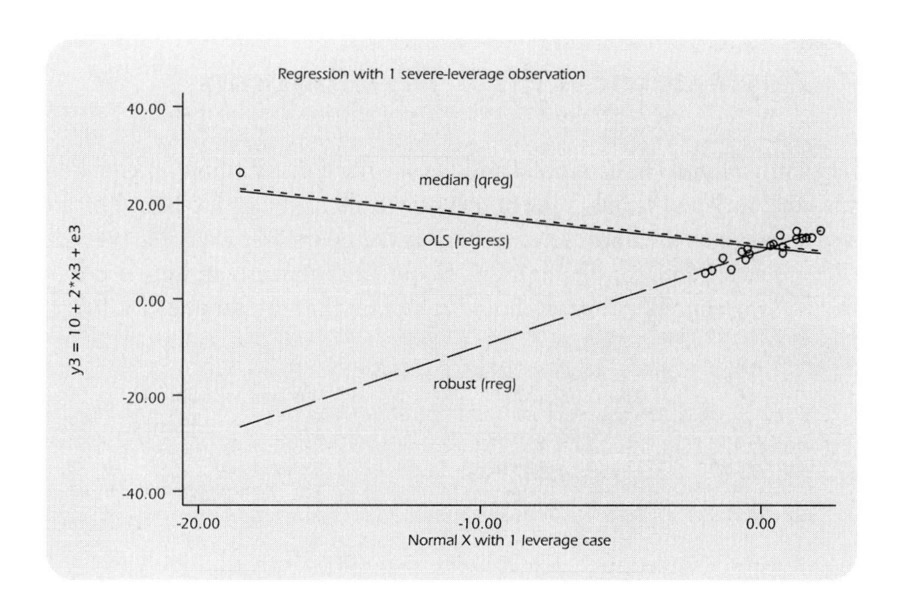


```
. rreg y3 x3, nolog
Robust regression estimates                          Number of obs =      19
                                                     F( 1,     17) =   63.01
                                                     Prob > F      =  0.0000

------------------------------------------------------------------------------
      y3 |      Coef.   Std. Err.       t     P>|t|     [95% Conf. Interval]
---------+--------------------------------------------------------------------
      x3 |   1.979015   .2493146      7.938   0.000     1.453007    2.505023
   _cons |   10.00897   .3071265     32.589   0.000     9.360986    10.65695
------------------------------------------------------------------------------
```

Setting aside high-influence observations, as done by **rreg**, provides a simple but not foolproof way to deal with leverage. More comprehensive methods termed bounded-influence regression also exist and should soon be available within Stata.

The examples in Figures 9.2 and 9.3 involve single outliers, but robust procedures can handle more. Too many severe outliers, or a cluster of similar outliers, can cause them to break down. But in such situations, which often are noticeable in diagnostic graphs, the analyst must question whether fitting a linear model makes sense.

Monte Carlo research shows that estimators like **rreg** and **qreg** generally remain unbiased, with better-than-OLS efficiency, when applied to heavy-tailed (outlier-prone) but symmetrical error distributions. The next section illustrates what can happen when errors have asymmetrical distributions.

Asymmetrical Error Distributions

The variable *e4* in *robust.dta* has a skewed and outlier-filled distribution: *e4* equals *e1* (a standard normal variable) raised to the fourth power, then adjusted to have 0 mean. These skewed errors, plus the linear relation with *x*, define the last *y* variable: $y4 = 10 + 2x + e4$. Regardless of an error distribution's shape, OLS remains an unbiased estimator—over the long run, its estimates should center on the true parameter values:

```
. regress y4 x

  Source |       SS       df       MS                  Number of obs =      20
---------+------------------------------               F(  1,    18) =    6.97
   Model |  155.870383     1  155.870383               Prob > F      =  0.0166
Residual |  402.341909    18  22.3523283               R-squared     =  0.2792
---------+------------------------------               Adj R-squared =  0.2392
   Total |  558.212291    19  29.3795943               Root MSE      =  4.7278

------------------------------------------------------------------------------
      y4 |      Coef.   Std. Err.       t     P>|t|       [95% Conf. Interval]
---------+--------------------------------------------------------------------
       x |   2.208388   .8362862      2.641   0.017       .4514157     3.96536
   _cons |   9.975681   1.062046      9.393   0.000       7.744406    12.20696
------------------------------------------------------------------------------
```

The same is not true for most robust estimators. Unless errors are symmetrical, the median line estimated by **qreg**, or the biweight line estimated by **rreg**, do not theoretically coincide with the expected-*y* line estimated by **regress**. So long as the errors' skew reflects only a small fraction of their distribution, **rreg** might exhibit little bias. But when the entire distribution is skewed, as with *e4*, **rreg** will downweight mostly one side, resulting in noticeably biased *y*-intercept estimates:

```
. rreg y4 x, nolog
Robust regression estimates                            Number of obs =      20
                                                       F(  1,    18) = 1319.29
                                                       Prob > F      =  0.0000

------------------------------------------------------------------------------
      y4 |      Coef.   Std. Err.       t     P>|t|       [95% Conf. Interval]
---------+--------------------------------------------------------------------
       x |   1.952073   .0537435     36.322   0.000       1.839163    2.064984
   _cons |   7.476669   .0682518    109.545   0.000       7.333278    7.620061
------------------------------------------------------------------------------
```

Although the **rreg** estimated *y*-intercept in Figure 9.4 is too low, the slope remains parallel to the OLS line and the true model. Being less affected by outliers, the **rreg** slope (dashed line) actually is closer to the true slope (dotted line) and obtains a much smaller standard error than **regress**. This illustrates the tradeoff of using **rreg** or similar estimators with skewed errors: We risk getting biased estimates of the *y*-intercept, but can still expect unbiased and relatively precise estimates of other

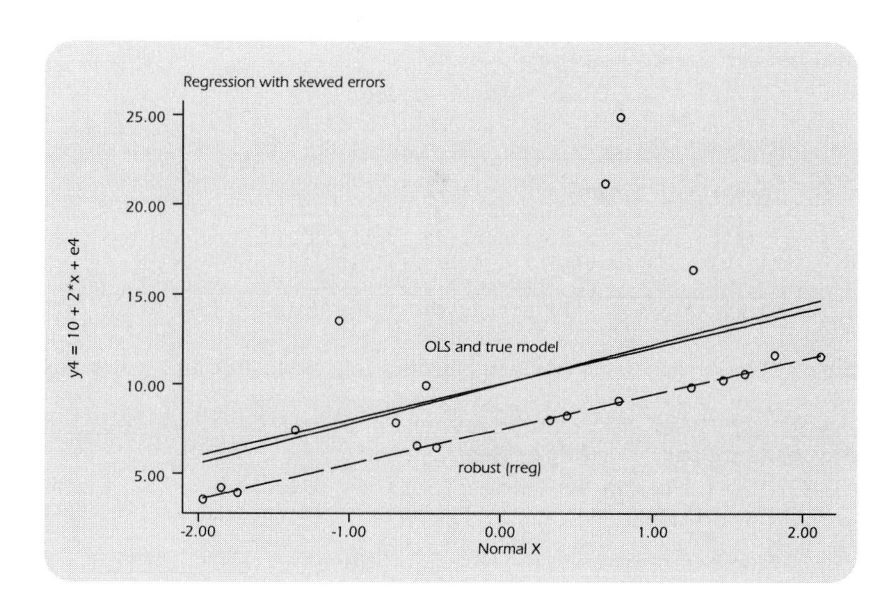

Figure 9.4

regression coefficients. In many applications such coefficients are substantively more interesting than the y-intercept, so this tradeoff is worthwhile. Moreover, the robust t and F tests, unlike those of OLS, do not assume normal errors.

Robust Analysis of Variance

rreg can also perform robust analysis of variance or covariance, once they are cast in regression form. For illustration, consider these data (*faculty1.dta*) on college faculty salaries:

```
Contains data from c:\data\faculty1.dta
  obs:          226                           College faculty salaries
 vars:            6                           23 Dec 1996 16:29
 size:        2,938 (96.1% of memory free)
---------------------------------------------------------------------------
   1. rank        byte   %8.0g      rank    Academic rank
   2. gender      byte   %8.0g      sex     Gender (dummy variable)
   3. female      byte   %8.0g              Gender (effect coded)
   4. assoc       byte   %8.0g              Assoc Professor (effect coded)
   5. full        byte   %8.0g              Full Professor (effect coded)
   6. pay         float  %9.0g              1989-90 salary
---------------------------------------------------------------------------
Sorted by:
```

Faculty salaries increase with rank. Men have higher average salaries:

```
. tabulate gender rank, sum(pay) means
                        Means of 1989-90 salary

Gender     | Academic rank
(dummy     |
variable)  |    Assist       Assoc        Full |      Total
-----------+-----------------------------------+-----------
      Male |     29280   38622.222   52084.902 |  41035.772
    Female | 28711.034   38019.048       47190 |  35228.052
-----------+-----------------------------------+-----------
     Total | 29022.188   38380.952   51569.649 |  39057.035
```

An ordinary (OLS) analysis of variance indicates that both *rank* and *gender* significantly affect salary. Their interaction is not significant:

```
. anova pay rank gender rank*gender
                        Number of obs =     226     R-squared     =  0.7305
                        Root MSE      = 5108.21     Adj R-squared =  0.7244

            Source |  Partial SS    df       MS           F     Prob > F
        -----------+----------------------------------------------------
             Model |  1.5560e+10     5  3.1120e+09     119.26     0.0000
                   |
              rank |  7.6124e+09     2  3.8062e+09     145.87     0.0000
            gender |   127361829     1   127361829       4.88     0.0282
       rank*gender |  87997720.1     2  43998860.1       1.69     0.1876
                   |
          Residual |  5.7406e+09   220  26093824.5
        -----------+----------------------------------------------------
             Total |  2.1300e+10   225  94668810.3
```

But salary is not normally distributed, and the senior-rank averages reflect the influence of a few highly paid outliers. Suppose we want to check these results by performing a robust analysis of variance. We need effect-coded versions of the *rank* and *gender* variables, which this dataset also contains:

```
. tab gender female
Gender     | Gender (effect coded)
(dummy     |
variable)  |        -1          1 |     Total
-----------+----------------------+----------
      Male |       149          0 |       149
    Female |         0         77 |        77
-----------+----------------------+----------
     Total |       149         77 |       226

. tab rank assoc
Academic   | Assoc Professor (effect coded)
rank       |        -1          0          1 |     Total
-----------+---------------------------------+----------
    Assist |        64          0          0 |        64
     Assoc |         0          0        105 |       105
      Full |         0         57          0 |        57
-----------+---------------------------------+----------
     Total |        64         57        105 |       226
```

```
. tab rank full

Academic   | Full Professor (effect coded)
rank       |        -1           0          1 |     Total
-----------+---------------------------------+----------
    Assist |        64           0          0 |        64
     Assoc |         0         105          0 |       105
      Full |         0           0         57 |        57
-----------+---------------------------------+----------
     Total |        64         105         57 |       226
```

If *faculty1.dta* did not already have these effect-coded variables (*female*, *assoc*, and *full*), we could create them from *gender* and *rank* using a series of **generate** and **replace** statements. We also need two interaction terms, representing female associate professors and female full professors:

```
. generate femassoc = female*assoc

. generate femfull = female*full
```

Males and assistant professors are "omitted categories" in this example. Now we can duplicate the previous ANOVA using regression:

```
. regress pay assoc full female femassoc femfull

  Source |       SS       df       MS                  Number of obs =     226
---------+------------------------------               F(  5,   220) =  119.26
   Model |  1.5560e+10     5  3.1120e+09               Prob > F      =  0.0000
Residual |  5.7406e+09   220  26093824.5               R-square      =  0.7305
---------+------------------------------               Adj R-square  =  0.7244
   Total |  2.1300e+10   225  94668810.3               Root MSE      =  5108.2

------------------------------------------------------------------------------
     pay |      Coef.   Std. Err.        t     P>|t|      [95% Conf. Interval]
---------+--------------------------------------------------------------------
   assoc |  -663.8995   543.8499     -1.221    0.223     -1735.722    407.9229
    full |   10652.92   783.9227     13.589    0.000      9107.957    12197.88
  female |  -1011.174   457.6938     -2.209    0.028     -1913.199   -109.1483
femassoc |   709.5864   543.8499      1.305    0.193      -362.236    1781.409
 femfull |  -1436.277   783.9227     -1.832    0.068     -2981.237    108.6819
   _cons |   38984.53   457.6938     85.176    0.000      38082.51    39886.56
------------------------------------------------------------------------------

. test assoc full

 ( 1)  assoc = 0.0
 ( 2)  full = 0.0
       F(  2,   220) =  145.87

            Prob > F =    0.0000

. test female

 ( 1)  female = 0.0
       F(  1,   220) =    4.88

            Prob > F =    0.0282
```

```
. test femassoc femfull

 ( 1)  femassoc = 0.0
 ( 2)  femfull = 0.0
       F(  2,   220) =    1.69

            Prob > F =    0.1876
```

regress followed by appropriate **test** commands obtains exactly the same R^2 and F test results we found earlier using **anova**. Predicted values from this regression equal the mean salaries.

```
. predict predpay1

. label variable predpay1 "OLS predicted salary"

. tabulate gender rank, summ(predpay1) means

                      Means of OLS predicted salary

Gender     | Academic rank
(dummy     |
variable)  |    Assist       Assoc        Full |     Total
-----------+-----------------------------------+----------
      Male |     29280   38622.223   52084.902 | 41035.772
    Female | 28711.035   38019.047       47190 | 35228.052
-----------+-----------------------------------+----------
     Total | 29022.188   38380.952   51569.649 | 39057.036
```

Predicted values (means), R^2, and F tests would also be the same regardless of which categories we chose to omit from the regression. Our "omitted categories," males and assistant professors, are not really absent. Their information is implied by the included categories: If a faculty member is not female, he must be male, and so forth.

To perform a robust analysis of variance, apply **rreg** to this model:

```
. rreg pay assoc full female femassoc femfull, nolog

Robust regression estimates                          Number of obs =     226
                                                     F(  5,   220) =  138.25
                                                     Prob > F      =  0.0000

------------------------------------------------------------------------------
     pay |      Coef.   Std. Err.       t     P>|t|      [95% Conf. Interval]
---------+--------------------------------------------------------------------
   assoc |  -315.6463   458.1588     -0.689   0.492     -1218.588    587.2957
    full |   9765.296   660.4048     14.787   0.000      8463.766    11066.83
  female |  -749.4949   385.5778     -1.944   0.053     -1509.394    10.40398
femassoc |   197.7833   458.1588      0.432   0.666     -705.1587    1100.725
 femfull |   -913.348   660.4048     -1.383   0.168     -2214.878    388.1816
   _cons |   38331.87   385.5778     99.414   0.000      37571.97    39091.77
------------------------------------------------------------------------------
```

```
. test assoc full

 ( 1)  assoc = 0.0
 ( 2)  full = 0.0

       F(  2,   220) =  182.67
            Prob > F =    0.0000

. test female

 ( 1)  female = 0.0

       F(  1,   220) =    3.78
            Prob > F =    0.0532

. test femassoc femfull

 ( 1)  femassoc = 0.0
 ( 2)  femfull = 0.0

       F(  2,   220) =    1.16
            Prob > F =    0.3144
```

rreg downweights several outliers, mainly high-paid male full professors. To see the robust means, again use predicted values:

```
. predict predpay2

. label variable predpay2 "Robust predicted salary"

. tabulate gender rank, summ(predpay2) means
                       Means of Robust predicted salary

Gender      | Academic rank
(dummy      |
variable)   |    Assist      Assoc        Full |     Total
 -----------+----------------------------------+----------
       Male | 28916.148  38567.934  49760.008 |  40131.58
     Female | 28848.289  37464.512   46434.32 | 34918.387
 -----------+----------------------------------+----------
      Total |   28885.4  38126.565  49409.935 | 38355.404
```

The male-female salary gap among assistant and full professors appears smaller if we use robust means. It does not entirely vanish, however, and the gender gap among associate professors slightly widens.

Further Robust Applications

Diagnostic statistics and graphs (Chapter 7) and nonlinear transformations (Chapter 8) safeguard and extend the usefulness of **rreg**, as they do ordinary regression. With transformed variables, **rreg** or **qreg** fit robust curvilinear regression models. **rreg** can also robustly perform simpler types of analysis. To obtain a 90% confidence interval for the mean of a single variable, *y*, we could type the usual confidence-interval command **ci**:

```
. ci y, level(90)
```

Or we could get exactly the same interval by regression:

```
. regress y, level(90)
```

Similarly, we can obtain a robust mean with 90% confidence interval by typing:

```
. rreg y, level(90)
```

In all three commands, the **level( )** option specifies the desired degree of confidence. If we omit this option, Stata automatically displays a 95% confidence interval.

To compare two means, analysts typically employ a two-sample *t* test (**ttest**) or one-way analysis of variance (**oneway** or **anova**). As seen earlier, we can perform equivalent tests (yielding identical *t* and *F* statistics) with regression, for example by regressing the measurement variable on a dummy variable (here called *group*) representing the two categories:

```
. regress y group
```

A robust version of this test results from typing the following:

```
. rreg y group
```

qreg performs median regression by default, but it is actually a more general tool: It can fit a linear model for any quantile of *y*, not just the median (.5 quantile). For example, commands like this can analyze how the first quartile (.25 quantile) of *y* changes with *x*:

```
. qreg y x, quant(.25)
```

Assuming constant error variance, the slopes of the .25 and .75 quantile lines should be the same. **qreg** could thus perform a check for heteroscedasticity or subtle kinds of nonlinearity.

10

Logistic Regression

The regression and ANOVA methods described in Chapters 5 through 9 require measured *y* variables. Stata also offers many techniques for modeling categorical, ordinal, and censored dependent variables. A partial list of relevant commands follows. For more details about any of these, type **`help`** *`command`*.

`clogit`	Maximum-likelihood conditional logistic regression.
`clogitp`	Obtain predicted probabilities after **`clogit`**.
`cnreg`	Censored-normal regression, assuming that *y* follows a Gaussian distribution but is censored at a point that can vary from observation to observation.
`dprobit`	Probit regression giving changes in probabilities instead of coefficients.
`glm`	Generalized linear models. Includes options to model logistic, probit, or complementary log-log links. Allows response variable to be binary or proportional for grouped data. Diagnostic statistics also available.
`logistic`	Logistic regression, giving odds ratios and diagnostics.
`logit`	Logistic regression, giving estimated coefficients.
`loglin`	Loglinear modeling.
`lpredict`	Predicted probabilities and diagnostic statistics after **`logistic`**.
`mlogit`	Multinomial logistic regression, with polytomous *y* variable.

ologit	Logistic regression with ordinal *y* variable.
ologitp	Obtain predicted probabilities after **ologit**.
oprobit	Probit regression with ordinal *y* variable.
oprobitp	Obtain predicted probabilities after **oprobit**.
probit	Probit regression, with dichotomous *y* variable.
tobit	Tobit regression, assuming *y* follows a Gaussian distribution but is censored at a known, fixed point (see **cnreg** for a more general version).

Examples of several of these commands appear in the next section.

The remainder of this chapter concentrates on one particular, widely used family of techniques called logit or logistic regression. We review logit methods for dichotomous, ordinal, and polytomous dependent variables.

Example Commands

. logistic *y x1 x2 x3*

Performs logistic regression of {0,1} variable *y* on predictors *x1*, *x2*, and *x3*.

. lfit

Presents a Pearson chi-square goodness-of-fit test for the estimated logistic model: observed versus expected frequencies of *y* = 1, using cells defined by the covariate (*x*-variable) patterns. When a large number of *x* patterns exist, we might group them according to estimated probabilities. **lfit, group(10)** would perform the test with 10 approximately equal-size groups.

. lstat

Presents classification statistics and classification table. **lstat**, **lroc**, and **lsens** (see later) are particularly useful when the point of analysis is classification. These commands all refer to the previously estimated **logistic** model.

. lroc

Graphs the receiver operating characteristic (ROC) curve and calculates area under the curve.

. lsens

Graphs both sensitivity and specificity versus the probability cutoff.

. lpredict ***phat***

Generates a new variable (arbitrarily named *phat*) equal to predicted probabilities that $y = 1$ based on the most recent **logistic** model.

. lpredict ***dX2*****, dx2**

Generates a new variable (arbitrarily named *dX2*), the diagnostic statistic measuring "change in Pearson chi-square," from the most recent **logistic** analysis.

. mlogit ***y x1 x2 x3*****, base(3) rrr nolog**

Performs multinomial logistic regression of multiple-category variable *y* on three *x* variables. Uses $y = 3$ as the base category for comparison; gives relative risk ratios instead of regression coefficients; does not show the list of maximum-likelihood iterations.

. predict ***P2*****, outcome(2)**

Generates a new variable (arbitrarily named *P2*) representing the predicted probability that $y = 2$, based on the most recent **mlogit** analysis.

. glm ***success x1 x2 x3*****, family(binomial** ***trials*****) eform**

Performs a logistic regression via generalized linear modeling using tabulated rather than individual-observation data. The variable *success* gives the number of times that the outcome of interest occurred, and *trials* gives the number of times it could have occurred, for each combination of the predictors *x1*, *x2*, and *x3*. That is, *success/trials* would equal the proportion of times that an outcome such as "patient recovers" occurred. The **eform** option asks for results in the form of odds ratios ("exponentiated form") rather than logit coefficients.

. cnreg ***y x1 x2 x3*****, censored(*****cen*****)**

Performs censored-normal regression of measurement variable *y* on three predictors *x1*, *x2*, and *x3*. If an observation's true *y* value is unknown because of left or right censoring, it is replaced for this regression by the nearest *y* value at which censoring occurs. The censoring variable *cen* is a {0,1} indicator of whether each observation's value of *y* has been censored and hence replaced in this fashion.

. loglin ***count x1 x2 x3*****, fit(*****x1 x2*****,** ***x1 x3*****)**

Fits a Poisson maximum-likelihood log-linear model to cross-tabulated data. (If you have data organized by individual observations rather than a cross-tabulation, the **poisson** command works better; see Chapter 11.) Variable *count* gives the cell frequency at each level of *x1*, *x2*, and *x3*. The option **fit(*****x1 x2*****,** ***x1 x3*****)** specifies a model that fits the *x1x2* and also the *x1x3* marginals. An "independence" model is **fit(*****x1*****,** ***x2*****,** ***x3*****)**; a saturated model is **fit(*****x1 x2 x3*****)**.

Space Shuttle Data

Our main example for this chapter, *shuttle.dta*, involves data covering the first 25 flights of the U.S. space shuttles. These data contain evidence that, if properly analyzed, might have warned in advance that *Challenger* (the 25th flight, designated STS 51-L) should not be launched (*Report of the Presidential Commission on the Space Shuttle Challenger Accident*, 1986).

```
. use shuttle, clear
(Space Shuttle data)

. describe
Contains data from c:\data\shuttle.dta
  obs:          25                                   Space Shuttle data
 vars:           6                                   24 Dec 1996 10:14
 size:         275 (96.3% of memory free)
-----------------------------------------------------------------------------
   1. flight       byte    %8.0g       flbl      Flight
   2. month        byte    %8.0g                 Month of launch
   3. day          byte    %8.0g                 Day of launch
   4. year         int     %8.0g                 Year of launch
   5. distress     byte    %8.0g       dlbl      Thermal distress incidents
   6. temp         byte    %8.0g                 Joint temperature, degrees F
-----------------------------------------------------------------------------
Sorted by:  year  month  day

. list
       flight      month        day       year  distress        temp
  1.     STS-1         4         12       1981      none          66
  2.     STS-2        11         12       1981    1 or 2          70
  3.     STS-3         3         22       1982      none          69
  4.     STS-4         6         27       1982         .          80
  5.     STS-5        11         11       1982      none          68
  6.     STS-6         4          4       1983    1 or 2          67
  7.     STS-7         6         18       1983      none          72
  8.     STS-8         8         30       1983      none          73
  9.     STS-9        11         28       1983      none          70
 10.  STS_41-B         2          3       1984    1 or 2          57
 11.  STS_41-C         4          6       1984    3 plus          63
 12.  STS_41-D         8         30       1984    3 plus          70
 13.  STS_41-G        10          5       1984      none          78
 14.  STS_51-A        11          8       1984      none          67
 15.  STS_51-C         1         24       1985    3 plus          53
 16.  STS_51-D         4         12       1985    3 plus          67
 17.  STS_51-B         4         29       1985    3 plus          75
 18.  STS_51-G         6         17       1985    3 plus          70
 19.  STS_51-F         7         29       1985    1 or 2          81
 20.  STS_51-I         8         27       1985    1 or 2          76
 21.  STS_51-J        10          3       1985      none          79
 22.  STS_61-A        10         30       1985    3 plus          75
 23.  STS_61-B        11         26       1985    1 or 2          76
 24.  STS_61-C         1         12       1986    3 plus          58
 25.  STS_51-L         1         28       1986         .          31
```

This chapter studies three variables:

distress The number of "thermal distress incidents," in which hot gas blow-through or charring damaged joint seals of a flight's booster rockets. Burn-through of a booster joint seal precipitated the *Challenger* disaster. Many previous flights had experienced similar but less severe damage, so the joint seals were known to be a source of possible danger.

temp The calculated joint temperature at launch time, in degrees Fahrenheit. Temperature depends largely on weather. Rubber O-rings sealing the booster rocket joints become less flexible when cold.

date Date, measured in days elapsed since January 1, 1960 (an arbitrary starting point). *date* is generated from the month, day, and year of launch using the **mdy** (month-day-year to elapsed time; see **help dates**) function:

```
. generate date = mdy(month, day, year)

. label variable date "Date (days since 1/1/60)"
```

Launch date matters because several changes over the course of the shuttle program might have made it riskier. Booster rocket walls were thinned to save weight and increase payloads, and joint seals were subjected to higher-pressure testing. Furthermore, the reusable shuttle hardware was aging. So we might ask, did the probability of booster joint damage (one or more distress incidents) increase with launch date?

distress is a labeled numeric variable:

```
. tabulate distress

Thermal     |
distress    |
incidents   |      Freq.     Percent        Cum.
------------+-----------------------------------
       none |          9       39.13       39.13
     1 or 2 |          6       26.09       65.22
     3 plus |          8       34.78      100.00
------------+-----------------------------------
      Total |         23      100.00
```

Ordinarily, **tabulate** displays the labels, but the **nolabel** option reveals that the underlying numerical codes are 0 = "none", 1 = "1 or 2", and 2 = "3 plus":

```
. tabulate distress, nolabel

Thermal     |
distress    |
incidents   |      Freq.     Percent        Cum.
------------+-----------------------------------
          0 |          9       39.13       39.13
          1 |          6       26.09       65.22
          2 |          8       34.78      100.00
------------+-----------------------------------
      Total |         23      100.00
```

We can use these codes to create a new dummy variable, *any*, coded 0 for no distress and 1 for one or more distress incidents:

```
. generate any = distress
(2 missing values generated)

. replace any = 1 if distress == 2
(8 real changes made)

. label variable any "Any thermal distress"
```

To see what this accomplished,

```
. tabulate distress any

Thermal     | Any thermal distress
distress    |
incidents   |         0          1 |     Total
------------+----------------------+ ---------
       none |         9          0 |         9
     1 or 2 |         0          6 |         6
     3 plus |         0          8 |         8
------------+----------------------+ ---------
      Total |         9         14 |        23
```

Logistic regression models how a {0,1} dichotomy like *any* depends on one or more *x* variables. The syntax of **logit** resembles that of **regress**, listing dependent variable first:

```
. logit any date

Iteration 0:  Log Likelihood =-15.394543
Iteration 1:  Log Likelihood = -13.01923
Iteration 2:  Log Likelihood =-12.991146
Iteration 3:  Log Likelihood =-12.991096

Logit Estimates                                     Number of obs =     23
                                                    chi2(1)       =   4.81
                                                    Prob > chi2   = 0.0283
Log Likelihood = -12.991096                         Pseudo R2     = 0.1561

 ------------------------------------------------------------------------------
     any |      Coef.   Std. Err.       z     P>|z|      [95% Conf. Interval]
---------+ --------------------------------------------------------------------
    date |   .0020907   .0010703      1.953   0.051     -6.93e-06    .0041884
   _cons |  -18.13116   9.517217     -1.905   0.057     -36.78456    .5222396
 ------------------------------------------------------------------------------
```

The **`logit`** iterative estimation procedure maximizes the logarithm of the likelihood function, as shown at the output's top. At iteration 0, the log likelihood describes the fit of a model including only the constant. The last log likelihood describes the fit of the final model:

$$L = -18.13116 + .0020907 \times date \qquad [10.1]$$

where L represents the predicted logit, or log odds, of any distress incidents:

$$L = \ln[P(any{=}1) \,/\, P(any{=}0)] \qquad [10.2]$$

An overall χ^2 test at the upper right evaluates the null hypothesis that all coefficients in the model, except the constant, equal zero:

$$\chi^2 = -2(\ln \mathcal{L}_{\mathrm{i}} - \ln \mathcal{L}_{\mathrm{f}}) \qquad [10.3]$$

where $\ln \mathcal{L}_{\mathrm{i}}$ is the initial or iteration 0 (model with constant only) log likelihood, and $\ln \mathcal{L}_{\mathrm{f}}$ is the final iteration's log likelihood. Here,

$$\begin{aligned} \chi^2 &= -2[-15.394543 - (-12.991096)] \\ &= 4.81 \end{aligned}$$

The probability of a greater χ^2, with 1 degree of freedom (the difference in complexity between initial and final models), is low enough (.0283) to reject the null hypothesis in this example—so *date* does have a significant effect.

Less accurate, though convenient, tests are provided by the asymptotic z (standard normal) statistics displayed with **`logit`** results. With one predictor variable, that predictor's z statistic and the overall χ^2 statistic test equivalent hypotheses, analogous to the usual t and F statistics in simple OLS regression. Unlike their OLS counterparts, the logit z approximation and χ^2 tests sometimes disagree (they do here). The χ^2 test has more general validity.

Like Stata's other maximum-likelihood estimation procedures, **`logit`** displays a pseudo R^2 with its output:

$$\text{pseudo } R^2 = 1 - \ln \mathcal{L}_{\mathrm{f}} \,/\, \ln \mathcal{L}_{\mathrm{i}} \qquad [10.4]$$

For this example:

$$\begin{aligned} \text{pseudo } R^2 &= 1 - (-12.991096) \,/\, (-15.394543) \\ &= .1561 \end{aligned}$$

Although they provide a quick way to describe or compare the fit of different models for the same dependent variable, pseudo R^2 statistics lack the straightforward explained-variance interpretation of true R^2 in OLS regression.

After **`logit`**, the **`predict`** command (with no options) obtains predicted probabilities:

$$Phat = 1 \,/\, (1 + e^{-L}) \qquad [10.5]$$

Figure 10.1

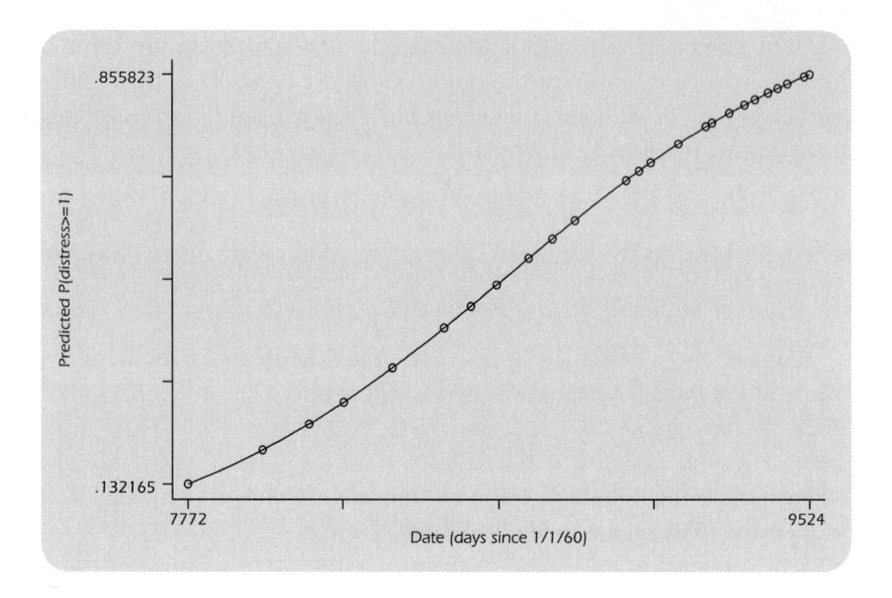

Graphed against *date*, these probabilities follow an S-shaped logistic curve as seen in Figure 10.1:

```
. predict Phat
. label variable Phat "Predicted P(distress>=1)"
. graph Phat date, connect(s)
```

The coefficient given by **logit** (.0020907) describes the effect of *date* on the logit or log odds of any thermal distress incidents. Each additional day increased the predicted log odds of thermal distress incidents by .0020907. Equivalently, we could say that each additional day multiplied predicted odds of thermal distress by $e^{.0020907}$ = 1.0020929; each 100 days therefore multiplied the odds by $(e^{.0020907})^{100} = 1.23$. ($e \approx 2.71828$, the base number for natural logarithms.) Stata can make these calculations using the _b[*varname*] coefficients stored after any estimation:

```
. display exp(_b[date])
1.0020929

. display exp(_b[date])^100
1.2325359
```

Or we could simply add an **or** (odds ratio) option to the **logit** command line. A better approach, though, employs the **logistic** command described in the next section. **logistic** performs the same estimation as **logit**, but it displays a table containing odds ratios rather than coefficients and provides several useful diagnostic tools.

Using Logistic Regression

Here is the same regression seen earlier, but using **logistic** instead of **logit**:

```
. logistic any date

Logit Estimates                                   Number of obs =       23
                                                  chi2(1)       =     4.81
                                                  Prob > chi2   =   0.0283
Log Likelihood = -12.991096                       Pseudo R2     =   0.1561

------------------------------------------------------------------------------
     any | Odds Ratio   Std. Err.       z     P>|z|     [95% Conf. Interval]
---------+--------------------------------------------------------------------
    date |   1.002093   .0010725      1.953   0.051     .9999931    1.004197
------------------------------------------------------------------------------
```

Note the identical log likelihoods and χ^2 statistics. Instead of coefficients (b), **logistic** displays odds ratios (e^b). The numbers in the "Odds Ratio" column of the **logistic** output are amounts by which the odds favoring $y = 1$ are multiplied, with each 1-unit increase in that x variable (if other x variables' values stay the same).

After any **logistic** regression, we can obtain a classification table and related statistics by typing **lstat**.

```
. lstat

Logistic model for any

              ----------True -----------
Classified |          D            ~D        Total
-----------+--------------------------+-----------
     +     |         12             4 |         16
     -     |          2             5 |          7
-----------+--------------------------+-----------
   Total   |         14             9 |         23

Classified + if predicted Pr(D) >= .5
True D defined as any ~= 0
--------------------------------------------------
Sensitivity                     Pr( +| D)   85.71%
Specificity                     Pr( -|~D)   55.56%
Positive predictive value       Pr( D| +)   75.00%
Negative predictive value       Pr(~D| -)   71.43%
--------------------------------------------------
False + rate for true ~D        Pr( +|~D)   44.44%
False - rate for true D         Pr( -| D)   14.29%
False + rate for classified +   Pr(~D| +)   25.00%
False - rate for classified -   Pr( D| -)   28.57%
--------------------------------------------------
Correctly classified                        73.91%
--------------------------------------------------
```

By default, **lstat** employs a probability of .5 as its cutoff (although we can change this by adding a **cutoff( )** option). Symbols in the classification table have the following meanings:

D	The event of interest did occur (that is, $y = 1$) for that observation. In this example, D indicates that any thermal distress occurred.
~D	The event of interest did not occur (that is, $y = 0$) for that observation. In this example, ~D corresponds to flights having no thermal distress.
+	The logistic model's predicted probability is greater than or equal to the cutoff point. Because we used the default cutoff, + here indicates that the model predicts a .5 or higher probability of thermal distress.
–	The predicted probability is less than the cutoff. Here, – means a predicted probability of thermal distress is below .5.

Thus for 12 flights, classifications are accurate in the sense that the model estimated at least a .5 probability of thermal distress, and distress did in fact occur. For 5 other flights, the model predicted less than a .5 probability, and distress did not occur. The overall "correctly classified" rate is therefore 12 + 5 = 17 out of 23, or 73.91%. The table also gives conditional probabilities such as "sensitivity" or the percentage of observations with P > .5 given that thermal distress occurred (12 out of 14 or 85.71%).

Another post-**logistic** command, **lpredict**, calculates a variety of fit and diagnostic statistics. For definitions of these statistics, see Hosmer and Lemeshow (1989).

lpredict ***newvar***	Predicted probability that $y = 1$.
lpredict ***newvar*, dbeta**	ΔB influence statistic, analogous to Cook's D.
lpredict ***newvar*, deviance**	Deviance residual for jth x pattern, d_j.
lpredict ***newvar*, dx2**	Change in Pearson χ^2, written as $\Delta\chi^2$ or $\Delta\chi^2_{\mathrm{P}}$.
lpredict ***newvar*, ddeviance**	Change in deviance χ^2, written as ΔD or $\Delta\chi^2_{\mathrm{D}}$.
lpredict ***newvar*, hat**	Leverage of the jth x pattern, h_j.
lpredict ***newvar*, number**	Assigns numbers to x patterns, $j = 1,2,3\ldots J$.
lpredict ***newvar*, resid**	Pearson residual for jth x pattern, r_j.
lpredict ***newvar*, rstandard**	Standardized Pearson residual.

Statistics obtained by the **dbeta**, **dx2**, **ddeviance**, and **hat** options do not measure the influence of individual observations, as their counterparts in ordinary regression do. Rather, these logit influence statistics measure the influence of "covariate patterns," that is, the consequences of dropping all observations with that particular combination of *x* values. See Hosmer and Lemeshow (1989) for details. A later section of this chapter shows these statistics in use.

Does booster joint temperature also affect the probability of any distress incidents? We could investigate by including *temp* as a second predictor variable:

```
. logistic any date temp

Logit Estimates                                   Number of obs =       23
                                                  chi2(2)       =     8.09
                                                  Prob > chi2   =   0.0175
Log Likelihood = -11.350748                       Pseudo R2     =   0.2627

 --------------------------------------------------------------------------
      any | Odds Ratio   Std. Err.       z     P>|z|     [95% Conf. Interval]
 --------+ ----------------------------------------------------------------
     date |    1.00297   .0013675      2.175   0.030     1.000293    1.005653
     temp |   .8408309   .0987887     -1.476   0.140     .6678848    1.058561
 --------------------------------------------------------------------------
```

The classification table indicates that including temperature as a predictor improved our correct classification rate to 78.26%.

```
. lstat

Logistic model for any

            ----------True-----------
Classified |          D            ~D        Total
 ----------+-------------------------+-----------
     +     |         12             3  |        15
     -     |          2             6  |         8
 ----------+-------------------------+-----------
   Total   |         14             9  |        23

Classified + if predicted Pr(D) >= .5
True D defined as any ~= 0
--------------------------------------------------
Sensitivity                     Pr( +| D)   85.71%
Specificity                     Pr( -|~D)   66.67%
Positive predictive value       Pr( D| +)   80.00%
Negative predictive value       Pr(~D| -)   75.00%
--------------------------------------------------
False + rate for true ~D        Pr( +|~D)   33.33%
False - rate for true D         Pr( -| D)   14.29%
False + rate for classified +   Pr(~D| +)   20.00%
False - rate for classified -   Pr( D| -)   25.00%
--------------------------------------------------
Correctly classified                        78.26%
--------------------------------------------------
```

According to the estimated model, each 1-degree increase in joint temperature multiplies the odds of booster joint damage by .84 (that is, each 1-degree warming reduces the odds of damage by about 16%). Although this effect seems strong enough to cause concern, the asymptotic z test says that it is not statistically significant ($z = -1.476$, $P = .140$). A more definitive test, however, employs the likelihood-ratio χ^2. The **lrtest** command compares nested models estimated by maximum likelihood. First, estimate a "complete" model containing all variables of interest, as done earlier with the **logistic *any date temp*** command. Next, type the command

```
. lrtest, saving(0)
```

Now estimate a reduced model, including only a subset of the x variables from the complete model. (Such reduced models are said to be "nested.") Finally, type **lrtest** again. For example (using the **quietly** prefix, because we already saw this output once):

```
. quietly logistic any date

. lrtest
Logistic:   likelihood-ratio test                    chi2(1)     =      3.28
                                                     Prob > chi2 =    0.0701
```

This second **lrtest** command tests the recent (presumably nested) model against the model previously saved by **lrtest, saving(0)**. It employs a general test statistic for nested maximum-likelihood models:

$$\chi^2 = -2(\ln\mathcal{L}_1 - \ln\mathcal{L}_0) \qquad [10.6]$$

$\ln\mathcal{L}_0$ is the log likelihood for the first model (with all x variables), and $\ln\mathcal{L}_1$ the log likelihood for the second model (with a subset of those x variables). Compare the resulting test statistic to a χ^2 distribution with degrees of freedom equal to the difference in complexity (number of x variables dropped) between models 0 and 1. Type **help lrtest** for more about this command, which works with any of Stata's maximum-likelihood estimation procedures (**logit**, **probit**, **stcox** and so on). The overall χ^2 statistic routinely given by **logit** or **logistic** output (equation [10.3] is a special case of [10.6]).

The previous **lrtest** example performed this calculation:

$$\chi^2 = -2[-12.991096 - (-11.350748)]$$
$$= 3.28$$

The calculation had 1 degree of freedom, yielding P = `chiprob(1,3.28)` = .0701; the effect of *temp* is significant at $\alpha = .10$. Given the small sample and fatal consequences of a Type II error, $\alpha = .10$ seems a more prudent cutoff than the usual $\alpha = .05$.

Conditional Effect Plots

Conditional effect plots help in understanding what a logit or logistic model implies for probabilities. For example, **summarize *date*, detail** tells us that the 25th percentile of *date* equals 8569. To find the predicted probability of any distress incidents, as a function of *temp*, with *date* fixed at its 25th percentile:

```
. quietly logit any date temp

. generate L1 = _b[_cons] + _b[date]*8569 + _b[temp]*temp

. generate Phat1 = 1/(1 + exp(-L1))

. label variable Phat1 "P(distress>=1|date=8569)"
```

L1 is the predicted logit, and *Phat1* equals the corresponding predicted probability, calculated according to equation [10.5]. Similar steps find the predicted probability of any distress with date fixed at its 75th percentile (9341):

```
. generate L2 = _b[_cons] + _b[date]*9341 + _b[temp]*temp

. generate Phat2 = 1/(1 + exp(-L2))

. label variable Phat2 "P(distress>=1|date=9341)"
```

We can now graph the relation between *temp* and the probability of any distress, for the two levels of *date* (Figure 10.2):

```
. graph Phat1 Phat2 temp, connect(ss) ylabel xlabel
      l1(Probability of distress>=1)
```

Among earlier flights (*date* = 8569, left curve), probability of thermal distress goes from very low, at around 80° F, to near 1, below 50° F. Among later flights (*date* = 9341, right curve), however, the probability of any distress exceeds .5 even in warm weather and climbs toward 1 on flights below 70° F. Note that *Challenger*'s launch temperature, 31° F, places it at top left in Figure 10.2. This analysis predicts almost certain booster joint damage for the *Challenger* flight.

Diagnostic Statistics and Graphs

As mentioned earlier, the influence and diagnostic statistics obtained by **lpredict** (after **logistic**) refer not to individual observations, as do the regression diagnostics of Chapter 7. Rather, the **lpredict** diagnostics refer to *x* patterns. With the space shuttle data, however, each *x* pattern is unique—no two flights share the

Figure 10.2

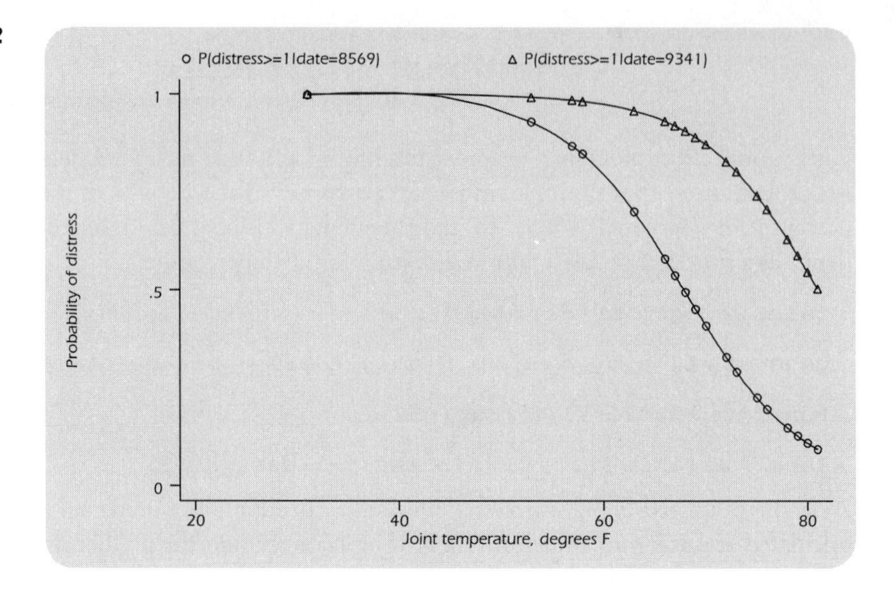

same combination of *date* and *temp* (naturally, because no two were launched the same day).

Hosmer and Lemeshow (1989) suggest plots that help in reading these diagnostics. To obtain these, we calculate and label several new variables. Because **lpredict** uses results from the most recent **logistic** estimation, we begin by quietly repeating this **logistic** command, to be sure it is what we think:

```
. quietly logistic any date temp

. lpredict Phat3

. label variable Phat3 "Predicted probability"

. lpredict dX2, dx2

. label variable dX2 "Change in Pearson chi-square"

. lpredict dB, dbeta

. label variable dB "Influence"

. lpredict dD, ddeviance

. label variable dD "Change in deviance"
```

To graph change in Pearson χ^2 versus probability of distress (Figure 10.3), type:

```
. graph dX2 Phat3, ylabel xlabel
```

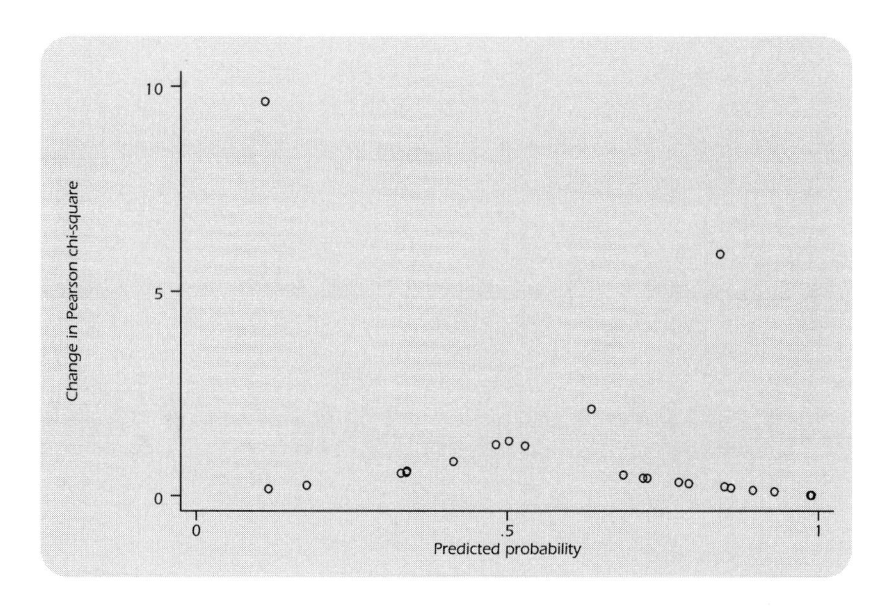

Figure 10.3

Two poorly fit *x* patterns, at upper right and left, stand out. We can identify these two flights (STS-2 and STS 51-A) by drawing the graph with flight numbers as plotting symbols. This is done by adding to the **graph** command the options **symbol([*flight*]) psize(150)**, with the results shown in Figure 10.4.

```
. graph dX2 Phat3, ylabel xlabel symbol([flight]) psize(120)

. list flight any date temp dX2 Phat3 if dX2 > 5

         flight        any         date        temp          dX2       Phat3
  2.      STS-2          1         7986          70     9.630337    .1091805
  4.      STS-4          .         8213          80            .           .
 14.   STS_51-A          0         9078          67     5.899742    .8400974
 25.   STS_51-L          .         9524          31            .           .
```

Flight STS 51-A experienced no thermal distress, despite late launch date and cool temperature (see Figure 10.2). The model predicts a .84 probability of distress for this flight. All points along the up-to-right curve in Figure 10.4 have *any* = 0, meaning no thermal distress. Atop the up-to-left (*any* = 1) curve, flight STS-2 experienced thermal distress despite being one of the earliest flights and being launched in slightly milder weather. The model predicts only a .109 probability of distress. (Stata considers missing values as "high" numbers, and thus lists two missing-values flights, including *Challenger*, among those with ***dX2*** > 5.)

Similar findings result from plotting ***dD*** versus predicted probability (Figure 10.5):

Figure 10.4

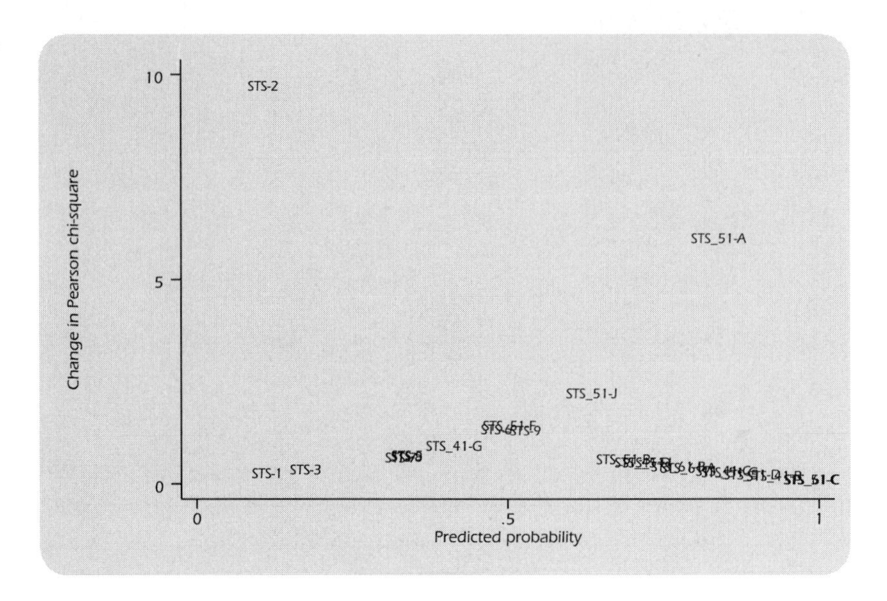

Figure 10.5

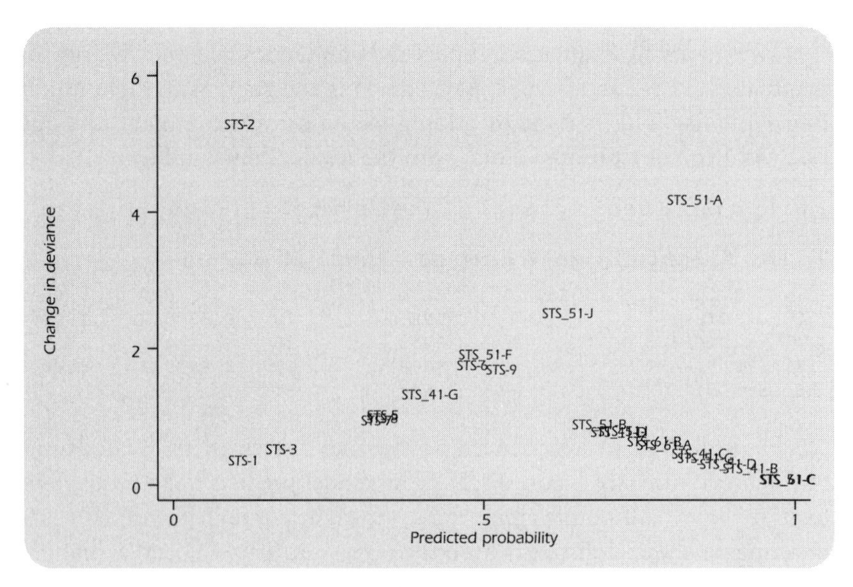


```
. graph dD Phat3, ylabel xlabel symbol([flight])
```

Again, flights STS-2 (top left) and STS 51-A (top right) stand out as poorly fit.

dB measures an *x* pattern's influence in logistic regression, as Cook's *D* measures an individual observation's influence in OLS. For a logistic-regression analogue to the OLS diagnostic plot in Figure 7.9, we can make the plotting symbols proportional to influence (Figure 10.6):

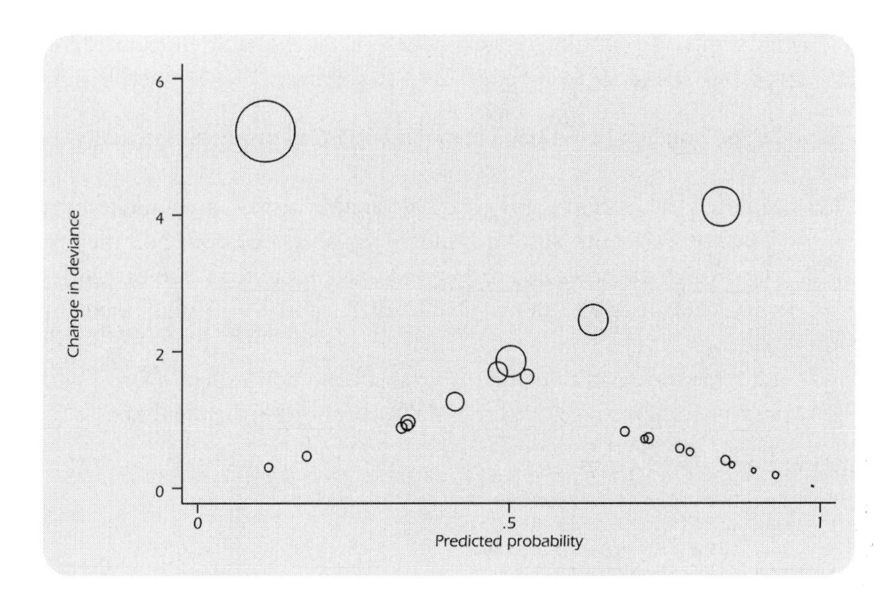

Figure 10.6

```
. graph dD Phat3 [iweight = dB], ylabel xlabel
```

Figure 10.6 reveals that the two worst-fit observations are also the most influential.

Observations poorly fit and influential deserve special attention because they both contradict the main pattern of the data and pull model estimates in their contrary direction. Of course, simply removing such outliers allows a "better fit" with the remaining data—but this is circular reasoning. A more thoughtful reaction would be to investigate what makes the outliers unusual. Why did shuttle flight STS-2, but not STS 51-A, experience booster joint damage? Seeking an answer might lead investigators to previously overlooked variables or to otherwise respecify the model.

Logistic Regression with Ordered-Category y

`logit` and `logistic` estimate models with two-category {0,1} *y* variables. We need other methods to estimate models in which *y* takes on more than two categories. For example:

`ologit` Ordered logistic regression, where *y* is an ordinal (ordered-category) variable. The numerical values representing the categories do not matter, except that higher numbers mean "more." For example, the *y* categories might be {1 = "poor," 2 = "fair," 3 = "excellent"}.

mlogit Multinomial logistic regression, where *y* has multiple but unordered categories like {1 = "Democrat," 2 = "Republican," 3 = "undeclared"}.

If *y* is {0,1}, **logit** (or **logistic**), **ologit**, and **mlogit** all produce essentially the same estimates.

We earlier simplified the three-category ordinal variable *distress* into a dichotomy, *any*. **logit** and **logistic** require {0,1} dependent variables. **ologit**, on the other hand, can handle ordinal variables like *distress* that have more than two categories. Recall that *distress* has categories 0 = "none," 1 = "1 or 2," and 2 = "3 plus" incidents of booster-joint distress.

Ordered-logit regression indicates that *date* and *temp* both affect *distress*, with the same signs (positive for *date*, negative for *temp*) seen in our earlier analyses:

```
. ologit distress date temp

Iteration 0:  Log Likelihood =-24.955257
Iteration 1:  Log Likelihood =-19.093131
Iteration 2:  Log Likelihood =-18.805891
Iteration 3:  Log Likelihood =-18.797073
Iteration 4:  Log Likelihood = -18.79706

Ordered Logit Estimates                              Number of obs =      23
                                                     chi2(2)       =   12.32
                                                     Prob > chi2   =  0.0021
Log Likelihood =   -18.79706                         Pseudo R2     =  0.2468

 ------------------------------------------------------------------------------
distress |      Coef.   Std. Err.       z     P>|z|     [95% Conf. Interval]
---------+--------------------------------------------------------------------
    date |    .003286   .0012662     2.595   0.009     .0008043     .0057677
    temp |  -.1733752   .0834473    -2.078   0.038     -.336929    -.0098215
---------+--------------------------------------------------------------------
   _cut1 |   16.42813   9.554813             (Ancillary parameters)
   _cut2 |   18.12227   9.722293
 ------------------------------------------------------------------------------
```

Likelihood-ratio tests are more accurate than the asymptotic *z* tests shown. First, have **lrtest** save results from the "full" model (including both predictors) just estimated:

```
. lrtest, saving(0)
```

Next, estimate a simpler model without *temp:*

```
. quietly ologit distress date

. lrtest
Ologit:  likelihood-ratio test                     chi2(1)     =       6.12
                                                   Prob > chi2 =     0.0133
```

The likelihood-ratio test indicates that *temp*'s effect is significant. Similar steps find that *date* also has a significant effect:

```
. quietly ologit distress temp

. lrtest
Ologit:   likelihood-ratio test                          chi2(1)     =      10.33
                                                         Prob > chi2 =     0.0013
```

The ordered-logit model estimates a score, *S*, as a linear function of *date* and *temp:*

$$S = .003286 \times date - .1733752 \times temp$$

Predicted probabilities depend on the value of *S* (plus a logistically distributed disturbance *u*) relative to the estimated cut points:

$P(distress\text{=“none”}) = P(S+u \leq \_\text{cut1}) = P(S+u \leq 16.42813)$

$P(distress\text{=“1 or 2”}) = P(\_\text{cut1} < S+u \leq \_\text{cut2}) = P(16.42813 < S+u \leq 18.12227)$

$P(distress\text{=“3 plus”}) = P(\_\text{cut2} < S+u) = P(18.12227 < S+u)$

After **ologit**, the command **ologitp** calculates predicted probabilities for each category of the dependent variable. We supply **ologitp** with names for these probabilities. For example: *none* could denote the probability of no distress incidents (first category of *distress*); *onetwo* the probability of 1 or 2 incidents (second category of *distress*); and *threeplu* the probability of 3 or more incidents (third and last category of *distress*):

```
. quietly ologit distress date temp

. ologitp none onetwo threeplu
```

This creates three new variables:

```
. describe none onetwo threeplu
  18. none          float  %9.0g               Pr(xb+u<_cut1)
  19. onetwo        float  %9.0g               Pr(_cut1<xb+u<_cut2)
  20. threeplu      float  %9.0g               Pr(_cut2<xb+u)
```

Predicted probabilities for *Challenger*'s last flight, the 25th in these data, are unsettling:

```
. list flight none onetwo threeplu if flight == 25
        flight        none      onetwo   threeplu
 25.  STS_51-L    .0000754    .0003346     .99959
```

Our model, based on analysis of 23 pre-*Challenger* shuttle flights, predicts little chance ($P = .000075$) of *Challenger* experiencing no booster joint damage, a scarcely greater likelihood of one or two incidents (P = .0003), but virtual certainty (P = .9996) of three or more damage incidents.

See Greene (1990) for more about ordered logistic regression, or consult the *Stata Reference Manuals* regarding Stata's implementation and options.

Multinomial Logistic Regression

When the dependent variable's categories have no natural ordering, we resort to multinomial logistic regression, also called polytomous logistic regression. Dataset *dover.dta* contains information from an environmental-issues survey of registered voters in Dover, New Hampshire:

```
Contains data from c:\data\dover.dta
  obs:           150                          5/85 Dover water survey
 vars:             9                          24 Dec 1996 10:17
 size:         1,950 (96.2% of memory free)
-----------------------------------------------------------------------
   1. age          byte    %8.0g              Years of age
   2. educ         byte    %8.0g              Respondent's education in years
   3. gender       byte    %8.0g    sexrlbl   Respondent's gender
   4. kids         byte    %8.0g    anyklbl   Have kids <18 in household?
   5. party        byte    %8.0g    partylb   Political party registration
   6. zoning       byte    %8.0g    zonlbl    Pass aquifer zoning
   7. superf       byte    %8.0g    superlbl  For Superfund landfill cleanup
   8. mapping      byte    %8.0g    maplbl    Detailed mapping of groundwater
   9. cleanup      byte    %8.0g    hiplbl    Participated in toxics pickup
-----------------------------------------------------------------------
Sorted by:
```

The variable *party* records respondents' political affiliation:

```
. tabulate party

Political    |
party        |
registration|      Freq.     Percent        Cum.
------------+-----------------------------------
   Democrat |         45       30.00       30.00
   Republic |         62       41.33       71.33
   undeclar |         43       28.67      100.00
------------+-----------------------------------
      Total |        150      100.00
```

party is actually a labeled numeric variable, as revealed by typing:

```
. tabulate party, nolabel

Political    |
party        |
registration |      Freq.     Percent        Cum.
-------------+-----------------------------------
           1 |         45       30.00       30.00
           2 |         62       41.33       71.33
           3 |         43       28.67      100.00
-------------+-----------------------------------
       Total |        150      100.00
```

The coding {1 = "Democrat," 2 = "Republican," 3 = "undeclared"} has no numerical meaning. We could equally well have used {1 = "Republican," 5 = "undeclared," 29 = "Democrat"}, or any other scheme.

Is there a "gender gap" among Dover voters? A Pearson χ^2 test finds no association between *party* and *gender:*

```
. tabulate party gender, chi2

Political    | Respondent's gender
party        |
registration |
             |      male     female |     Total
-------------+----------------------+----------
   Democrat  |        18         27 |        45
   Republic  |        33         29 |        62
   undeclar  |        18         25 |        43
-------------+----------------------+----------
      Total  |        69         81 |       150
          Pearson chi2(2) =   2.2520   Pr = 0.324
```

Multinomial logistic regression can replicate this simple analysis, obtaining a similar χ^2 value from its likelihood-ratio test:

```
. mlogit party gender, rrr nolog

Multinomial regression                            Number of obs =      150
                                                  chi2(2)       =     2.25
                                                  Prob > chi2   =   0.3240
Log Likelihood = -161.55453                       Pseudo R2     =   0.0069

------------------------------------------------------------------------------
   party |        RRR   Std. Err.        z     P>|z|     [95% Conf. Interval]
---------+--------------------------------------------------------------------
Democrat |
  gender |   1.706897   .6771427     1.348    0.178     .7843811    3.714388
---------+--------------------------------------------------------------------
undeclar |
  gender |    1.58046    .632857     1.143    0.253     .7210084    3.464388
------------------------------------------------------------------------------
(Outcome party==Republic is the comparison group)
```

Note the output's bottom line: *party* = "Republican" is the comparison group or base category. Unless we tell it otherwise, **mlogit** automatically chooses the most frequent category (in this instance, Republicans) as a base. The **rrr** option instructs

mlogit to show relative risk ratios, which resemble the odds ratios given by **logistic**. Another option, **nolog**, suppresses printing the iteration log.

Using the **tabulate** output from the previous page, we can calculate that *among males* the odds favoring Democrat over Republican are

$$P(\text{Democrat}) \,/\, P(\text{Republican}) = (18/69) \,/\, (33/69) = .5454545$$

Among females the odds favoring Democrat over Republican are

$$P(\text{Democrat}) \,/\, P(\text{Republican}) = (27/81) \,/\, (29/81) = .9310344$$

Thus the odds favoring Democrat over Republican are

$$.9310344 \,/\, .5454545 = 1.706897$$

times higher for females (*gender* = 1) than for males (*gender* = 0). This multiplier, a ratio of two odds, equals the relative risk ratio (1.706897) displayed by **mlogit**.

In general, the relative risk ratio for category j of y, and predictor x_k, equals the amount by which predicted odds favoring $y = j$ (compared with y = base) are multiplied, per 1-unit increase in x_k, other things being equal. In other words, the relative risk ratio rrr_{jk} is a multiplier such that, if all x variables except x_k stay the same,

$$\text{rrr}_{jk} \times \frac{P(y=j \mid x_k)}{P(y=\text{base} \mid x_k)} = \frac{P(y=j \mid x_k+1)}{P(y=\text{base} \mid x_k+1)} \qquad [10.7]$$

We can override the default base category by specifying **base( )** as an option. Because "undeclared" is category 3 of *party*, **base(3)** makes "undeclared" the **mlogit** base category.

```
. mlogit party gender, rrr base(3) nolog
Multinomial regression                              Number of obs =       150
                                                    chi2(2)       =      2.25
                                                    Prob > chi2   =    0.3240
Log Likelihood = -161.55453                         Pseudo R2     =    0.0069

------------------------------------------------------------------------------
   party |        RRR   Std. Err.        z     P>|z|      [95% Conf. Interval]
---------+--------------------------------------------------------------------
Democrat |
  gender |       1.08    .4684613      0.177   0.859      .4615366     2.52721
---------+--------------------------------------------------------------------
Republic |
  gender |   .6327273    .2533607     -1.143   0.253      .2886509    1.386948
------------------------------------------------------------------------------
(Outcome party==undeclar is the comparison group)
```

Changing the base category changes relative risk ratios but has no effect on the model fit, predictions or χ^2 statistic.

Both **tabulate** and **mlogit** (with any base category) agree that gender does not predict political party affiliation. Age and education exhibit stronger effects:

```
. mlogit party age educ, rrr base(3) nolog

Multinomial regression                            Number of obs =      150
                                                  chi2(4)       =    34.58
                                                  Prob > chi2   =   0.0000
Log Likelihood = -145.39136                       Pseudo R2     =   0.1063

------------------------------------------------------------------------------
   party |        RRR   Std. Err.       z     P>|z|     [95% Conf. Interval]
---------+--------------------------------------------------------------------
Democrat |
     age |   1.077405   .0197447     4.068   0.000      1.039393    1.116807
    educ |   1.266085   .1322149     2.259   0.024      1.031747    1.553646
---------+--------------------------------------------------------------------
Republic |
     age |   1.088564   .0193815     4.766   0.000      1.051232    1.127222
    educ |   1.248008   .1252581     2.207   0.027      1.025145    1.519321
------------------------------------------------------------------------------
(Outcome party==undeclar is the comparison group)
```

The *z* tests suggest that both *age* and *educ* have significant effects. To perform a likelihood-ratio test of the effect of education, compare the full model just estimated with a reduced model, lacking *educ* but otherwise the same:

```
. lrtest, saving(0)

. quietly mlogit party age

. lrtest
Mlogit:  likelihood-ratio test                   chi2(2)      =       6.56
                                                 Prob > chi2 =      0.0376
```

We can reject the null hypothesis that *educ* has no effect. (The **base( ) rrr nolog** options do not matter here because they do not change likelihood ratios.) We can also test the effect of *age*, by comparing the full model with a reduced model without *age:*

```
. quietly mlogit party educ

. lrtest
Mlogit:  likelihood-ratio test                   chi2(2)      =      32.89
                                                 Prob > chi2 =      0.0000
```

Both *age* and *educ* significantly predict *party*. What else can we say? To interpret these effects, recall that "undeclared" is now the base category. The relative risk ratios tell us that:

> Odds of "Democrat" rather than "undeclared" are multiplied by 1.077 (increase about 8%) with each 1-year increase in age, controlling for education.

Odds of "Democrat" rather than "undeclared" are multiplied by 1.266 (increase about 27%) with each 1-year increase in education, controlling for age.

Odds of "Republican" rather than "undeclared" are multiplied by 1.089 (increase about 9%) with each 1-year increase in age, controlling for education.

Odds of "Republican" rather than "undeclared" are multiplied by 1.248 (increase about 25%) with each 1-year increase in education, controlling for age.

Thus older or better educated voters are more likely to register for one of the political parties, rather than as "undeclared."

age and *educ* predict "Democrat" versus "undeclared," and "Republican" versus "undeclared." But neither *age* nor *educ* has much effect on the odds of "Republican" over "Democrat," as we see by changing the base category:

```
. mlogit party age educ, rrr base(1) nolog
Multinomial regression                                Number of obs =     150
                                                      chi2(4)       =   34.58
                                                      Prob > chi2   = 0.0000
Log Likelihood = -145.39136                           Pseudo R2     = 0.1063

------------------------------------------------------------------------------
   party |       RRR   Std. Err.       z     P>|z|     [95% Conf. Interval]
---------+--------------------------------------------------------------------
Republic |
     age |  1.010358   .0144412      0.721   0.471     .9824461    1.039062
    educ |  .9857225    .085046     -0.167   0.868      .832367    1.167332
---------+--------------------------------------------------------------------
undeclar |
     age |  .9281564   .0170096     -4.068   0.000     .8954098    .9621004
    educ |  .7898367   .0824812     -2.259   0.024     .6436472    .9692297
------------------------------------------------------------------------------
(Outcome party==Democrat is the comparison group)
```

Nor do *age* or *educ* affect the odds of "Democrat" over "Republican":

```
. mlogit party age educ, rrr base(2) nolog
Multinomial regression                                Number of obs =     150
                                                      chi2(4)       =   34.58
                                                      Prob > chi2   = 0.0000
Log Likelihood = -145.39136                           Pseudo R2     = 0.1063

------------------------------------------------------------------------------
   party |       RRR   Std. Err.       z     P>|z|     [95% Conf. Interval]
---------+--------------------------------------------------------------------
Democrat |
     age |  .9897486   .0141467     -0.721   0.471     .9624065    1.017868
    educ |  1.014484   .0875275      0.167   0.868     .8566541    1.201393
---------+--------------------------------------------------------------------
undeclar |
     age |  .9186415   .0163561     -4.766   0.000      .887137    .9512648
    educ |  .8012769   .0804213     -2.207   0.027     .6581889    .9754717
------------------------------------------------------------------------------
(Outcome party==Republic is the comparison group)
```

predict can calculate predicted probabilities from **mlogit**. The **outcome ( )** option specifies for which *y* category we want probabilities. For example, to get predicted probabilities that *party* = "Republican" (category 2):

```
. quietly mlogit party age educ

. predict PRepub, outcome(2)

. label variable PRepub "Predicted P(party=Republican)"
```

Predicted probabilities of registering Republican (*PRepub*) tend to be higher for actual Republicans than they are for undeclared voters. For Republicans and Democrats, though, the predicted probabilities of registering Republican are almost the same—because age and education did not much discriminate between Dover's voters from these two parties:

```
. tabulate party, summ(PRepub)

Political    | Summary of Predicted P(party=Republican)
party        |
registration|        Mean   Std. Dev.       Freq.
 -----------+ ----------------------------------
   Democrat |   .44265034   .12437918          45
   Republic |   .46324686   .12532096          62
   undeclar |    .3106844   .13358532          43
 -----------+ ----------------------------------
      Total |   .41333333   .14270703         150
```

mlogit involves much more complex internal code than do either **logit** or **ologit**. Consequently, if you apply **mlogit** to very large datasets, be prepared to wait a while—even with a fast computer.

The *Stata Reference Manual* provides more details. An article by Hamilton and Seyfrit (1994) describes how to construct conditional effect plots as an aid to interpreting multinomial logit analysis. Good introductions to logit regression topics appear in Aldrich and Nelson (1984), Greene (1990), and particularly Hosmer and Lemeshow (1989). An excellent recent book by Long (1997) presents logit regression and its variants in a broader context, and also consideres probit, tobit, Poisson, negative binomial and generalized linear models.

11

Survival Analysis and Event-Count Models

This chapter presents methods you can use to analyze event data. *Survival analysis* encompasses several related techniques that focus on times until the event of interest occurs. Although the event could be good or bad, by convention we refer to that event as a "failure." The time until failure is "survival time." Survival analysis is particularly relevant in biomedical research, but it can equally well apply to other fields from engineering to social science—for example, in modeling the time until an unemployed person gets a job, or a single person gets married.

We also look briefly at Poisson regression and its relatives. These methods focus not on survival times but, rather, on the rates or counts of events over a specified interval of time. Event-count methods include Poisson regression and negative binomial regression. Such models can be fit either through specialized commands, or through a broad approach called generalized linear modeling (GLM).

Selvin (1995) provides well-illustrated introductions to survival analysis and Poisson regression. I have borrowed (with permission) several of his examples. Other good introductions to survival analysis include a chapter in Rosner (1995) and the more comprehensive treatment by Lee (1992). McCullagh and Nelder (1989) describe generalized linear models; Long (1997) covers both event counts and GLM.

Example Commands

```
. stset time failure
```

Identifies single-record survival time data. Variable *time* indicates the time elapsed before either a particular event (called a "failure") occurred or the period of observation ended ("censoring"). Variable *failure* indicates whether a failure (*failure* = 1) or censoring (*failure* = 0) occurred at *time*. The dataset contains only one record per individual. Data must be **stset** before any further **st*** commands will work. If we subsequently **save** the dataset, however, the **stset** definitions are saved as well.

. stset *time failure*, id(*patient*) t0(*start*)

Identifies multiple-record survival time data. *time* indicates elapsed time before failure or censoring; *failure* indicates whether failure (1) or censoring (0) occurred at this time. *patient* is an identification number. The same individual can contribute more than one record to the data, but always has the same identification number. *start* contains the value of "time 0," the time when each observation entered the data (became at risk).

. stdes

Describes survival-time data, listing the definitions set by **stset** and other characteristics of the data.

. stsum

Obtains summary statistics: The total time at risk, incidence rate, number of subjects, and percentiles of survival time.

. ctset *time nfail ncensor nenter*, by(*ethnic sex*)

Identifies count-time data. *time* is a measure of time; *nfail* is the number of failures occurring at *time*. Optionally, we can also specify *ncensor* (number of censored observations at *time*) or *nenter* (number entering at *time*). *ethnic* and *sex* are other categorical variables defining observations in these data.

. cttost

Converts count-time data, previously identified by a **ctset** command, into survival-time form that can be analyzed by **st*** commands.

. sts graph

Graphs the Kaplan-Meier survivor function. To compare visually two or more survivor functions, such as one for each value of the categorical variable *sex*, use the **by( )** option:

. sts graph, by(*sex*)

To adjust, through Cox regression, for the effects of a continuous independent variable such as *age*, use the **adjustfor( )** option:

. sts graph, by(*sex*) adjustfor(*age*)

Note: The **by()** and **adjustfor()** options work similarly with other **sts** commands: **sts list**, **sts generate**, and **sts test**.

. sts list

Lists the estimated Kaplan-Meier survivor (failure) function.

. sts test *sex*

Tests the equality of the Kaplan-Meier survivor function across categories of *sex*.

. sts generate *survfunc* = S

Creates a new variable arbitrarily named *survfunc*, containing the estimated Kaplan-Meier survivor function.

. stcox *x1 x2 x3*

Estimates a Cox proportional hazard model, regressing time to failure on continuous or dummy variable predictors *x1–x3*. Requires **stset** data. Note: Stata also offers another Cox regression command that does not require **stset** data: **cox**. It likewise offers alternatives to the **stweib** and **stereg** commands (described later). The alternatives can produce the same results as corresponding **st*** commands, but they have different syntax and can cause confusion if used interchangeably. The **st*** commands are easier to use.

. stcox *x1 x2 x3*, strata(*x4*) basehaz(*hazard*) robust

Estimates Cox proportional hazard model, stratified by *x4*. Store the group-specific baseline hazard estimate as a new variable named *hazard*. (Baseline survivor function estimates could be obtained through a **basesur(*survive*)** option.) Obtains robust estimates of variances.

. stweib *x1 x2*

Estimates Weibull-distribution model regression of the time to failure on continuous or dummy variable predictors *x1* and *x2*. Requires **stset** data.

. stereg *x1 x2 x3 x4*

Estimates exponential-distribution model regression of time to failure on continuous or dummy predictors *x1–x4*. Requires **stset** data.

. poisson *count x1 x2 x3*, irr exposure(*x4*)

Performs Poisson regression of event-count variable *count* (assumed to follow a Poisson distribution) on continuous or dummy independent variables *x1–x3*. Independent-variable effects will be reported as incident rate ratios (**irr**). The **exposure( )** option identifies a variable indicating the amount of exposure, if this is not the same for all observations.
Note: A Poisson model assumes that the event probability remains constant, regardless of how many times an event occurs for each observation. If the probability does not remain constant, we should consider using **nbreg** (negative binomial regression) or **gnbreg** (generalized negative binomial regression) instead.

```
. glm count x1 x2 x3, link(log) family(poisson)
      lnoffset(x4) eform
```

Performs the same regression specified in the **poisson** example, but as a generalized linear model (GLM). **glm** can fit Poisson, negative binomial, logit, and many other types of models, depending on what **link( )** (link function) and **family( )** (distribution family) options we employ.

Survival-Time Data

Survival-time data contain, at a minimum, one variable measuring how much time elapsed before a certain event occurred to each observation. The literature often terms this event of interest a "failure," regardless of its substantive meaning. When failure has not occurred to an observation, by the time data collection ends, that observation is said to be "censored." The **stset** command sets up a dataset for survival-time analysis by identifying which variable measures time and (if necessary) which variable is a dummy indicating whether the observation failed or was censored. The dataset can also contain any number of other measurement or categorical variables, and individuals (for example, medical patients) can be represented by more than one observation.

To illustrate the use of **stset**, we will begin with an example from Selvin (1995:453) concerning 51 individuals diagnosed with HIV. The data initially reside in a raw-data file (*aids.raw*) that looks like this:

```
   1     1     1    34
   2    17     1    42
   3    37     0    47
     (rows 4–50 omitted)
  51    81     0    29
```

The first column values are case numbers (1, 2, 3, ... , 51). The second column tells how many months elapsed after the diagnosis, before that person either developed symptoms of AIDS or the study ended (1, 17, 37, ...). The third column holds a 1 if the individual developed AIDS symptoms (failure), or a 0 if no symptoms had appeared by the end of the study (censoring). The last column reports the individual's age at the time of diagnosis.

We can read the raw data into memory using **infile**, then label the variables and data and save them in Stata format as file *aids0.dta*:

```
. infile case time aids age using aids.raw, clear
(51 observations read)

. label variable case "Case ID number"
```

```
. label variable time "Months since HIV diagnosis"

. label variable aids "Developed AIDS symptoms"

. label variable age "Age in years"

. label data "AIDS (Selvin 1995:453)"

. compress
case was float now byte
time was float now byte
aids was float now byte
age was float now byte

. save aids0
file c:\data\aids0.dta saved
```

The next step is to identify which variable measures time, and which indicates failure/censoring. Although not necessary with these single-record data, we can also note which variable holds individual case identification numbers. In an **stset** command, the first-named variable measures time and the second is a dummy for failure (1) or censored (0). After using **stset**, we save the data again to preserve this information.

```
. stset time aids, id(case)
note:  making entry-time variable t0
       (within case, t0 will be 0 for the 1st observation and the
       lagged value of exit time time thereafter)

   data set name:  c:\data\aids0.dta
              id:  case
      entry time:  t0
       exit time:  time
 failure/censor:  aids

. save, replace
file c:\data\aids0.dta saved
```

stdes yields a brief description of how our survival-time data are structured. In this simple example, we have only one record per subject, so some of this information is unneeded.

```
. stdes

    failure time:  time
      entry time:  t0
  failure/censor:  aids
              id:  case

                                  |---------------per subject --------------|
Category                   total        mean         min     median        max
------------------------------------------------------------------------------
no. of subjects               51
no. of records                51           1           1          1          1

(first) entry time                         0           0          0          0
(final) exit time                   62.03922           1         67         97

subjects with gap              0
time on gap if gap             0           .           .          .          .
time at risk                3164    62.03922           1         67         97

failures                      25    .4901961           0          0          1
------------------------------------------------------------------------------
```

The **stsum** command obtains summary statistics. We have 25 failures out of 3,164 person-months, giving an incidence rate of 25/3164 = .0079014. The percentiles of survival time derive from a Kaplan-Meier survivor function (next section). This function estimates about a 25% chance of developing AIDS within 41 months after diagnosis, and 50% within 81 months. Over the observed range of the data (to 97 months) the probability of AIDS does not reach 75%, so there is no 75th percentile given.

```
. stsum

    failure time:  time
      entry time:  t0
  failure/censor:  aids
              id:  case

         |               incidence       no. of     |----Survival time ---|
         | time at risk     rate        subjects        25%       50%       75%
---------+---------------------------------------------------------------------
   total |         3164   .0079014            51         41        81         .
```

If the data happened to include a grouping or categorical variable such as *sex* (0 = male, 1 = female), we could obtain summary statistics on survival time separately for each group by a command of the form:

```
. stsum, by(sex)
```

Later sections describe more formal methods for comparing survival times from two or more groups.

Count-Time Data

Survival-time (**st**) datasets like *aids0.dta* contain information on individual people or things, with variables indicating the time at which failure or censoring occurred for each individual. A different type of dataset called count-time (**ct**) contains aggregate data, with variables counting the number of individuals that failed or were censored at time *t*. For example, *diskdriv.dta* contains hypothetical test information on 25 disk drives. All but five drives failed before testing ended at 1,200 hours.

```
. use diskdriv, clear
(Count-time data on disk drives)

. list
        hours  failures   censored
  1.      200         2          0
  2.      400         3          0
  3.      600         4          0
  4.      800         8          0
  5.     1000         3          0
  6.     1200         0          5
```

To set up a count-time dataset, we specify the time variable, the number-of-failures variable, and the number-censored variable, in that order. After **ctset**, the **cttost** command automatically converts our count-time data to survival-time format.

```
. ctset hours failures censored
  data set name:  c:\data\diskdriv.dta
           time:  hours
       no. fail:  failures
       no. lost:  censored
      no. enter:  -                 (meaning all enter at time 0)

. cttost
(data is now st)
             id:  -                 (meaning each record a unique subject)
     entry time:  -                 (meaning all entered at time 0)
      exit time:  hours
 failure/censor:  failures
         weight:  [fweight=w]

. list
        hours  failures          w
  1.     1200         0          5
  2.      200         1          2
  3.      400         1          3
  4.      600         1          4
  5.      800         1          8
  6.     1000         1          3
```

```
. stdes

    failure time:  hours
  failure/censor:  failures
          weight:  [fweight=w]

                                     |-------------- per subject ------------|
                          unweighted   unweighted            unweighted
Category                     total        mean         min      median      max
 -----------------------------------------------------------------------------
no. of subjects                  6
no. of records                   6          1            1          1          1

(first) entry time                          0            0          0          0
(final) exit time                         700          200        700       1200

subjects with gap                0
time on gap if gap               0
time at risk                  4200        700          200        700       1200

failures                         5    .8333333           0          1          1
 -----------------------------------------------------------------------------
```

The **cttost** command defines a set of frequency weights, *w*, in the resulting **st**-format dataset. **st*** commands automatically recognize and use these weights in any survival-time analysis, so the data are now viewed as containing 25 observations (25 disk drives) instead of the previous 6 (6 time periods):

```
. stsum

    failure time:  hours
  failure/censor:  failures
          weight:  [fweight=w]

       |                 incidence        no. of     |-----Survival time ----|
       | time at risk       rate          subjects      25%       50%       75%
-------+---------------------------------------------------------------------
 total |        19400    .0010309              25       600       800      1000
```

Kaplan-Meier Survivor Functions

Let n_t represent the number of observations that have not failed and are not censored at the beginning of time period *t*. d_t represents the number of failures that occur to these observations during time period *t*. The Kaplan-Meier estimator of surviving beyond time *t* is the product of survival probabilities in *t* and the preceding periods:

$$S(t) = \prod_{j=t0}^{t} \{ (n_j - d_j) / n_j \} \qquad [11.1]$$

For example, in the AIDS data seen earlier, 1 of the 51 individuals developed symptoms only one month after diagnosis. No observations were censored this early, so the probability of "surviving" (meaning, not developing AIDS) beyond *time* = 1 is:

$$S(1) = (51 - 1) / 51 = .9804$$

A second patient developed symptoms at *time* = 2, and a third at *time* = 9:

$$S(2) = .9804 \times (50 - 1) / 50 = .9608$$

$$S(9) = .9608 \times (49 - 1) / 49 = .9412$$

Graphing *S(t)* against *t* produces a Kaplan-Meier survivor curve, like the one seen in Figure 11.1. Stata draws such graphs automatically with the **sts graph** command. For example:

```
. use aids2, clear
(AIDS (Selvin 1995:453))

. sts graph
```

Figure 11.1

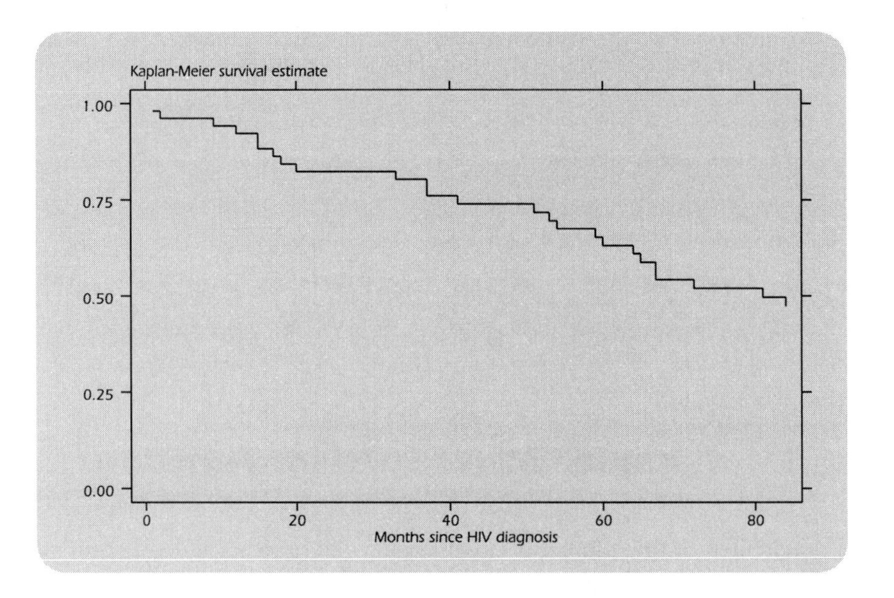

For a second example of survivor functions, we turn to data from Rosner (1995). The observations are 234 former smokers, attempting to quit. Most did not succeed. Variable *days* records how many days elapsed between quitting and starting up again. The study lasted one year, and variable *smoking* indicates whether an individual resumed smoking before the end of this study (*smoking* = 1, "failure") or not (*smoking* =

0, "censored"). With new data, we should begin by using **stset** to set them up for survival-time analysis:

```
. use smoking, clear
(Smoking (Rosner 1995:607)

. describe
Contains data from c:\data\smoking.dta
  obs:           234                        Smoking (Rosner 1995:607)
 vars:             8                        7 Dec 1996 11:38
 size:         3,744 (96.2% of memory free)
-------------------------------------------------------------------------------
   1. id           int    %9.0g             Case ID number
   2. days         int    %9.0g             Days abstinent
   3. smoking      byte   %9.0g             Resumed smoking
   4. age          byte   %9.0g             Age in years
   5. sex          byte   %8.0g      sex    Sex (female)
   6. cigs         byte   %9.0g             Cigarettes per day
   7. co           int    %9.0g             Carbon monoxide x 10
   8. minutes      int    %9.0g             Minutes elapsed since last cig
-------------------------------------------------------------------------------
Sorted by:

. stset days smoking
   data set name:  c:\data\smoking.dta
              id:  -                (meaning each record a unique subject)
      entry time:  -                (meaning all entered at time 0)
       exit time:  days
 failure/censor:  smoking

. save smoking0
file c:\data\smoking0.dta saved
```

The study involved 110 men and 124 women. Incidence rates for both sexes appear similar:

```
. stsum, by(sex)
    failure time:  days
  failure/censor:  smoking

       |              incidence        no. of     |-----Survival time ---|
sex    | time at risk    rate          subjects       25%        50%        75%
 ------+-----------------------------------------------------------------------
  Male |        8813   .0105526          110          4         15         68
Female |       10133   .0106582          124          4         15         91
 ------+-----------------------------------------------------------------------
 total |       18946   .0106091          234          4         15         73
```

Figure 11.2 confirms this similarity, showing little difference between the survivor functions of men and women. That is, both sexes returned to smoking at about the same rate. The survival probabilities of nonsmokers decline very steeply during

Figure 11.2

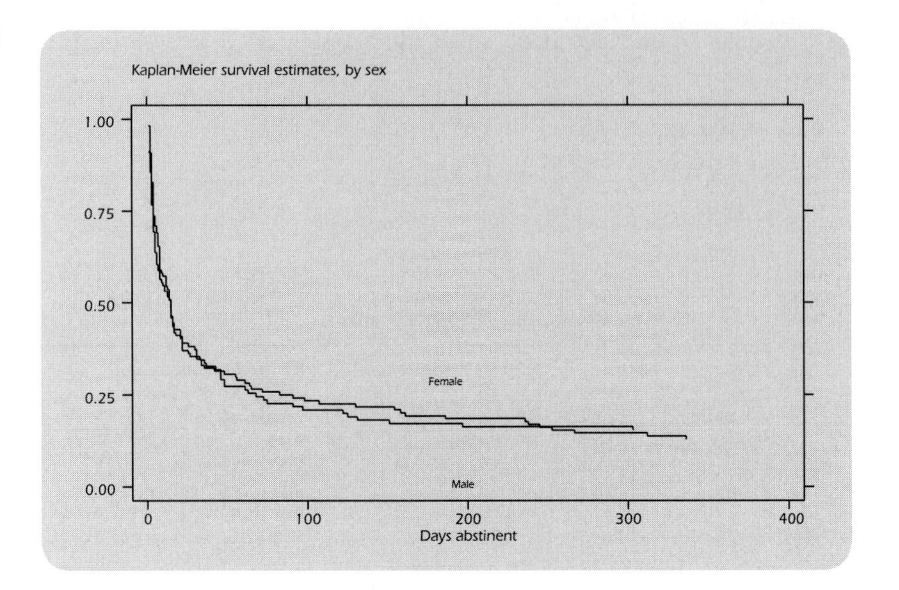

the first 30 days after quitting. For either sex, there is less than a 15% chance of surviving beyond a full year.

```
. sts graph, by(sex)
```

We can also formally test for the equality of survivor functions using a log-rank test. Unsurprisingly, this test finds no significant difference ($P = .6772$) between the smoking recidivism of men and women.

```
. sts test sex

    failure time:  days
  failure/censor:  smoking


Log-rank test for equality of survivor functions
------------------------------------------------

       | Events
sex    | observed        expected
------+------------------------
Male   |       93           95.88
Female |      108          105.12
------+------------------------
Total  |      201          201.00

              chi2(1) =       0.17
              Pr>chi2 =     0.6772
```

Cox Proportional Hazard Models

Regression methods allow us to take survival analysis further and examine the effects of multiple continuous or categorical predictors. One widely used method known as Cox regression employs a proportional hazard model. The hazard rate for failure at time t is defined as:

$$h(t) = \frac{\text{probability of failing between times } t \text{ and } t + \Delta t}{(\Delta t)\ (\text{probability of failing after time } t)} \qquad [11.2]$$

We model this hazard rate as a function of the baseline hazard (h_0) at time t, and the effects of one or more x variables:

$$h(t) = h_0(t) \exp(\beta_1 x_1 + \beta_2 x_2 + \ldots + \beta_k x_k) \qquad [11.3a]$$

or, equivalently:

$$\ln[h(t)] = \ln[h_0(t)] + \beta_1 x_1 + \beta_2 x_2 + \ldots + \beta_k x_k \qquad [11.3b]$$

"Baseline hazard" means the hazard for an observation with all x variables equal to 0. Cox regression estimates this hazard nonparametrically and obtains maximum-likelihood estimates of the β parameters in [11.3]. Stata's **stcox** procedure ordinarily reports hazard ratios, which are estimates of exp(β). These indicate proportional changes relative to the baseline hazard rate.

Does age affect the onset of AIDS symptoms? Dataset *aids2.dta* contains information that helps answer this question. Note that with **stcox**, unlike most other Stata model-fitting commands, we list only the independent variable(s). The survival-analysis dependent variable, a survivor or hazard function, is understood automatically with **stset** data.

```
. use aids2, clear
(AIDS (Selvin 1995:453))

. stcox age

     failure time:  time
       entry time:  t0
   failure/censor:  aids
               id:  case

Iteration 0:  Log Likelihood = -89.07617
Iteration 1:  Log Likelihood =-86.658023
Iteration 2:  Log Likelihood =-86.576308
Iteration 3:  Log Likelihood =-86.576295
Refining estimates:
Iteration 0:  Log Likelihood =-86.576295

Cox regression — entry time t0
```

```
No. of subjects =          51                  Log likelihood =   -86.576295
No. of failures =          25                  chi2(1)        =         5.00
Time at risk    =        3164                  Prob > chi2    =       0.0254

------------------------------------------------------------------------------
   time |
   aids | Haz. Ratio   Std. Err.       z     P>|z|      [95% Conf. Interval]
--------+---------------------------------------------------------------------
    age |   1.084557   .0378623     2.325   0.020       1.01283    1.161363
------------------------------------------------------------------------------
```

We might interpret the estimated hazard ratio, 1.084557, with reference to two HIV-positive individuals whose ages are α and $\alpha + 1$. The older person is 8.5% more likely to develop AIDS symptoms over a short period of time (that is, the ratio of their respective hazards is 1.084557). This ratio differs significantly ($P = .020$) from 1. If we wanted to state our findings for a five-year difference in age, we could raise the hazard ratio to the fifth power:

```
. display exp(_b[age])^5
1.5005865
```

Thus the hazard of AIDS onset is about 50% higher when the second person is five years older than the first. Alternatively, we could learn the same thing (and obtain the new confidence interval) by repeating the regression after temporarily rescaling *age* to have five-year units. The **nolog noshow** options that follow suppress display of the iteration log and the **st**-dataset description.

```
. replace age = age/5
age was byte now float
(51 real changes made)

. stcox age, nolog noshow
Cox regression -- entry time t0

No. of subjects =          51                  Log likelihood =   -86.576295
No. of failures =          25                  chi2(1)        =         5.00
Time at risk    =        3164                  Prob > chi2    =       0.0254

------------------------------------------------------------------------------
   time |
   aids | Haz. Ratio   Std. Err.       z     P>|z|      [95% Conf. Interval]
--------+---------------------------------------------------------------------
    age |   1.500587   .2619305     2.325   0.020      1.065815    2.112711
------------------------------------------------------------------------------
```

Like ordinary regression, Cox models can have more than one independent variable. Dataset *heart.dta* contains survival-time data from Selvin (1995), on 35 patients with very high cholesterol levels. Variable *time* gives the number of days each patient was under observation. *coronary* indicates whether a coronary event occurred during this time (*coronary* = 1) or not (*coronary* = 0). The data also include cholesterol levels and other factors thought to affect heart disease. The file *heart.dta* was previously set up for survival-time analysis by an **stset *time coronary*** command, so we can go directly to **st** analysis.

```
. use heart, clear
(Heart disease (Selvin 1995:436))

. describe
Contains data from c:\data\heart.dta
 obs:            35                         Heart disease (Selvin 1995:436)
vars:             8                         7 Dec 1996 14:58
size:           560 (96.3% of memory free)
-----------------------------------------------------------------------------
  1. patient     byte    %9.0g              Patient ID number
  2. time        int     %9.0g              Time in days
  3. coronary    byte    %9.0g              Coronary event (1) or none (0)
  4. weight      int     %9.0g              Weight in pounds
  5. sbp         int     %9.0g              Systolic blood pressure
  6. chol        int     %9.0g              Cholesterol level
  7. cigs        byte    %9.0g              Cigarettes smoked per day
  8. ab          byte    %9.0g              Type A (1) or B (0) personality
-----------------------------------------------------------------------------
Sorted by:  patient  time


. stdes
    failure time:  time
  failure/censor:  coronary

                                   |--------------per subject-------------|
Category                   total        mean        min     median        max
-----------------------------------------------------------------------------
no. of subjects               35
no. of records                35           1          1          1          1

(first) entry time                         0          0          0          0
(final) exit time                   2580.629        773       2875       3141

subjects with gap              0
time on gap if gap             0
time at risk               90322    2580.629        773       2875       3141

failures                       8    .2285714          0          0          1
-----------------------------------------------------------------------------
```

Cox regression finds that cholesterol level and cigarettes both significantly increase the hazard of a coronary event. Counterintuitively, weight appears to decrease the hazard. Systolic blood pressure and A/B personality do not have significant net effects.

```
. stcox weight sbp chol cigs ab, noshow nolog
Cox regression -- entry time 0
No. of subjects =         35                   Log likelihood =  -17.263231
No. of failures =          8                   chi2(5)        =       13.97
Time at risk    =      90322                   Prob > chi2    =      0.0158
```

```
------------------------------------------------------------------------------
    time |
coronary | Haz. Ratio   Std. Err.       z     P>|z|       [95% Conf. Interval]
---------+--------------------------------------------------------------------
  weight |   .9349336    .0305184    -2.061   0.039       .8769919    .9967034
     sbp |   1.012947    .0338061     0.385   0.700       .9488087    1.081421
    chol |   1.032142    .0139984     2.333   0.020       1.005067    1.059947
    cigs |   1.203335    .1071031     2.080   0.038       1.010707    1.432676
      ab |    3.04969    2.985616     1.139   0.255       .4476492    20.77655
------------------------------------------------------------------------------
```

After estimating the model, **stcox** can also generate new variables holding the estimated baseline hazard and survivor functions. Because "baseline" refers to a situation with all *x* variables equal to zero, however, we first need to rescale some variables so that 0 values make sense. A patient who weighs 0 pounds, or has 0 blood pressure, does not provide a useful comparison. Guided by the minimum values actually in our data, we might rescale *weight* so that 0 indicates 120 pounds, *sbp* so that 0 indicates 100, and *chol* so that 0 indicates 340:

```
. summ

Variable |     Obs        Mean    Std. Dev.        Min         Max
---------+-----------------------------------------------------
 patient |      35          18    10.24695           1          35
    time |      35    2580.629    616.0796         773        3141
coronary |      35    .2285714     .426043           0           1
  weight |      35    170.0857    23.55516         120         225
     sbp |      35    129.7143    14.28403         104         154
    chol |      35    369.2857    51.32284         343         645
    cigs |      35    17.14286    13.07702           0          40
      ab |      35    .5142857    .5070926           0           1

. replace weight = weight - 120
(35 real changes made)

. replace sbp = sbp - 100
(35 real changes made)

. replace chol = chol - 340
(35 real changes made)

. summ

Variable |     Obs        Mean    Std. Dev.        Min         Max
---------+-----------------------------------------------------
 patient |      35          18    10.24695           1          35
    time |      35    2580.629    616.0796         773        3141
coronary |      35    .2285714     .426043           0           1
  weight |      35    50.08571    23.55516           0         105
     sbp |      35    29.71429    14.28403           4          54
    chol |      35    29.28571    51.32284           3         305
    cigs |      35    17.14286    13.07702           0          40
      ab |      35    .5142857    .5070926           0           1
```

Zero values for all the *x* variables now make more substantive sense. To create new variables holding the baseline survivor and hazard function estimates, we repeat the regression with **basesurv()** and **basehaz()** options:

```
. stcox weight sbp chol cigs ab, noshow nolog
      basesurv(survivor) basehaz(hazard)
Cox regression — entry time 0
No. of subjects =          35              Log likelihood =  -17.263231
No. of failures =           8              chi2(5)        =       13.97
Time at risk    =       90322              Prob > chi2    =      0.0158

-------------------------------------------------------------------------
    time |
coronary | Haz. Ratio   Std. Err.       z     P>|z|     [95% Conf. Interval]
---------+---------------------------------------------------------------
  weight |   .9349336   .0305184    -2.061   0.039     .8769919    .9967034
     sbp |   1.012947   .0338061     0.385   0.700     .9488087    1.081421
    chol |   1.032142   .0139984     2.333   0.020     1.005067    1.059947
    cigs |   1.203335   .1071031     2.080   0.038     1.010707    1.432676
      ab |    3.04969   2.985616     1.139   0.255     .4476492    20.77655
-------------------------------------------------------------------------
```

Note that rescaling three *x* variables had no effect on the hazard ratios, standard errors, and so forth. The command created two new variables, arbitrarily named *survivor* and *hazard.* To graph the baseline survivor function, we plot *survivor* against *time* and connect data points with a step function (**connect(J)**) as seen in Figure 11.3:

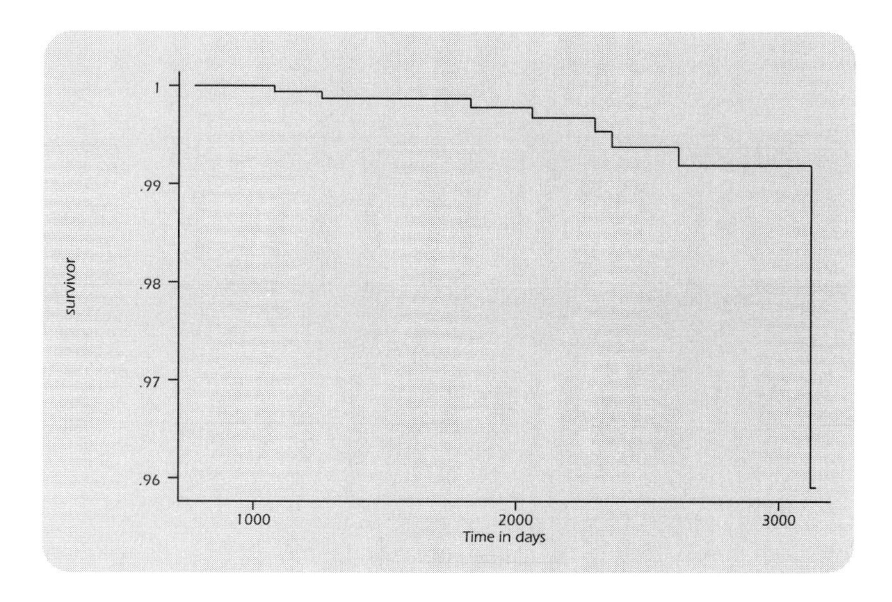

Figure 11.3

```
. graph survivor time, connect(J) symbol(.) sort ylabel xlabel
```

The baseline survivor function—which depicts survival probabilities for patients having "0" weight (120 pounds), "0" blood pressure (100), "0" cholesterol (340), 0 cigarettes per day, and a type B personality—declines with time. Although this decline looks precipitous at the right, notice that the probability really only falls from 1 to about .96. Given less favorable values of the predictor variables, the survival probabilities would fall much farther.

The same baseline survivor-function graph could have been obtained another way, without **stcox**. The alternative, shown in Figure 11.4, employs an **sts graph** command with **adjustfor()** option listing the predictor variables:

```
. sts graph, adjustfor(weight sbp chol cigs ab)
```

Figure 11.4, unlike Figure 11.3, follows the usual survivor-function convention of scaling the vertical axis from 0 to 1. Apart from this difference in scaling, Figures 11.4 and 11.3 depict the same curve.

Figure 11.5 graphs the estimated baseline hazard function against time, using the variable (*hazard*) generated by our **stcox** command. We have just 8 data points because only 8 patients "failed" (suffered coronary events). A hazard function is not estimated for time periods in which no failures occurred. This graph shows the baseline hazard increasing with time, from near 0 to .033.

```
. graph hazard time, ylabel xlabel
```

Figure 11.4

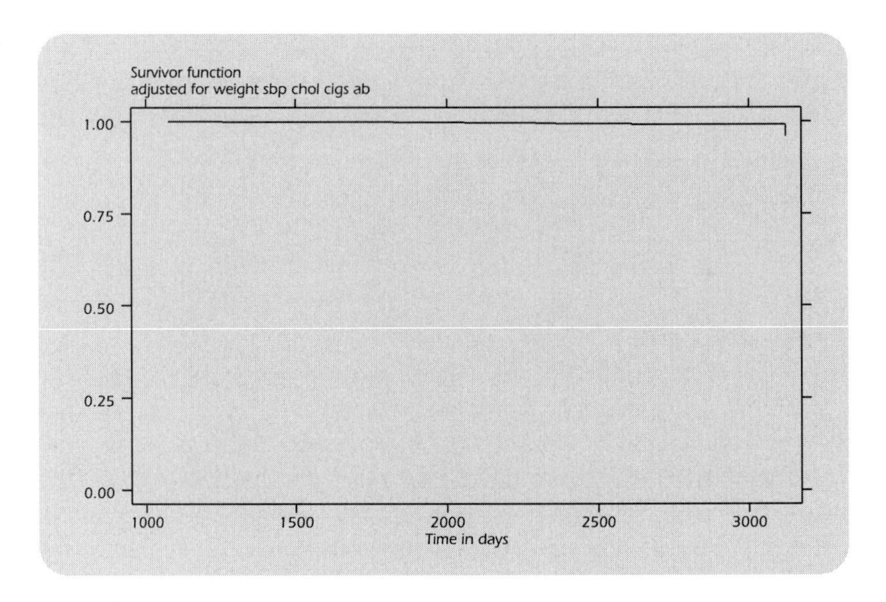

Figure 11.5

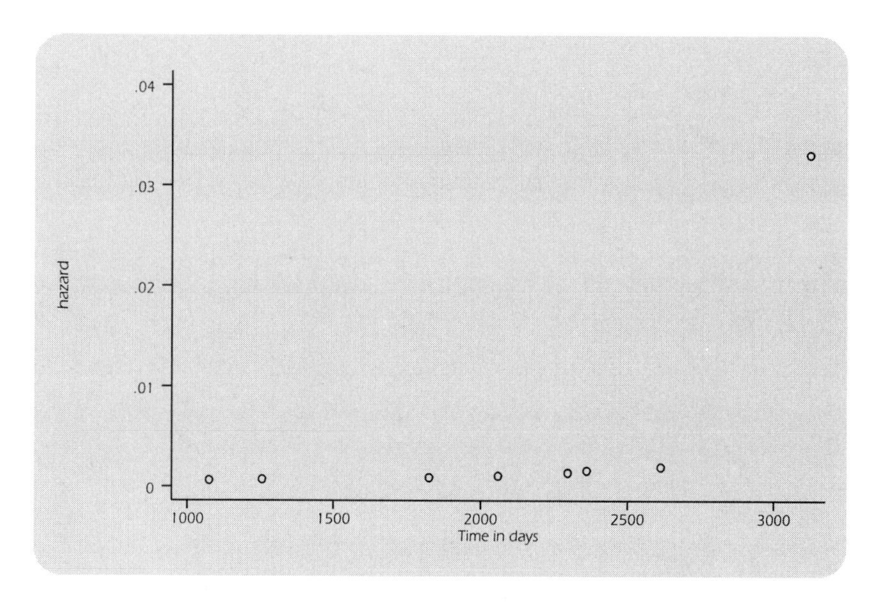

Exponential and Weibull Regression

Cox regression estimates the baseline survivor function empirically, without reference to any theoretical distribution. Two alternative approaches, exponential and Weibull regression, begin from the assumption that survival times do follow a known theoretical distribution. The basic exponential and Weibull models have the same general form as Cox regression (equations 11.2 and 11.3), but define the baseline hazard $h_0(t)$ differently.

If failures occur randomly, with a constant probability, then survival times follow an exponential distribution and could be analyzed by *exponential regression.* Constant probability means that the individuals studied do not "age," in the sense that they are no more or less likely to fail late in the period of observation than they were at its start. Over the long term, this assumption seems unjustified for machines or living organisms, but it might approximately hold if the period of observation covers a relatively small fraction of their life spans. An exponential model implies that logarithms of the survivor function, $\ln(S(t))$, are linearly related to t.

A second common parametric approach, *Weibull regression,* is based on the more general Weibull distribution. This does not require failure rates to remain constant but allows them to increase or decrease smoothly over time. The Weibull model implies that $\ln(-\ln(S(t)))$ is a linear function of $\ln(t)$.

Graphs provide a useful diagnostic for the appropriateness of exponential or Weibull models. For example, returning to *aids2.dta,* we can construct a graph

Figure 11.6

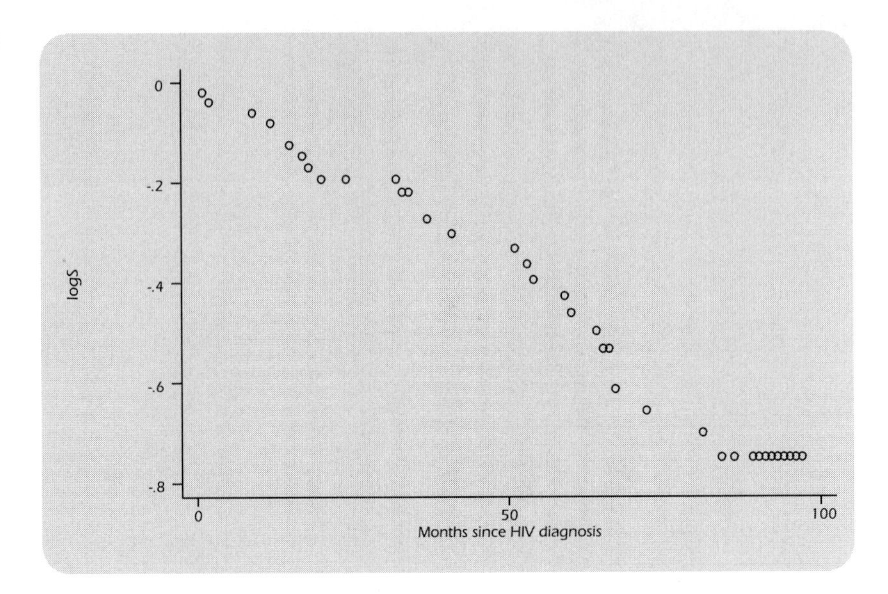

(Figure 11.6) of $\ln(S(t))$ versus time, after first generating Kaplan-Meier estimates of the survivor function $S(t)$:

```
. use aids2, clear
(AIDS (Selvin 1995:453))

. sts gen S = S

. generate logS = ln(S)

. graph logS time, ylabel xlabel
```

The pattern in Figure 11.6 appears somewhat linear, encouraging us to try an exponential regression:

```
. stereg age
    failure time:  time
      entry time:  t0
  failure/censor:  aids
              id:  case

Iteration 0:  Log Likelihood =   -62.16773
Iteration 1:  Log Likelihood =  -60.118195
Iteration 2:  Log Likelihood =  -59.997152
Iteration 3:  Log Likelihood =  -59.996976
Iteration 4:  Log Likelihood =  -59.996976

Exponential regression — entry time t0
log relative hazard form
```

```
No. of subjects =          51                Log likelihood =  -59.996976
No. of failures =          25                chi2(1)        =        4.34
Time at risk    =        3164                Prob > chi2    =      0.0372

------------------------------------------------------------------------------
   time | Haz. Ratio   Std. Err.        z     P>|z|      [95% Conf. Interval]
-------+----------------------------------------------------------------------
    age |   1.074414   .0349626      2.206   0.027      1.008028    1.145172
------------------------------------------------------------------------------
```

The hazard ratio (1.074) and standard error (.035) estimated by this exponential regression do not greatly differ from their counterparts (1.085 and .038) in our earlier Cox regression. The similarity reflects the degree of correspondence between empirical and exponential hazard functions.

A Weibull distribution can appear curvilinear in a plot of ln(*S*(*t*)) versus *t*, but it should be linear in a plot of ln(–ln(*S*(*t*))) versus ln(*t*) such as Figure 11.7. An exponential distribution, on the other hand, will appear linear in both plots and have a slope equal to 1 in the ln(–ln(*S*(*t*))) versus ln(*t*) plot. The data points in Figure 11.7 actually do fall close to a line with slope 1, suggesting that an exponential model is adequate.

```
. generate loglogS = ln(-ln(S))

. generate logtime = ln(time)

. graph loglogS logtime, ylabel xlabel
```

Although we do not need the additional complexity of a Weibull model with these data, results are given below for illustration. The many iteration steps, some

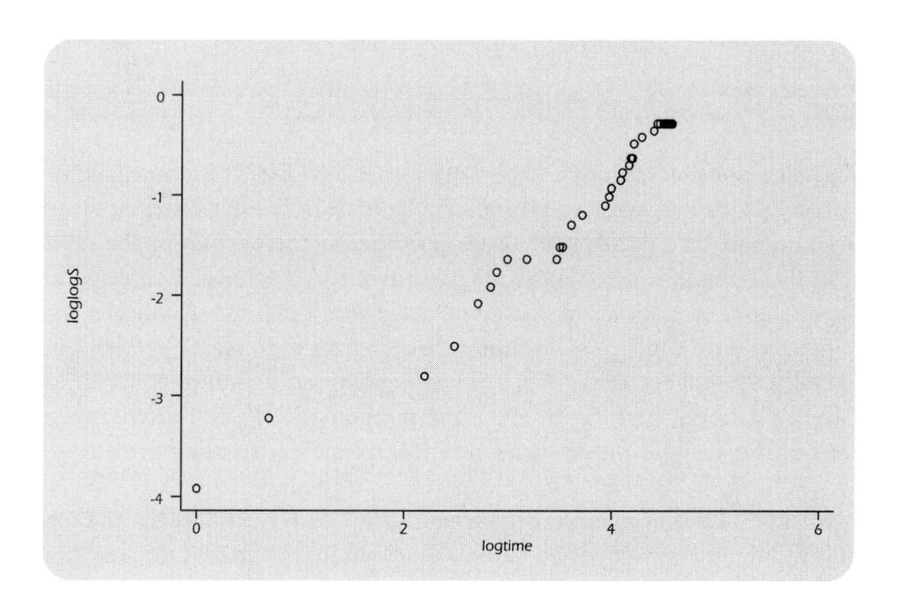

Figure 11.7

with disconcerting messages such as "unproductive step attempted" or "nonconcave function encountered" are typical of Weibull regression and no cause for alarm. (To suppress the iteration log and **st** setup information we could have typed **stweib *age*, nolog noshow** instead.)

```
. stweib age

    failure time:  time
      entry time:  t0
  failure/censor:  aids
              id:  case

Iteration 0:  Log Likelihood = -62.116475
(unproductive step attempted)
Iteration 1:  Log Likelihood =  -62.04218
(unproductive step attempted)
Iteration 2:  Log Likelihood = -61.970352
(unproductive step attempted)
Iteration 3:  Log Likelihood = -61.905587
Iteration 4:  Log Likelihood =  -60.54884
     (output omitted)
Iteration 17:  Log Likelihood =  -59.77826
Iteration 18:  Log Likelihood = -59.778259

Weibull regression — entry time t0
log relative hazard form

No. of subjects =          51                 Log likelihood =  -59.778259
No. of failures =          25                 chi2(1)        =        4.68
Time at risk    =        3164                 Prob > chi2    =      0.0306

----------------------------------------------------------------------------
    time | Haz. Ratio   Std. Err.        z     P>|z|      [95% Conf. Interval]
-------+ -------------------------------------------------------------------
     age |   1.079488   .0363546     2.271   0.023      1.010535    1.153146
----------------------------------------------------------------------------
    ln p |   .1230851   .1821982     0.676   0.499     -.2340167    .4801869
       p |   1.130981                                   .7913486    1.616376
     1/p |   .8841884                                   .6186678    1.263666
----------------------------------------------------------------------------
```

The Weibull regression obtains a hazard ratio estimate (1.079) intermediate between our previous Cox and exponential results. The most noticeable difference from those earlier models is the presence of three new lines at the bottom of the table. These refer to the Weibull distribution shape parameter p. A p value of 1 corresponds to an exponential model: The hazard does not change with time. $p > 1$ indicates that the hazard increases with time; $p < 1$ indicates that the hazard decreases. A 95% confidence interval for p ranges from .79 to 1.62, so we have no reason to reject an exponential ($p = 1$) model here. Different, but mathematically equivalent, parameterizations of the Weibull model focus on $\ln(p)$, p, or $1/p$, so Stata provides all three.

Exponential or Weibull regression are preferable to Cox regression when survival times actually follow an exponential or Weibull distribution. When they do not, these models are misspecified and can yield misleading results. Cox regression, which makes no *a priori* assumptions about distribution shape, remains useful in a wider variety of situations.

Poisson Regression

If events occur independently and with constant probability, then counts of events over a given period of time follow a Poisson distribution. Let r_j represent the incidence rate for the jth observation:

$$r_j = \frac{\text{count of events}}{\text{number of times event could have occurred}} \qquad [11.4]$$

The denominator in [11.4] is termed the "exposure" and often measured in units such as person-years. We model the logarithm of incidence rate as a linear function of one or more predictor (x) variables:

$$\ln(r_j) = \beta_0 + \beta_1 x_1 + \beta_2 x_2 + \ldots + \beta_k x_k \qquad [11.5a]$$

Equivalently, the model describes logs of expected event counts:

$$\ln(\textit{expected count}) = \ln(\textit{exposure}) + \beta_0 + \beta_1 x_1 + \beta_2 x_2 + \ldots + \beta_k x_k \qquad [11.5b]$$

Assuming that a Poisson process underlies the events of interest, Poisson regression iteratively finds maximum-likelihood estimates of the parameters.

Data on radiation exposure and cancer deaths among workers at Oak Ridge National Laboratory provide an example. The 56 observations in dataset *oakridge.dta* represent 56 age/radiation-exposure categories (7 categories of age × 8 categories of radiation). For each combination, we know the number of deaths and the number of person-years of exposure.

```
. use oakridge, clear
(Radiation (Selvin 1995:474))

. describe
Contains data from c:\data\oakridge.dta
  obs:            56                           Radiation (Selvin 1995:474)
 vars:             4                           6 Dec 1996 14:50
 size:           616 (96.2% of memory free)
-------------------------------------------------------------------------------
  1. age          byte    %9.0g       ageg     Age group
  2. rad          byte    %9.0g                Radiation exposure level
  3. deaths       byte    %9.0g                Number of deaths
  4. pyears       float   %9.0g                Person-years
-------------------------------------------------------------------------------
Sorted by:
```

```
. summ

Variable |     Obs        Mean   Std. Dev.       Min        Max
---------+-----------------------------------------------------
     age |      56           4     2.0181          1          7
     rad |      56         4.5   2.312024          1          8
  deaths |      56    1.839286   3.178203          0         16
  pyears |      56    3807.679   10455.91         23      71382

. list in 1/6

          age       rad    deaths     pyears
  1.     < 45         1         0      29901
  2.    45-49         1         1       6251
  3.    50-54         1         4       5251
  4.    55-59         1         3       4126
  5.    60-64         1         3       2778
  6.    65-69         1         1       1607
```

Does the death rate increase with exposure to radiation? Poisson regression finds a statistically significant effect:

```
. poisson deaths rad, nolog exposure(pyears) irr

Poisson regression, normalized by pyears          Number of obs   =       56
Goodness-of-fit chi2(54)    =   254.548           Model chi2(1)   =   14.872
Prob > chi2                 =    0.0000           Prob > chi2     =   0.0001
Log Likelihood              =  -169.736           Pseudo R2       =   0.0420

------------------------------------------------------------------------------
  deaths |       IRR   Std. Err.       z     P>|z|       [95% Conf. Interval]
---------+--------------------------------------------------------------------
     rad |  1.236469   .0603551     4.348   0.000       1.123657    1.360606
------------------------------------------------------------------------------
```

For the regression above, we specified the event count (*deaths*) as the dependent variable and radiation (*rad*) as the independent variable. The Poisson "exposure" variable is *pyears*, or person-years in each category of *rad*. The **irr** option calls for incident rate ratios rather than regression coefficients in the results table—that is, we get estimates of exp(β) instead of β, the default. According to this incident rate ratio, the death rate becomes 1.236 times higher (increased by 23.6%) with each increase in radiation category. Although that ratio is statistically significant, the overall model fits poorly. The pseudo R^2 (see equation [10.4]) is only .042, and the model's predictions differ significantly from the actual counts—as shown by the goodness-of-fit chi square statistic ($\chi^2 = 254$, $P < .00005$).

We can substantially improve the fit by including *age* as a second predictor variable. Pseudo R^2 rises to .5966, and the goodness-of-fit test no longer leads us to reject our model.

```
. poisson deaths rad age, nolog exposure(pyears) irr

Poisson regression, normalized by pyears          Number of obs   =       56
Goodness-of-fit chi2(53)    =    58.005           Model chi2(2)   =  211.415
Prob > chi2                 =    0.2960           Prob > chi2     =   0.0000
Log Likelihood              =   -71.465           Pseudo R2       =   0.5966
```

```
------------------------------------------------------------------------------
deaths |         IRR   Std. Err.        z     P>|z|      [95% Conf. Interval]
-------+----------------------------------------------------------------------
   rad |   1.176673    .0593446     3.226   0.001      1.065924    1.298929
   age |   1.960034    .0997536    13.223   0.000      1.773955    2.165631
------------------------------------------------------------------------------
```

For simplicity, to this point we have treated *rad* and *age* as if both were continuous variables, and we expect their effects on the log death rate to be linear. In fact, however, both independent variables are measured as ordered categories. *rad* = 1, for example, means 0 radiation exposure; *rad* = 2 means 0 to 19 milliseiverts; *rad* = 3 means 20 to 39 milliseiverts, and so forth. An alternative way to include radiation exposure categories in the regression, while watching for nonlinear effects, is as a set of dummy variables. Below we use the **gen( )** option of **tabulate** to create 8 dummy variables, *r1* to *r8*, representing each of the eight values of *rad*.

```
. tabulate rad, gen(r)
Radiation   |
exposure    |
level       |      Freq.     Percent        Cum.
------------+-----------------------------------
          1 |          7       12.50       12.50
          2 |          7       12.50       25.00
          3 |          7       12.50       37.50
          4 |          7       12.50       50.00
          5 |          7       12.50       62.50
          6 |          7       12.50       75.00
          7 |          7       12.50       87.50
          8 |          7       12.50      100.00
------------+-----------------------------------
      Total |         56      100.00

. describe
Contains data from c:\data\oakridge.dta
 obs:            56                           Radiation (Selvin 1995:474)
vars:            12                           6 Dec 1996 14:50
size:         1,064 (95.8% of memory free)
-----------------------------------------------------------------------------
  1. age            byte   %9.0g       ageg      Age group
  2. rad            byte   %9.0g                 Radiation exposure level
  3. deaths         byte   %9.0g                 Number of deaths
  4. pyears         float  %9.0g                 Person-years
  5. r1             byte   %8.0g                 rad==      1.0000
  6. r2             byte   %8.0g                 rad==      2.0000
  7. r3             byte   %8.0g                 rad==      3.0000
  8. r4             byte   %8.0g                 rad==      4.0000
  9. r5             byte   %8.0g                 rad==      5.0000
 10. r6             byte   %8.0g                 rad==      6.0000
 11. r7             byte   %8.0g                 rad==      7.0000
 12. r8             byte   %8.0g                 rad==      8.0000
-----------------------------------------------------------------------------
Sorted by:
     Note:  data has changed since last save
```

We now include seven of these dummies (omitting one to avoid multicollinearity) as regression predictors. The additional complexity of this dummy-variable model brings little improvement in fit. It does, however, add to our interpretation. The overall effect of radiation on death rate appears to come primarily from the two highest radiation levels (*r7* and *r8*, corresponding to 100 to 119 and 120 or more milliseiverts). At these levels, the incidence rates are about four times higher.

```
. poisson deaths r2-r8 age, nolog exposure(pyears) irr

Poisson regression, normalized by pyears            Number of obs   =        56
Goodness-of-fit chi2(47)     =     53.978           Model chi2(8)   =   215.442
Prob > chi2                  =     0.2251           Prob > chi2     =    0.0000
Log Likelihood               =    -69.452           Pseudo R2       =    0.6080
------------------------------------------------------------------------------
 deaths |        IRR   Std. Err.         z     P>|z|       [95% Conf. Interval]
-------+----------------------------------------------------------------------
     r2 |   1.473591    .4268979     1.338    0.181     .8351884    2.599975
     r3 |   1.630688    .6659256     1.197    0.231     .7324279    3.630587
     r4 |   2.375967    1.088835     1.888    0.059     .9677429    5.833388
     r5 |   .7278112    .7518254    -0.308    0.758     .0961019    5.511956
     r6 |   1.168477     1.20691     0.151    0.880     .1543196    8.847469
     r7 |   4.433726    3.337737     1.978    0.048     1.013862    19.38915
     r8 |    3.89188    1.640978     3.223    0.001     1.703168    8.893266
    age |   1.961907    .1000652    13.213    0.000     1.775267    2.168169
------------------------------------------------------------------------------
```

Radiation levels 7 and 8 seem to have similar effects, so we might simplify the model by combining them. First, we test whether their coefficients are significantly different. They are not:

```
. test r7 = r8

 ( 1)   r7 - r8 = 0.0

           chi2(  1) =     0.03
         Prob > chi2 =     0.8676
```

Next, generate a new dummy variable *r78*, which equals 1 if either *r7* or *r8* equal 1:

```
. gen r78 = (r7 | r8)
```

Finally, substitute the new predictor for *r7* and *r8* in the regression:

```
. poisson deaths r2-r6 r78 age, irr ex(pyears) nolog

Poisson regression, normalized by pyears            Number of obs   =        56
Goodness-of-fit chi2(48)     =     54.005           Model chi2(7)   =   215.415
Prob > chi2                  =     0.2557           Prob > chi2     =    0.0000
Log Likelihood               =    -69.465           Pseudo R2       =    0.6079

------------------------------------------------------------------------------
 deaths |        IRR   Std. Err.         z     P>|z|       [95% Conf. Interval]
-------+----------------------------------------------------------------------
     r2 |   1.473602    .4269012     1.338    0.181     .8351949    2.599995
     r3 |   1.630718    .6659381     1.197    0.231     .7324414    3.630655
     r4 |   2.376065     1.08888     1.889    0.059     .9677823    5.833629
     r5 |   .7278387    .7518538    -0.308    0.758     .0961055    5.512164
     r6 |   1.168507    1.206941     0.151    0.880     .1543236    8.847701
    r78 |   3.980326    1.580024     3.480    0.001     1.828214    8.665831
    age |   1.961722     .100043    13.213    0.000     1.775122    2.167937
------------------------------------------------------------------------------
```

We could simplify the model further in this fashion. At each step, we can use **test** to evaluate whether combining two dummy variables is justifiable.

Generalized Linear Models

Generalized linear models (GLM) have the form:

$$g[\mathrm{E}(y)] = \beta_0 + \beta_1 x_1 + \beta_2 x_2 + \ldots + \beta_k x_k, \qquad y \sim F \qquad [11.6]$$

where $g[\ ]$ is the *link function* and F the distribution family. This general formulation encompasses many specific models. For example, if $g[\ ]$ is the identity function and y follows a normal (Gaussian) distribution, we have a linear regression model:

$$\mathrm{E}(y) = \beta_0 + \beta_1 x_1 \beta_2 x_2 + \ldots + \beta_k x_k, \qquad y \sim \text{Normal} \qquad [11.7]$$

If $g[\ ]$ is the logit function and y follows a Bernoulli distribution, we have logit regression instead:

$$\text{logit}[\mathrm{E}(y)] = \beta_0 + \beta_1 x_1 + \beta_2 x_2 + \ldots + \beta_k x_k, \qquad y \sim \text{Bernoulli} \qquad [11.8]$$

Because of its broad applications, GLM could have been introduced at several different points in this book. Its relevance to this chapter comes from the ability to fit Poisson regression models. This requires that $g[\ ]$ is the natural log function and y follows a Poisson distribution:

$$\ln[\mathrm{E}(y)] = \beta_0 + \beta_1 x_1 + \beta_2 x_2 + \ldots + \beta_k x_k, \qquad y \sim \text{Poisson} \qquad [11.9]$$

Stata's GLM command can have the following general syntax:

```
. glm y x1 x2 x3, family(familyname) link(linkname)
      lnoffset(exposure) eform
```

Here **family( )** specifies the y distribution family, **link( )** the link function, and **lnoffset( )** the "exposure" variable needed for Poisson regression. The **eform** option asks for regression coefficients in exponentiated form, exp(β) rather than β. Type **help glmpred** for details about obtaining predicted values after fitting a GLM model.

Possible distribution families are

family(gaussian)	Gaussian or normal (default)
family(igaussian)	Inverse Gaussian
family(binomial)	Bernoulli binomial
family(poisson)	Poisson

family(nbinomial) Negative binomial

family(gamma) Gamma

We can also specify a number or variable indicating the binomial denominator *N* (number of trials), or a number indicating the negative binomial variance and deviance functions, by declaring them in the **family( )** option:

family(binomial # **)**

family(binomial *varname* **)**

family(nbinomial # **)**

Possible link functions are

link(identity) Identity (default)

link(log) Log

link(logit) Logit

link(probit) Probit

link(cloglog) Complementary log–log

link(opower # **)** Odds power

link(power # **)** Power

link(nbinomial) Negative binomial

For example, we could replicate the first simple regression in Chapter 6 as follows. In the output below, Stata advises us that the model we specified is ordinary regression, so it would make more sense to use **regress** or **fit** instead.

```
. use states90, clear
(U.S. states data 1990-91)

. glm csat expense, link(identity) family(gaussian)
Iteration 1 : deviance = 175306.2097

Residual df  =        49                          No. of obs =        51
Pearson X2   =  175306.2                          Deviance   =  175306.2
Dispersion   =  3577.678                          Dispersion =  3577.678
Gaussian (normal) distribution, identity link
------------------------------------------------------------------------------
    csat |      Coef.   Std. Err.       t     P>|t|     [95% Conf. Interval]
---------+--------------------------------------------------------------------
 expense |  -.0222756    .0060371    -3.690   0.001    -.0344077   -.0101436
   _cons |   1060.732     32.7009    32.437   0.000     995.0175    1126.447
------------------------------------------------------------------------------
(Model is ordinary regression, use fit or regress instead)
```

Alternatively, because the identity function and Gaussian distribution are defaults, we could just have typed

```
. glm csat expense
```

To repeat the first **logit** regression of Chapter 10 through **glm**:

```
. use shuttle2, clear
(Space Shuttle data)

. glm any date, link(logit) family(bernoulli)
Iteration 1 : deviance =   25.9905
Iteration 2 : deviance =   25.9822
Iteration 3 : deviance =   25.9822
Iteration 4 : deviance =   25.9822

Residual df  =          21                      No. of obs =          23
Pearson X2   =    22.88855                      Deviance   =    25.98219
Dispersion   =    1.089931                      Dispersion =    1.237247

Bernoulli distribution, logit link
------------------------------------------------------------------------------
     any |      Coef.   Std. Err.       z     P>|z|      [95% Conf. Interval]
---------+--------------------------------------------------------------------
    date |   .0020907    .0010703      1.953   0.051     -6.94e-06    .0041884
   _cons |  -18.13116    9.517253     -1.905   0.057     -36.78463    .5223109
------------------------------------------------------------------------------
```

Alternatively, we could obtain the odds ratio or exponential form of this logit model, corresponding to the first **logistic** regression of Chapter 10, by including an **eform** option:

```
. glm any date, link(logit) family(bernoulli) eform nolog
Residual df  =          21                      No. of obs =          23
Pearson X2   =    22.88855                      Deviance   =    25.98219
Dispersion   =    1.089931                      Dispersion =    1.237247
Bernoulli distribution, logit link
------------------------------------------------------------------------------
     any | Odds Ratio   Std. Err.       z     P>|z|      [95% Conf. Interval]
---------+--------------------------------------------------------------------
    date |   1.002093    .0010725      1.953   0.051       .9999931   1.004197
------------------------------------------------------------------------------
```

The final **poisson** regression of the present chapter corresponds to this **glm** model:

```
. use oakridg2, clear
(Radiation (Selvin 1995:474))

. glm deaths r2-r6 r78 age, link(log) family(poisson)
      lnoffset(pyears) eform nolog
Residual df  =          48                      No. of obs =          56
Pearson X2   =    53.93496                      Deviance   =     54.0054
Dispersion   =    1.123645                      Dispersion =    1.125113
```

```
Poisson distribution, log link, offset ln(pyears)
------------------------------------------------------------------------------
  deaths |        IRR   Std. Err.        z     P>|z|      [95% Conf. Interval]
---------+--------------------------------------------------------------------
      r2 |   1.473602    .4269013      1.338   0.181      .8351949    2.599996
      r3 |   1.630718    .6659382      1.197   0.231      .7324415    3.630655
      r4 |   2.376065     1.08888      1.889   0.059      .9677823    5.833629
      r5 |   .7278387    .7518466     -0.308   0.758      .0961073    5.512057
      r6 |   1.168507    1.206928      0.151   0.880       .154327    8.847507
     r78 |   3.980326    1.580024      3.480   0.001      1.828214    8.665832
     age |   1.961722     .100043     13.213   0.000      1.775122    2.167937
------------------------------------------------------------------------------
```

Where a specialized model-fitting command such as **fit**, **logistic**, or **poisson** already exists, the specialized command generally works faster and offers more useful options than **glm** applied to the same problem. The chief advantage of **glm** is that it can estimate models for which Stata has no specialized command. For Poisson regression, however, the **glm** deviance-based dispersion statistic (deviance/*df*) also provides information about the appropriateness of a Poisson model. Dispersion values higher than 1.5 indicate overdispersion, in which case a negative binomial model might be more appropriate. In this example we see dispersion = 1.125113, giving us no cause to reject the Poisson assumptions.

12

Principal Components and Factor Analysis

Principal components and factor analysis provide methods for simplification—combining many correlated variables into a smaller number of underlying dimensions. Along the way to achieving simplification, the analyst must choose from a daunting variety of options. If the data really do reflect distinct underlying dimensions, different options might nonetheless converge on similar results. In the absence of distinct underlying dimensions, however, different options often lead to divergent results. Experimenting with these options can tell us how stable a particular finding is, or how much it depends on arbitrary choices about the specific analytical technique.

Stata accomplishes principal components and factor analysis with four basic commands:

`factor`	Extracts principal components or factors of several different types.
`greigen`	Constructs a scree graph, or plot of the eigenvalues, from the recent **`factor`**.
`rotate`	Performs orthogonal (uncorrelated factors) or oblique (correlated factors) rotation, after **`factor`**.
`score`	Generates factor scores (composite variables) after **`factor`** or **`rotate`**.

The composite variables generated by **`score`** can subsequently be saved, listed, graphed, or analyzed like any other Stata variable.

Users who create composite variables by the older method of adding other variables together, without doing factor analysis, should assess their results by calculating an α reliability coefficient:

`alpha`	Cronbach's α reliability

Example Commands

. factor ***x1-x20*****, pcf mineigen(1)**

Obtains principal components of the variables *x1* through *x20*. Retains components having eigenvalues greater than 1.

. factor ***x1-x20*****, ml factor(5)**

Maximum likelihood factor analysis of the variables *x1* through *x20*. Retains only the first five factors.

. greigen

Graphs eigenvalues versus factor or component number from the most recent **factor** command (also known as a "scree graph").

. rotate, varimax factors(2)

Performs orthogonal (varimax) rotation of the first two factors or components from the most recent **factor** command.

. rotate, promax factors(3)

Performs oblique (promax) rotation of the first three factors or components from the most recent **factor** command.

. score ***f1 f2 f3***

Generates three new factor score variables named *f1*, *f2*, and *f3*, based on the most recent **factor** and **rotate** commands.

. alpha ***x1-x10***

Calculates Cronbach's α reliability coefficient, for a composite variable defined as the sum of *x1* through *x10*. The sense of items entering negatively is ordinarily reversed. Options can override this default or form a composite variable by adding either the original variables or their standardized values together.

Principal Components

To illustrate basic principal components and factor analysis commands, we will use a small dataset that describes the nine major planets of this solar system. The data include several variables in both raw and natural logarithm form. Logarithms are employed here to reduce skew and linearize relations among the variables.

```
. use planets, clear
(Solar system data)

. describe
Contains data from c:\data\planets.dta
  obs:             9                        Solar system data
 vars:            12                        27 Dec 1996 17:20
 size:           441 (95.9% of memory free)
-------------------------------------------------------------------------------
   1. planet      str7   %9s                 Planet
   2. dsun        float  %9.0g               Mean dist. sun, km*10^6
   3. radius      float  %9.0g               Equatorial radius in km
   4. rings       byte   %8.0g    ringlbl    Has rings?
   5. moons       byte   %8.0g               Number of known moons
   6. mass        float  %9.0g               Mass in kilograms
   7. density     float  %9.0g               Mean density, g/cm^3
   8. logdsun     float  %9.0g               natural log dsun
   9. lograd      float  %9.0g               natural log radius
  10. logmoons    float  %9.0g               natural log (moons + 1)
  11. logmass     float  %9.0g               natural log mass
  12. logdense    float  %9.0g               natural log dense
-------------------------------------------------------------------------------
Sorted by:  dsun
```

To extract initial factors or principal components, use the command **factor** followed by a variable list (variables in any order) and one of these options:

pc	Principal components
pcf	Principal components factoring
pf	Principal factoring (default)
ipf	Principal factoring with iterated communalities
ml	Maximum-likelihood factoring

For example, to obtain principal components factors:

```
. factor rings logdsun-logdense, pcf
(obs=9)

            (principal component factors; 2 factors retained)
  Factor     Eigenvalue     Difference    Proportion    Cumulative
-------------------------------------------------------------------------
      1        4.62365        3.45469       0.7706        0.7706
      2        1.16896        1.05664       0.1948        0.9654
      3        0.11232        0.05395       0.0187        0.9842
      4        0.05837        0.02174       0.0097        0.9939
      5        0.03663        0.03657       0.0061        1.0000
      6        0.00006              .       0.0000        1.0000
```

```
            Factor Loadings
 Variable |      1          2     Uniqueness
----------+--------------------------------------------------------------
    rings |   0.97917    0.07720    0.03526
  logdsun |   0.67105   -0.71093    0.04427
   lograd |   0.92287    0.37357    0.00875
 logmoons |   0.97647    0.00028    0.04651
  logmass |   0.83377    0.54463    0.00821
 logdense |  -0.84511    0.47053    0.06439
```

Only the first two components have eigenvalues greater than 1, and these two components explain more than 96% of the six variables' combined variance. The unimportant 3rd through 6th principal components might safely be disregarded in subsequent analysis.

Two **factor** options provide control over the number of factors extracted:

factors(#) Where # specifies the number of factors

mineigen(#) Where # specifies the minimum eigenvalue for retained factors

The principal components factoring (**pcf**) procedure automatically drops factors with eigenvalues below 1, so

```
. factor rings logdsun-logdense, pcf
```

is equivalent to

```
. factor rings logdsun-logdense, pcf mineigen(1)
```

In this example, we could also have obtained the same results by typing

```
. factor rings logdsun-logdense, pcf factors(2)
```

To see a scree graph (plot of eigenvalues versus component or factor number) after any **factor**, simply type

```
. greigen
```

graph options can be added to **greigen**, to draw a more presentable image (Figure 12.1):

```
. quietly factor rings logdsun-logdense, pcf

. greigen, yline(1) ylabel(0,1,2,3,4,5) xlabel(1,2,3,4,5,6)
      b2(Component Number)
```

Figure 12.1 again emphasizes the unimportance of components 3 through 6.

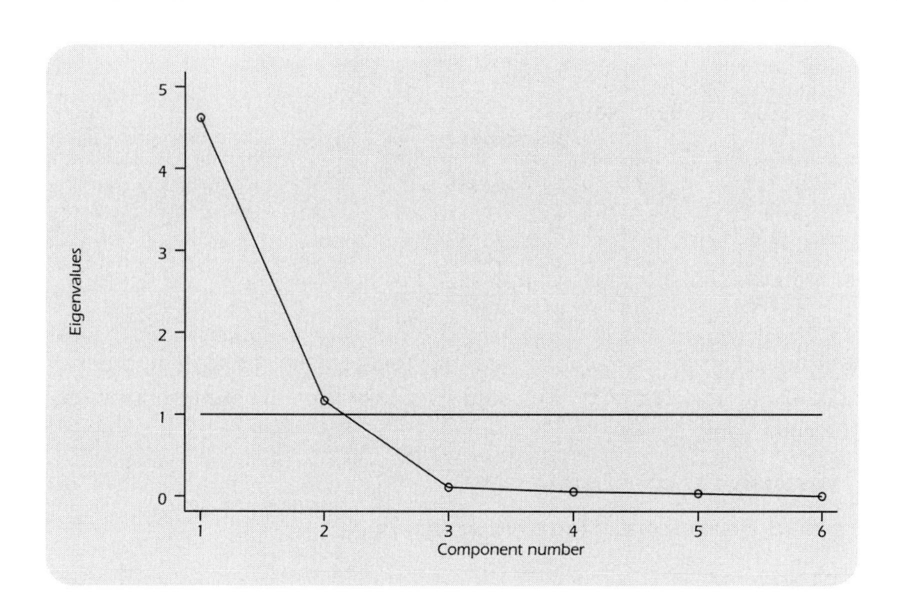

Figure 12.1

Rotation

Rotation further simplifies factor structure. After factoring, type **`rotate`** followed by either

`varimax` Varimax orthogonal rotation, for uncorrelated factors or components (default).

`promax( )` Promax oblique rotation, allowing correlated factors or components. Choose a number (promax power) ≤ 4; the higher the number, the greater the degree of interfactor correlation. **`promax(3)`** is the default.

Two additional **`rotate`** options are

`factors( )` As it does with **`factor`**, this option specifies how many factors to retain.

`horst` Horst modification to varimax and promax rotation.

Rotation (and factor scoring) can be performed following any factor analysis, whether it employed the **`pcf`**, **`pf`**, **`ipf`**, or **`ml`** options. In this section, we will continue with our **`pcf`** example. For orthogonal (default) rotation of the first two components found in the planetary data,

```
. rotate

              (varimax rotation)
              Rotated Factor Loadings
 Variable |       1           2     Uniqueness
---------+--------------------------------
    rings |  -0.74739     0.63729     0.03526
  logdsun |   0.02755     0.97723     0.04427
   lograd |  -0.91698     0.38781     0.00875
 logmoons |  -0.69112     0.68981     0.04651
  logmass |  -0.97481     0.20381     0.00821
 logdense |   0.26549    -0.93012     0.06439
```

This example accepts all the defaults: varimax rotation and the same number of factors retained in the last **factor**. We could have asked for the same rotation explicitly, by the command

```
. rotate, varimax factors(2)
```

For oblique promax rotation of the most recent factoring,

```
. rotate, promax

              (promax rotation)
              Rotated Factor Loadings
 Variable |       1           2     Uniqueness
---------+--------------------------------
    rings |  -0.70990     0.41625     0.03526
  logdsun |   0.23272     1.06946     0.04427
   lograd |  -0.95231     0.08350     0.00875
 logmoons |  -0.63571     0.49386     0.04651
  logmass |  -1.05540    -0.13721     0.00821
 logdense |   0.10691    -0.91073     0.06439
```

By default, this example used a promax power of 3. We could have asked for this explicitly:

```
. rotate, promax(3) factors(2)
```

promax(4) would permit further simplification of the loading matrix, at the cost of stronger interfactor correlations and hence less simplification.

After promax rotation, *rings*, *lograd*, *logmoons*, and *logmass* load most heavily (and all negatively) on factor 1. This appears to be a "smallness/few satellites" dimension. *logdsun* and *logdense* load higher on factor 2, forming a "far out/low density" dimension. The next section shows how to create new variables representing these dimensions.

Factor Scores

Factor scores are linear composites, formed by standardizing each variable to zero mean and unit variance, then weighting with factor score coefficients and summing for each factor. **score** performs these calculations automatically, using the most recent **rotate** or **factor** results:

```
. score f1 f2
              (based on rotated factors)
              Scoring Coefficients
 Variable |       1           2
----------+--------------------
    rings |  -0.21915     0.13667
  logdsun |   0.13871     0.46517
   lograd |  -0.32032    -0.01750
 logmoons |  -0.18953     0.17360
  logmass |  -0.36774    -0.11655
 logdense |  -0.01402    -0.37922
```

```
. label variable f1 "smallness/few satellites"
```

```
. label variable f2 "far out/low density"
```

We supply names for the new variables, unimaginatively called *f1* and *f2*.

```
. list planet f1 f2
        planet         f1          f2
  1.   Mercury   .8522902   -1.273661
  2.     Venus   .4416931   -1.182095
  3.     Earth   .3271711   -1.025886
  4.      Mars   .6580691     -.62064
  5.   Jupiter  -1.361802    .4573379
  6.    Saturn  -1.155303    .9708819
  7.    Uranus  -.7230982    .9527438
  8.   Neptune  -.6068392    .8303058
  9.     Pluto   1.567819    .8910129
```

Being standardized variables, the new factor scores have means (approximately) equal to zero and standard deviations equal to one:

```
. summ f1 f2
Variable |     Obs        Mean    Std. Dev.        Min          Max
---------+-----------------------------------------------------
      f1 |       9    3.31e-09           1   -1.361802    1.567819
      f2 |       9    3.31e-09           1   -1.273661    .9708819
```

Thus the factor scores are measured in units of standard deviations from their means. Mercury, for example, is about .85 standard deviations above average on the smallness/few satellites dimension, and 1.27 standard deviations below average on far out/low density dimension.

Promax rotation permits correlations between factor scores:

```
. correlate f1 f2
(obs=9)
        |       f1        f2
--------+------------------
     f1|   1.0000
     f2|  -0.4865    1.0000
```

Scores on factor 1 have a moderate negative correlation with scores on factor 2: smaller planets are less likely to be far out/low density.

If we employed varimax instead of promax rotation, we would get uncorrelated factor scores:

```
. quietly factor rings logdsun-logdense, pcf

. quietly rotate

. quietly score varimax1 varimax2

. correlate varimax1 varimax2
(obs=9)
        | varimax1 varimax2
--------+------------------
varimax1|   1.0000
varimax2|   0.0000   1.0000
```

Once created by **score**, factor scores can be treated like any other Stata variable—listed, analyzed, graphed, and so forth. Graphs of principal component factors sometimes help us identify multivariate outliers or clusters of observations that stand apart from the rest. For example, Figure 12.2 reveals three distinct types of planets:

```
. graph f1 f2, noaxis yline(0) xline(0) ylabel xlabel
      symbol([planet]) psize(120)
```

The **symbol([*planet*]) psize(120)** options cause values of *planet* (planet name) to be used as plotting symbols, at 120% of normal plotting-symbol size. The inner, rocky planets (high on "smallness/few satellites" factor 1; low on "far out/low density" factor 2) cluster together at upper left. The outer gas giants have opposite

Figure 12.2

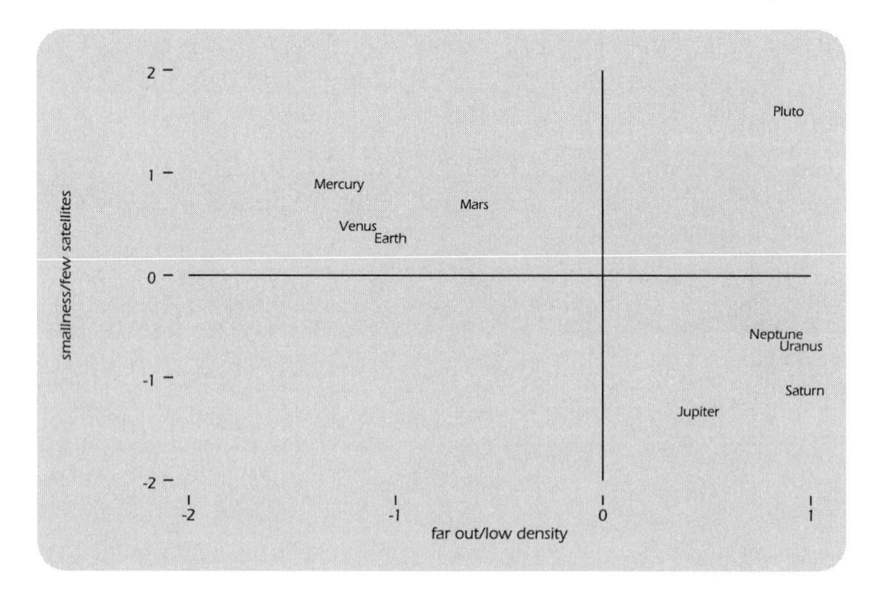

characteristics, and cluster together at lower right. Pluto, which physically resembles some outer-system moons, is unique among planets for being high on both "smallness/few satellites" and "far out/low density" dimensions.

Principal Factoring

Principal factoring extracts principal components from a modified correlation matrix, in which the main diagonal consists of communality estimates instead of 1's. The **factor** options **pf** and **ipf** both call for principal factoring. They differ in how communalities are estimated:

pf Communality estimates equal R^2 from regressing each variable on all the others.

ipf Iterative estimation of communalities.

Whereas principal components analysis focuses on explaining the variables' variance, principal factoring explains intervariable correlations.

Applying principal factoring with iterated communalities (**ipf**) to the planetary data:

```
. factor rings logdsun-logdense, ipf
(obs=9)

            (iterated principal factors; 5 factors retained)
  Factor     Eigenvalue     Difference     Proportion     Cumulative
------------------------------------------------------------------
     1        4.59663        3.46817        0.7903         0.7903
     2        1.12846        1.05107        0.1940         0.9843
     3        0.07739        0.06438        0.0133         0.9976
     4        0.01301        0.01176        0.0022         0.9998
     5        0.00125        0.00137        0.0002         1.0000
     6       -0.00012              .       -0.0000         1.0000

           Factor Loadings
Variable |      1          2          3          4          5    Uniqueness
---------+--------------------------------------------------------------
   rings |   0.97599    0.06649    0.11374   -0.02065   -0.02234    0.02916
 logdsun |   0.65708   -0.67054    0.14114    0.04471    0.00816    0.09663
  lograd |   0.92670    0.37001   -0.04504    0.04865    0.01662   -0.00036
logmoons |   0.96738   -0.01074    0.00781   -0.08593    0.01597    0.05636
 logmass |   0.83783    0.54576    0.00557    0.02824   -0.00714   -0.00069
logdense |  -0.84602    0.48941    0.20594   -0.00610    0.00997    0.00217
```

Only the first two factors have eigenvalues higher than 1. With **pcf** or **pf** factoring, we can simply disregard minor factors. Using **ipf**, however, we must decide how many factors to retain, then repeat the analysis asking for exactly that many factors. Here we should retain two factors:

```
. factor rings logdsun-logdense, ipf factor(2)
(obs=9)

          (iterated principal factors; 2 factors retained)
  Factor     Eigenvalue     Difference    Proportion    Cumulative
------------------------------------------------------------------
     1         4.57495        3.47412       0.8061        0.8061
     2         1.10083        1.07631       0.1940        1.0000
     3         0.02452        0.02013       0.0043        1.0043
     4         0.00439        0.00795       0.0008        1.0051
     5        -0.00356        0.02182      -0.0006        1.0045
     6        -0.02537              .      -0.0045        1.0000

            Factor Loadings
 Variable |      1          2     Uniqueness
----------+------------------------------
    rings |   0.97474    0.05374    0.04699
  logdsun |   0.65329   -0.67309    0.12016
   lograd |   0.92816    0.36047    0.00858
 logmoons |   0.96855   -0.02278    0.06139
  logmass |   0.84298    0.54616   -0.00890
 logdense |  -0.82938    0.46490    0.09599
```

After this final factor analysis, we can create composite variables by **rotate** and **score**. Rotation of the **ipf** factors produces results similar to those found earlier with **pcf**: a smallness/few satellites dimension, and a far out/low density dimension. When variables have a strong factor structure, as these do, the specific techniques we choose make less difference.

Maximum-Likelihood Factoring

The **ml** option calls for maximum-likelihood factoring:

```
. factor rings logdsun-logdense, ml factor(1)
(obs=9)
Iteration 0:  Log Likelihood =-2188.7586
Iteration 1:  Log Likelihood =-243.02794
Iteration 2:  Log Likelihood =-71.635796
Iteration 3:  Log Likelihood =-42.336298
Iteration 4:  Log Likelihood =-42.320907
Iteration 5:  Log Likelihood =-42.320558
Iteration 6:  Log Likelihood =-42.320537
Iteration 7:  Log Likelihood =-42.320535
Iteration 8:  Log Likelihood =-42.320535
Iteration 9:  Log Likelihood =-42.320535

          (maximum-likelihood factors; 1 factor retained)
  Factor     Variance       Difference    Proportion    Cumulative
------------------------------------------------------------------
     1        4.47257               .       1.0000        1.0000

Test:  1 vs. no   factors.  Chi2(   6) =   62.02, Prob > chi2 =  0.0000
Test:  1 vs. more factors.  Chi2(   9) =   51.73, Prob > chi2 =  0.0000
```

```
          Factor Loadings
 Variable |      1    Uniqueness
----------+--------------------
    rings |   0.98724    0.02535
  logdsun |   0.59219    0.64931
   lograd |   0.93655    0.12288
 logmoons |   0.95890    0.08052
  logmass |   0.86919    0.24451
 logdense |  -0.77145    0.40487
```

Unlike principal components or principal factoring, maximum-likelihood factor analysis supports formal tests for the appropriate number of factors. The output includes two χ^2 tests:

`J vs. no factors`

This tests whether the current model, with *J* factors, fits the observed correlation matrix significantly better than a no-factor model. A low probability indicates that the current model is a significant improvement over no factors.

`J vs. more factors`

This tests whether the current *J*-factor model fits significantly worse than a more complicated, perfect-fit model. A low *P*-value suggests that the current model does not have enough factors.

The previous 1-factor example yields these results:

`1 vs. no factors`

Probability of a greater $\chi^2 = 0.0000$ (actually, meaning $P < .00005$). The 1-factor model is significantly better than a no-factor model.

`1 vs. more factors`

Probability of a greater $\chi^2 = 0.0000$ ($P < .00005$). The 1-factor model is significantly worse than a perfect-fit model.

Perhaps a 2-factor model will do better:

```
. factor rings logdsun-logdense, ml factor(2)
(obs=9)
Iteration 0:  Log Likelihood =-12.540107
Iteration 1:  Log Likelihood =-11.770319
Iteration 2:  Log Likelihood = -8.456887
Iteration 3:  Log Likelihood =-6.2593382

            (maximum-likelihood factors; 2 factors retained)
  Factor     Variance       Difference     Proportion    Cumulative
----------------------------------------------------------------------
     1        3.64200         1.67115        0.6489        0.6489
     2        1.97085               .        0.3511        1.0000

Test:  2 vs. no    factors.  Chi2(  12) =  134.14, Prob > chi2 =  0.0000
Test:  2 vs. more factors.  Chi2(   4) =    6.72, Prob > chi2 =  0.1513
```

```
                   Factor Loadings
Variable |         1            2     Uniqueness
---------+--------------------------------
   rings |    0.86551    -0.41545     0.07829
 logdsun |    0.20920    -0.85593     0.22361
  lograd |    0.98438    -0.17528     0.00028
logmoons |    0.81560    -0.49982     0.08497
 logmass |    0.99965     0.02639     0.00000
logdense |   -0.46434     0.88565     0.00000
```

Now we find the following:

`2 vs. no factors`

Probability of a greater χ^2 = 0.0000 (actually, meaning $P < .00005$). The 2-factor model is significantly better than a no-factor model.

`2 vs. more factors`

Probability of a greater χ^2 = 0.1513. The 2-factor model is not significantly worse than a perfect-fit model.

These tests suggest that two factors provide an adequate model.

Computational routines performing maximum-likelihood factor analysis often yield "improper solutions"—unrealistic results like negative variance or zero uniqueness. When this happens (as it did in the 2-factor **`ml`** example), the χ^2 tests lack formal justification. Used descriptively, the tests can still provide informal guidance regarding the appropriate number of factors.

13

Introduction to Programming

As noted in Chapter 2 we can create simple programs by writing any sequence of Stata commands in an ASCII file. Then we enter Stata and type a command with the form **`do`** ***`filename`*** that tells Stata to read *filename.do* and run whatever commands it contains. More sophisticated programs are possible as well, using Stata's built-in programming language. Many of the commands used in previous chapters actually involve programs written in Stata. These programs might have originated either from Stata Corporation or from Stata users, who wanted something beyond Stata's built-in features to accomplish a particular task.

Stata programs can access all the existing features of Stata, call other programs that call other programs in turn, and use model-fitting aids including matrix algebra and maximum likelihood estimation. Whether our purposes are broad, such as adding a new statistical technique, or narrowly specialized, such as managing a particular dataset, our ability to write programs in Stata greatly extends what we can do.

The topic of Stata programming deserves a book in itself. This chapter only introduces some of the basic tools and gives examples showing some ways these tools can be used.

Basic Concepts and Tools

A few elementary concepts, combined with the Stata capabilities described in earlier chapters, suffice to get started.

Do-files

Do-files are ASCII (text) files, created by any word processor or editor and typically saved with a *.do* extension. The file can contain any sequence of legitimate Stata commands. In Stata, typing the following command instructs Stata to read *filename.do* and run the commands it contains:

```
. do filename
```

Each command in *filename.do*, including the last, must end with a hard return—unless we have reset the delimiter to some other character, through a **#delimit** command. For example:

```
#delimit ;
```

This sets a semicolon as the end-of-line delimiter, so that Stata does not consider a line finished until it encounters a semicolon. This permits a single command to extend over more than one physical line. Later, the following command resets "carriage return" as the usual end-of-line delimiter:

```
#delimit cr
```

Ado-files

Ado (automatic do) files are ASCII files containing sequences of Stata commands, much like do-files. The difference is that we need not type the **do *filename*** command to run an ado-file. Suppose we type the command

```
. clear
```

As with any command, Stata reads this and checks whether any intrinsic command by this name exists. If a **clear** command does not exist (and in fact, it does not) then Stata next searches in its usual "ado" directories, trying to find a file named *clear.ado.* If Stata finds such a file (as it should), it then runs whatever commands the file contains. Ado-files have the extension *.ado.* User-written programs usually go in the \ado directory (type **help adopath** for advice about changing this), whereas official Stata ado-files get installed in \stata\ado. Stata normally places hundreds of ado-files in the numerous subdirectories of \stata\ado.

Programs

Both do-files and ado-files might be viewed as types of programs, but Stata uses the word "program" in a narrower sense, to mean a sequence of commands stored in memory and executed by typing a particular program name. Do-files, ado-files, or commands typed interactively can define such programs. The definition begins with a statement that names the program. For example, to create a program named *new,* we start with

```
program define new
```

After the lines that actually define the program comes an **end** command, followed by one last hard return:

```
end
```

Once Stata has read the program-definition commands, it retains that definition of the program in memory and will run it again any time we type the program's name as a command:

```
. new
```

Programs effectively make new commands available within Stata, so most users do not need to know whether a given command comes from Stata itself or an ado-file-defined program.

Local Macros

Macros are names (with as many as 7 characters) that can stand for strings, program-defined results, or user-defined values. A *local macro* exists only within the program that defines it and cannot be passed to another program. To create a local macro named *iterate*, standing for the number "0":

```
local iterate = 0
```

To refer to the contents of a local macro ("0" in this example), place the macro name within left and right single quotes. For example:

```
display `iterate´
0
```

Thus to increase the value of *iterate* by one, we write

```
local iterate = `iterate´ + 1
```

Global Macros

Global macros are similar to local macros, but once defined they remain in memory and can be used by other programs. To refer to a global macro's contents, we preface the macro name with a dollar sign (instead of enclosing the name in left and right single quotes as done with local macros):

```
global foo = 73
```

```
display $foo * 2
146
```

Version

Stata's capabilities and features have changed over the years. Consequently, programs written for an older version of Stata might not run directly under the current version. The **version** command works around this problem so that old programs remain usable. Once we tell Stata for what version the program was written, Stata makes the necessary adjustments, so that an older program can run under a new version of Stata.

For example, if we begin our program with the following statement, Stata interprets all the program's commands as it would have in version 3.1:

```
version 3.1
```

Comments

Stata does not attempt to execute any line that begins with an asterisk. Such lines can therefore be used to insert comments and explanations into a program. For example:

```
* This entire line is a comment.
```

Alternatively, we can include a comment within an executable line by bracketing it by `/*` and `*/`. For example:

```
summarize var1 var2 var3     /* this part is the comment */
```

Looping

A section of program with the following general form will repeatedly run the commands within curly `{ }` brackets, as long as *expression* evaluates to "true" :

```
while expression {
          command A
          command B
          . . . .
          }
command Z
```

When *expression* evaluates to "false," the looping stops, and Stata goes on to run *command Z*. For example, here is simple program named *count5*, that uses a loop to display onscreen the iteration numbers from 1 through 5:

```
* Program that counts from one to five
program define count5
version 5.0
local iterate = 1
while `iterate' <= 5 {
       display `iterate'
       local iterate = `iterate' + 1
       }
end
```

By typing these commands, we define program *count5*. Alternatively, we could use a text editor or word processor to save the same series of commands as an ASCII file named *count5.do*. Then, typing the following causes Stata to read the file:

```
. do count5
```

Either way, by defining program *count5* we make this available as a new command:

```
. count5
1
2
3
4
5
```

Parse

The **parse** command, used only for programming, "grammatically" analyzes strings such as a typed-in command line. Like other programming commands, **parse** can accomplish complicated tasks. Here is a very simple example, a program that does nothing except display whatever variable list we type after a **listvars** command. Our program requires that some variable list exist.

```
program define listvars
    local varlist "required existing"
    parse "`*'"
    display "the variables are (`varlist')"
end
```

With a word processor or editor, we could type and save the previous commands as an ASCII file named *listvars.do.* Typing the following command causes Stata to read and run the file, thereby defining this program in memory:

```
. do listvars
```

To see the program in action, retrieve any dataset. Then issue a **listvars** command, followed by a few variable names. For example:

```
. use grnland
(Greenland municipalities)

. describe
Contains data from c:\data\grnland.dta
  obs:            19                          Greenland municipalities
 vars:             9                          1 Jan 1997 12:24
 size:           836 (96.4% of memory free)
---------------------------------------------------------------------------
   1. munic        str16  %16s                Municipality
   2. pop88        int    %9.0g               Total population 1988
   3. pop94        int    %9.0g               Total population 1994
   4. unemp94      float  %9.0g               % unemployed native-towns 94
   5. femper94     float  %9.0g               % female natives 20-39 '94
   6. inc94        float  %9.0g               Av. taxpayers income 94
   7. dfish        float  %9.0g               Increase fish value 88-94
```

```
   8. dpop         int    %9.0g                 Increase population 88-94
   9. dinc         int    %9.0g                 Increase income 89-94
-------------------------------------------------------------------------------
Sorted by:
```

```
. listvars dfish dpop dinc crimrate
the variables are (dfish dpop dinc crimrate)
```

Many general-purpose tools, mentioned in previous chapters, also facilitate Stata programming. These include the `_result` functions (which are actually global macros) that store results after statistical analysis, explicit subscripting, and system variables such as `_n` (observation number) and `_N` (total number of observations). The **`display`** command, used previously for on-screen calculations, also permits detailed control of column position and color for designing results tables.

Two Example Programs

This section contains two simple examples of Stata programs. Hundreds of others, many more advanced, can be found among Stata's ado-files.

One such Stata-supplied ado-file named *clear.ado* defines the program **`clear`**. Typing **`clear`** causes Stata to read this ado-file and perform the commands it contains to drop whatever variables, labels, matrices, scalars, constraints, equations, and automatically loaded programs happen to be in memory. Because Stata does all this automatically, a casual user has no reason to know that **`clear`** is not an intrinsic Stata command. File *clear.ado* illustrates the basic structure of a program-defining ado-file. Note that the `*! version 3.1.1` comment in *clear.ado* refers to the version of *clear.ado*, not to the version of Stata (which is set at 5.0 two lines later).

```
*! version 3.1.1  22aug1996
program define clear
     version 5.0
     drop _all
     label drop _all
     matrix drop _all
     scalar drop _all
     constraint drop _all
     eq drop _all
     discard
end
```

Our second example, file *tmean.ado*, defines a new statistical procedure not supplied with Stata: calculating 10% trimmed means. These are means of variables found after setting aside the lowest 10% and highest 10% of values. *tmean.ado* applies some of the programming tools discussed earlier: command-line parsing, local macros for variables needed within the program, looping, calculations using `_result()` values, a results table formatted by the **display** command, and global macros storing program results (the trimmed mean) for possible later use by another program. Comment lines, beginning with asterisks, explain each step.

```
* 23nov1996
* Program tmean calculates 10% trimmed means.
program define tmean
   version 5.0
   * A list of variables is required to exist.
   local varlist "required existing"
   * The next two lines parse the command line.
   parse "`*'"
   parse "`varlist'",parse(" ")
   * Set local macro I equal to one.
   local I = 1
   * Specify how results are to be displayed.
   display _n in green _col(1) "Variable" _col(20) "Trimmed Mean"
   display in green _col(1) "----" _col(20) "------"
   * Loop all existing variables to be calculated, starting with
   * the first listed on the command line. The loop ends when
   * there are no more variables.
   while "`1'" != "" {
      * Calculate summary statistics for variable 1.
      quietly summ `1', detail
      * Recalculate summaries, excluding values below the 10th
      * percentile (_result(8) after summarize, detail) or
      * above the 90th percentile (_result(12)).
      quietly summ `1' if `1' > _result(8) & `1' < _result(12)
      * Display variable in green, trimmed mean result in yellow.
      display in green "`1'" _col(20) in yellow %10.3f _result(3)
      * Save each trimmed mean in global macro: S_1, S_2, etc.
      global S_`I' = _result(3)
      * Reset I to the next variable.
      local I = `I'+ 1
      macro shift
   }
end
```

File *tmean.ado* is supplied with this book but not with Stata. For it to work, copy *tmean.ado* into one of the directories where Stata looks for ado files. You should create your own personal ado-file directory, such as *c:\ado*, to hold "unofficial" programs such as *tmean.ado* or items from the *Stata Technical Bulletin*. (Type **help ado** if you need advice about how to change Stata's ado-file directories.) Once we have *tmean.ado* in a recognized ado-file directory, Stata will automatically find and run it when we type the command **tmean**.

The data on Greenland municipalities seen earlier include one variable, *dinc* (increase in average income, 1988 to 1994) with both high and low outliers. The trimmed mean of this variable is somewhat higher (closer to 0) than the arithmetic mean found by **summarize**:

```
. use grnland, clear
(Greenland municipalities)

. summarize dinc

Variable |     Obs        Mean   Std. Dev.        Min         Max
---------+-----------------------------------------------------
    dinc |      18   -969.6667    10588.8      -30253       19704

. tmean dinc

Variable            Trimmed Mean
--------            ------------
dinc                  -389.286
```

Statistical Computing Environments for Social Research (edited by Stine and Fox, 1997) compares programming for statistical analysis using Stata, SAS, GAUSS, and other packages. Its Stata chapter includes a step-by-step development of **tmean**.

Matrix Algebra

For more advanced statistical modeling, matrix algebra provides essential tools. Stata's matrix commands are too diverse to describe them adequately here; the subject occupies more than 60 pages in the *Stata Reference Manual.* A few examples help give some of the flavor of these commands, however.

The intrinsic hardcoded Stata command **regress** performs ordinary least squares (OLS) regression, among other things. But suppose that for some reason we wanted to write an OLS program ourselves. *linreg1.do* (following) defines such a program. Comments bracketed by /* */ explain what each command does.

```
* Minimal linear regression program.
program define linreg1
     local varlist "require existing min(2)"
     tempname YXX XX Xy yy XXinv b
     parse "`*'"
     mat accum `YXX' = `varlist'          /* (y X)'(y X) */
     matrix `XX' = `YXX'[2...,2...]       /* extract X'X */
     matrix `Xy' = `YXX'[2...,1]          /* extract X'y */
```

```
    matrix `XXinv' = syminv(`XX')        /* X'X inverse */
    matrix `b'  = `XXinv' * `Xy'         /* coefficients */
    matrix list `b'
end
```

linreg1.do is a crude program that does nothing except calculate the vector of estimated regression coefficients according to the familiar OLS equation:

$$b = (X'X)^{-1}X'y$$

linreg1.do makes no provision for **in** or **if** qualifiers, syntax errors, or options, and it does not calculate standard errors, confidence intervals, or the other ancillary statistics we usually want with regression.

We can use a small dataset on nuclear reactors to illustrate. As using **regress** shows, the cost of decommissioning an old reactor increases with its generating capacity and with the number of years in operation.

```
. use c:\data\reactor, clear
(Reactor Decommissioning Costs)

. describe
Contains data from c:\data\reactor.dta
  obs:           5                          Reactor Decommissioning Costs
 vars:           6                          23 Nov 1996 13:46
 size:         130 (95.8% of memory free)
-----------------------------------------------------------------------------
   1. site        str14  %14s               Reactor site
   2. decom       byte   %8.0g              Decommissioning cost, millions
   3. capacity    int    %8.0g              Generating capacity, megawatts
   4. years       byte   %9.0g              Years in operation
   5. start       int    %8.0g              Year operations started
   6. close       int    %8.0g              Year operations closed
-----------------------------------------------------------------------------
Sorted by:  start

. regress decom capacity years

  Source |       SS       df       MS                  Number of obs =       5
---------+------------------------------               F(  2,     2) =  189.42
   Model |  4666.16571     2  2333.08286               Prob > F      =  0.0053
Residual |  24.6342883     2  12.3171442               R-squared     =  0.9947
---------+------------------------------               Adj R-squared =  0.9895
   Total |     4690.80     4     1172.70               Root MSE      =  3.5096

-----------------------------------------------------------------------------
   decom |      Coef.   Std. Err.       t     P>|t|     [95% Conf. Interval]
---------+-------------------------------------------------------------------
capacity |   .1758739   .0247774      7.098   0.019     .0692653    .2824825
   years |   3.899314   .2643087     14.753   0.005     2.762085    5.036543
   _cons |  -11.39963   4.330311     -2.633   0.119    -30.03146     7.23219
-----------------------------------------------------------------------------
```

We can replicate this regression using our home-brewed program *linreg1.do*, which yields exactly the same regression coefficients:

```
. do linreg1

. linreg1 decom capacity years
(obs=5)
__000005[3,1]
               decom
capacity    .1758739
   years   3.8993139
   _cons  -11.399633
```

A more refined linear regression program, written by Joe Hilbe, appears below. This program, *linreg2*, displays tests and other regression results in a nicely formatted table, with Stata's usual color conventions.

```
* Linear regression program.
* Hilbe 12-12-96
program define linreg2
  local options "Level(integer $S_level)"
  if "`*'" == "" | substr("`1'",1,1)=="," {     /*allows output redisplay*/
      if "$S_E_cmd" != "linreg2" {             /*by typing only linreg2*/
              error 301
      }
      parse "`*'"
  }
  else {
    quietly {
      local varlist "req ex min(2)"
      parse "`*'"
      parse "`varlist'",parse(" ")
      tempname YXX XX Xy yy XXinv b pv s2 V y my touse
      preserve               /* store original data set */
      local i= 1
      while "``i''"!=""  {  /* temp drop cases with missing values */
         drop if ``i''==.
         local i=`i'+1
      }
      local y `1'
      summ `y'
      global my = _result(3)                /* response mean */
      matrix accum `YXX' = `varlist'        /* (y X)'(y X) */
      local n = _result(1)                  /* cases in regression */
      local k 0
      local i 1
      matrix `XX' = `YXX'[2...,2...]       /* extract X'X */
      matrix `Xy' = `YXX'[2...,1]          /* extract X'y */
      scalar `yy' = `YXX'[1,1]             /* sum of y-squared */
      matrix `XXinv' = syminv(`XX')        /* X'X inverse */
```

```
        while `i'<=rowsof(`XXinv') {
                local k = `k' + (`XXinv'[`i',`i'] ~= 0)
                local i = `i'+1
        }
        matrix `b'  = `XXinv' * `Xy'            /* coefs     */
        matrix `pv'    = `b' ' * `Xy'           /* needed for s2 */
        scalar `s2' = (`yy'-`pv'[1,1])/(`n'-`k')     /* scale */
        matrix `V'  = `s2' * `XXinv'            /* variance */
        local df    = `n' - `k'
        global ess  = `s2'*(`n'-`k')            /* error sum squares */
        global tss  = `yy'-($my^2*`n')          /* total sum squares */
        global mss  = $tss - $ess               /* model sum squares */
        global mdf  = `n'-`df'-1                /* model df */
        global edf  = `n'-$mdf-1                /* error df */
        global tdf  = `n'-1                     /* total df */
        global   n  = `n'                       /* number of obs */
        global mse  = $ess/(`n'-`k')            /* mean squared error */
        global   f  = ($mss/$mdf)/$mse          /* f statistic  */
        global rsq  = 1-$ess/$tss               /* R-square */
        global ars  = 1-((1-$rsq)*($tdf)/$edf)  /* adj R-square */
        mat `b'     = `b' '                     /* transpose b */
        mat post `b' `V', dof(`df') obs(`n') depn(`y')
        global S_E_cmd "linreg2"
        restore                                 /* return original data set */
     }
  }
*
* Display the ANOVA table, F test and R squared:
  #delimit ;
  display _n in green "  Source          SS           df         MS"
    in green _col(56) "Number of obs" _col(70) "=" in yellow
    _col(72) %6.0f $n ;
  display in green "-----+----------------"
    in green _col(56) "F( $mdf, $edf)" _col(70) "=" in yellow
     _col(72) %6.2f $f ;
  display in green "   Model |" in yellow _col(9) %11.0g $mss _col(26)
     %5.0g $mdf _col(32) %11.0g $mss/$mdf in green _col(56) "Prob > F"
     _col(70) "=" in yellow _col(72) %6.4f fprob($mdf,$edf,$f);
  display in green "Residual |" in yellow _col(9) %11.0g $ess _col(26)
     %5.0g $edf _col(32) %11.0g $ess/$edf in green _col(56) "R-squared"
     _col(70) "=" in yellow _col(72) %6.4f $rsq ;
  display in green "-----+----------------"
     in green _col(56) "Adj R-squared" _col(70) "=" in yellow _col(72)
     %6.4f $ars ;
  display in green "   Total |" in yellow _col(9) %11.0g $tss _col(26)
```

```
    %5.0g $tdf _col(32) %11.0g $tss/$tdf in green _col(56) "Root MSE"
    _col(70) "=" in yellow _col(71) %7.0g sqrt($mse) _n ;
#delimit cr
*
* Display table of coefficients, SE's, t tests, confidence intervals:
  matrix mlout
end
```

Applied to the reactor dataset, *linreg2* estimates the same regression model as did *linreg1* and **regress**:

```
. quietly do linreg2

. linreg2 decom capacity years
  Source          SS          df          MS                Number of obs =       5
 --------+-----------------------------                     F(  2,      2) =  189.42
   Model | 4666.16571          2   2333.08286               Prob > F      =  0.0053
Residual | 24.6342883          2   12.3171442               R-squared     =  0.9947
 --------+-----------------------------                     Adj R-squared =  0.9895
   Total |      4690.8          4       1172.7              Root MSE      =  3.5096

------------------------------------------------------------------------------
   decom |      Coef.   Std. Err.          t     P>|t|      [95% Conf. Interval]
 --------+--------------------------------------------------------------------
capacity |   .1758739   .0247774       7.098    0.019      .0692653    .2824825
   years |   3.899314   .2643087      14.753    0.005      2.762085    5.036543
   _cons |  -11.39963   4.330311      -2.633    0.119     -30.03146     7.23219
------------------------------------------------------------------------------
```

Matrix algebra, combined with other capabilities such as maximum-likelihood estimation (type **help ml** for an overview), opens a vast scope for creativity. Stata's expansion in recent years occurred largely by this route: Stata's authors and users have built up a library of ado-files implementing hundreds of new statistical and data-management procedures.

Bootstrapping

Bootstrapping refers to a process of repeatedly drawing random samples, with replacement, from the data at hand. Instead of trusting theory to describe the sampling distribution of an estimator, we approximate that distribution empirically. Drawing *k* bootstrap samples of size *n* (from an original sample also size *n)* yields *k* new estimates. The distribution of these bootstrap estimates provides an empirical basis for estimating standard errors or confidence intervals (Efron and Tibshirani, 1986; for an introduction see Stine in Fox and Long, 1990). Bootstrapping seems most attractive in situations where the statistic of interest is theoretically intractable or where the usual theory regarding that statistic rests on untenable assumptions.

Unlike Monte Carlo simulations, which fabricate their data, bootstrapping typically works from real data. For illustration, we turn to *islands.dta*, which contains area and biodiversity measures for eight Pacific Island groups (from Cox and Moore, 1993).

```
. use islands
(Island biodiversity (Cox 93))

. describe
Contains data from c:\data\islands.dta
  obs:              8                           Island biodiversity (Cox 93)
 vars:              4                           1 Jan 1997 13:04
 size:            208 (96.5% of memory free)
-------------------------------------------------------------------------------
   1. island       str15  %15s                  Pacific island
   2. area         float  %9.0g                 Land area, km^2
   3. birds        byte   %8.0g                 Number of bird genera
   4. plants       int    %8.0g                 Number flowering plant genera
-------------------------------------------------------------------------------
Sorted by:
```

Suppose we want to form a confidence interval for the mean number of bird genera. The **ci** command accomplishes this by the usual *t*-distribution method:

```
. ci birds
Variable |      Obs         Mean    Std. Err.       [95% Conf. Interval]
---------+-------------------------------------------------------------------
   birds |        8       47.625    13.38034        15.98552    79.26448
```

Such confidence intervals, however, derive from a normality assumption. We might hesitate to trust this assumption, given the small sample size (n = 8) and skewed distribution of *birds:*

```
. summ birds, detail
                         Number of bird genera
-------------------------------------------------------------
      Percentiles      Smallest
 1%           10             10
 5%           10             17
10%           10             18       Obs                   8
25%         17.5             33       Sum of Wgt.           8

50%         43.5                      Mean             47.625
                        Largest       Std. Dev.      37.84531
75%         61.5             54
90%          126             59       Variance       1432.268
95%          126             64       Skewness       1.060604
99%          126            126       Kurtosis       3.304637
```

After any **summarize *varname*, detail** command, Stata unobtrusively saves the following results in memory:

`_result(1)` Number of observations

`_result(2)` Sum of weight (equals n unless weights are used)

`_result(3)`	Mean
`_result(4)`	Variance
`_result(5)`	Minimum
`_result(6)`	Maximum
`_result(7)`	5th percentile
`_result(8)`	10th percentile
`_result(9)`	25th percentile
`_result(10)`	50th percentile
`_result(11)`	75th percentile
`_result(12)`	90th percentile
`_result(13)`	95th percentile
`_result(14)`	Skewness
`_result(15)`	Kurtosis
`_result(16)`	1st percentile
`_result(17)`	99th percentile
`_result(18)`	Sum of variable

Many other analytical commands also save their results in memory. Consult the "Saved Results" section of each command's entry in the *Stata Reference Manual* for details.

Saved results have many uses in programming, as illustrated by *tmean.ado* earlier in this chapter. They make bootstrapping straightforward as well. The following **bs** command bootstraps the mean of *birds*, which equals `_result(3)` after **summ *birds*, detail**.

```
. bs "summ birds, detail" "_result(3)", rep(200)
      saving(boot01)

command:      summ birds, detail
statistic:    _result(3)
(obs=8)

Bootstrap statistics
Variable |  Reps  Observed       Bias   Std. Err.    [95% Conf. Interval]
---------+--------------------------------------------------------------
     bs1 |  200     47.625   -.014375   11.76591    24.42313  70.82687  (N)
         |                                             25.25   71.0625  (P)
         |                                             25.75      72.5 (BC)
------------------------------------------------------------------------
                                 N = normal, P = percentile, BC = bias corrected
```

The syntax of **bs** requires stating what analysis is to be bootstrapped, in double quotes (**"summ *birds*, detail"**). Following this comes the statistic(s) to be bootstrapped, likewise in its own double quotes (**"_result(3)"**). More than one statistic could be listed, each separated by a space. The previous example specifies two options:

rep(200)	Calls for 200 repetitions, or drawing 200 bootstrap samples.
saving(*boot01*)	Saves the 200 bootstrap means in a new dataset named *boot01.dta.*

The **bs** results table shows the number of repetitions performed and the "observed" (original-sample) value of the statistic being bootstrapped—in this case the mean *birds* value 47.625. The table also shows estimates of bias, standard error, and three types of confidence intervals. "Bias" here refers to the mean of the *k* bootstrap values of our statistic (for example, the mean of the 200 bootstrap means of *birds*), minus the observed statistic. In this example, the bias is near zero. The estimated standard error equals the standard deviation of the *k* bootstrap statistic values (for example, the standard deviation of the 200 bootstrap means of *birds*). This bootstrap standard error (11.76) is less than the theoretical standard error (13.38).

Normal-approximation (N) confidence intervals are as follows:

$$\text{observed sample statistic } \pm t \times \text{bootstrap standard error}$$

where t is chosen from the theoretical t distribution with $k - 1$ degrees of freedom. Their use is recommended when the bootstrap distribution appears unbiased and approximately normal.

Percentile (P) confidence intervals simply use percentiles of the bootstrap distribution (for a 95% interval, the 2.5th and 97.5th percentiles) as lower and upper bounds. These might be appropriate when the bootstrap distribution appears unbiased but nonnormal.

The bias-corrected (BC) interval also employs percentiles of the bootstrap distribution, but chooses these percentiles following a normal-theory adjustment for the proportion of bootstrap values less than or equal to the observed statistic. When substantial bias exists (by one rule of thumb, when bias exceeds 25% of one standard error) these intervals might be best.

We saved the bootstrap results in a file named *boot01.dta*, so we can retrieve this and examine the bootstrap distribution more closely if desired. The **saving(*filename*)** option with **bs** creates a dataset with **rep(#)** observations, and a variable named *bs1* (also *bs2*, *bs3* ... if we bootstrap more than one statistic).

```
. use boot01
(bs: summ birds, detail)
```

```
. describe

Contains data from c:\data\boot01.dta
  obs:           200                              bs: summ birds, detail
 vars:             1                              1 Jan 1997 13:08
 size:         1,600 (96.2% of memory free)
-------------------------------------------------------------------------------
  1. bs1          float  %9.0g                    _result(3)
-------------------------------------------------------------------------------
Sorted by:

. summarize

Variable |     Obs        Mean   Std. Dev.       Min        Max
---------+-----------------------------------------------------
     bs1 |     200    47.61063   11.76591       21.5     84.125
```

Note that the standard deviation of these 200 bootstrap means equals the standard error (11.76591) given earlier with the **bs** results table. The mean of the 200 means, minus the observed (original sample) mean, equals the bias

$$47.61063 - 47.625 = -.01437$$

Figure 13.1 shows the distribution of these 200 sample means. The distribution exhibits slight positive skew but is not far from a theoretical normal curve.

```
. graph bs1, bin(13) xline(47.6) ylabel xlabel norm
```

Figure 13.1

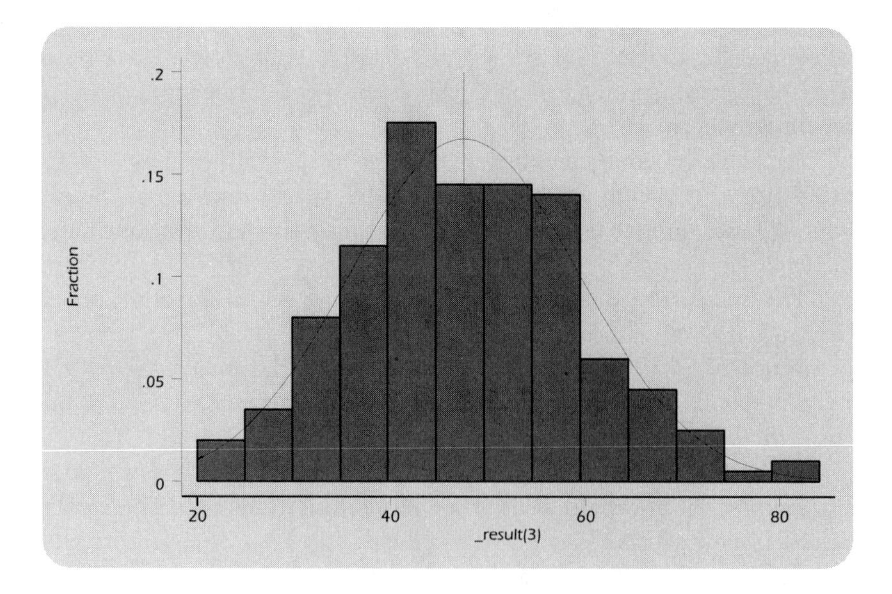

Biologists have observed that biodiversity, or the number of different kinds of plants and animals, tends to increase with island size. *Islands.dta* allows us to test this proposition with respect to flowering plants and birds. For example, we find a strong linear relation between *birds* and *area:*

```
. use islands, clear
(Island biodiversity (Cox 93))

. regress birds area
  Source |       SS       df       MS                  Number of obs =       8
---------+------------------------------               F(  1,     6) =  162.96
   Model |  9669.83255     1  9669.83255               Prob > F      =  0.0000
Residual |  356.042449     6  59.3404082               R-squared     =  0.9645
---------+------------------------------               Adj R-squared =  0.9586
   Total |   10025.875     7  1432.26786               Root MSE      =  7.7033

------------------------------------------------------------------------------
   birds |      Coef.   Std. Err.       t     P>|t|       [95% Conf. Interval]
---------+--------------------------------------------------------------------
    area |   .0026512   .0002077     12.765   0.000       .002143    .0031594
   _cons |   13.97169    3.79046      3.686   0.010      4.696772    23.24662
------------------------------------------------------------------------------
```

After any multiple or simple regression command such as **regress *birds area***, Stata stores the regression coefficients as global macros:

`_b[area]`	Regression coefficient on the predictor (x) variable *area*
`_b[_cons]`	Regression constant or y-intercept

It also stores standard errors, using a similar notation:

`_se[area]`	Estimated standard error of `_b[area]`
`_se[_cons]`	Estimated standard error of `_b[_cons]`

Finally, **regress** stores the following general results:

`_result(1)`	Number of observations
`_result(2)`	Model sum of squares
`_result(3)`	Model degrees of freedom
`_result(4)`	Residual sum of squares
`_result(5)`	Residual degrees of freedom
`_result(6)`	F statistic
`_result(7)`	R^2
`_result(8)`	Adjusted R^2
`_result(9)`	Root mean squared error [square root of (residual SS/residual df)]

Like means or other summary statistics, regression results can be bootstrapped through a **bs** command. We place the analysis to be bootstrapped in double quotes (**"regress *birds area*"**), then the statistics of interest inside a second set of double quotes (the regression slope and intercept, **"_b[*area*] _b[_cons]"**). This example calls for 500 repetitions and saves the results as *boot02.dta.* The bootstrap regression coefficients are referred to as *bs1* and regression constants as *bs2.*

```
. bs "regress birds area" "_b[area] _b[_cons]", rep(500)
      saving(boot02)
command:       regress birds area
statistics:   _b[area] _b[_cons]
(obs=8)
Bootstrap statistics

Variable |   Reps   Observed        Bias    Std. Err.    [95% Conf. Interval]
---------+------------------------------------------------------------------
     bs1 |   500    .0026512    -.000026    .0004734     .0017211  .0035812  (N)
         |                                               .0019587  .0030405  (P)
         |                                               .0018886  .0029085 (BC)
---------+------------------------------------------------------------------
     bs2 |   500    13.97169    .2600524    3.723821     6.655393  21.28799  (N)
         |                                               7.432551  21.81098  (P)
         |                                               7.716486  21.91253 (BC)
----------------------------------------------------------------------------
                                  N = normal, P = percentile, BC = bias corrected
```

Figure 13.2 shows the severely skewed distribution of these 500 bootstrap regression coefficients. Whereas the bootstrap distribution of means (Figure 13.1) appeared approximately normal, and produced bootstrap confidence intervals narrower than the theoretical confidence interval, in this regression example (Figure 13.2) bootstrapping obtains larger standard errors and wider confidence intervals.

```
. use boot02
(bs: regress birds area)
```

Figure 13.2

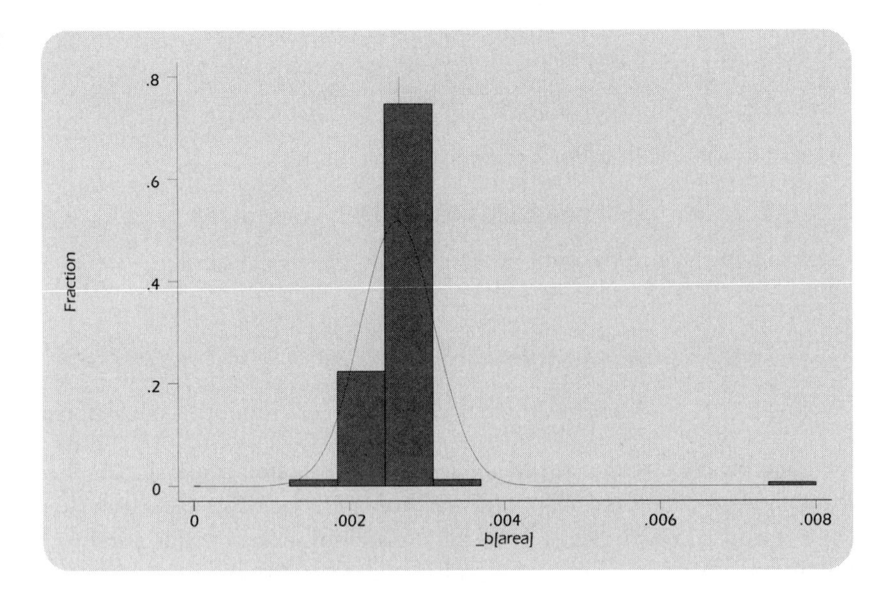

```
. graph bs1, bin(13) xline(.00265) ylabel xlabel norm
```

In this regression example, **bs** performs "data resampling" (resampling intact observations). An alternative procedure called "residual resampling" (resampling only the residuals) requires a bit more programming work. Two additional commands not described here, **bstrap** and **bsample**, make do-it-yourself bootstrap programming easier.

Monte Carlo Simulation

Monte Carlo simulations generate and analyze many samples of artificial data, allowing researchers to investigate the long-run behavior of their statistical techniques. The **simul** command makes designing a simulation straightforward, so that it only requires a small amount of additional programming. This section gives two examples.

Simulation commands typically take the form

```
. simul progname, reps(20) dots
```

This **simul** command asks Stata to find a program named *progname*, and run that program 20 times. With each repetition, a dot appears on screen.

To begin a simulation, we need to define a program that generates one sample of random data, analyzes it, and stores the results of interest in memory. Here is a very simple program that randomly generates 100 values of variable *x*, from a standard normal distribution, then finds the mean of *x* and stores this value as a variable named *xmean*:

```
program define simple
   version 5.0                  /* program written for Stata 5.0 */
   if "`1'" == "?" {
      global S_1 "xmean"        /* names the variable to hold results */
      exit
   }
   drop _all                    /* clears any data from memory */
   set obs 100                  /* sample will have n = 100 obs. */
   generate x = invnorm(uniform())  /* creates 100 values of x */
   summ x                       /* find the mean of x */
   post `1' _result(3)          /* posts mean to first-named variable */
end
```

We could type these lines interactively or else write them as an ASCII file named *simple.do*. In the latter case, typing **do *simple.do*** would cause Stata to read the file, thereby defining program *simple*. Once *simple* has been defined, we call it using the **simul** command:

```
. simul simple, reps(20) dots
```

This command called for 20 repetitions of program *simple*, with the 20 means of *x* (`_result(3)`) posted to a new file as variable *xmean*. This file of 20 means remains in memory when the simulation ends.

For a more interesting simulation, program *midsim* (defined below) generates both a standard normal *x* and a second variable, *w*, that is "contaminated normal": distributed as N(0,1) with probability .95, and as N(0,10) with probability .05. Contaminated distributions are often used to simulate variables that have occasional wild outliers or "bad" values (for example, large measurement errors). For each variable, *midsim* calculates both the mean and the median (`_result(10)` after **`summarize, detail`**).

```
program define midsim
* Performs one iteration of a Monte Carlo simulation comparing
* mean and median.  Generates one n = 100 sample containing two
* variables, x and w:
*
*      x ~ N(0,1)
*
*      w ~ N(0,1) with p = .95
*      w ~ N(0,10) with p = .05
*
      version 5.0
      if "`1'" == "?" {
            global S_1 "xmean xmedian wmean wmedian"
            exit
      }
      drop _all
      set obs 100
      generate x = invnorm(uniform())
      generate w = invnorm(uniform())
      replace w = 10*w if uniform() < .05
      summ x, detail
      local XMEAN = _result(3)
      local XMED = _result(10)
      summ w, detail
      post `1'  (`XMEAN') (`XMED') _result(3) _result(10)
* Note:  macro names should ordinarily be in (parentheses) within
* a post command, unless they are part of an expression.
end
```

The following commands repeat this simulation 200 times. We then summarize the resulting dataset containing means and medians from the 200 samples. Figure 13.3 shows these means and medians as boxplots.

Figure 13.3

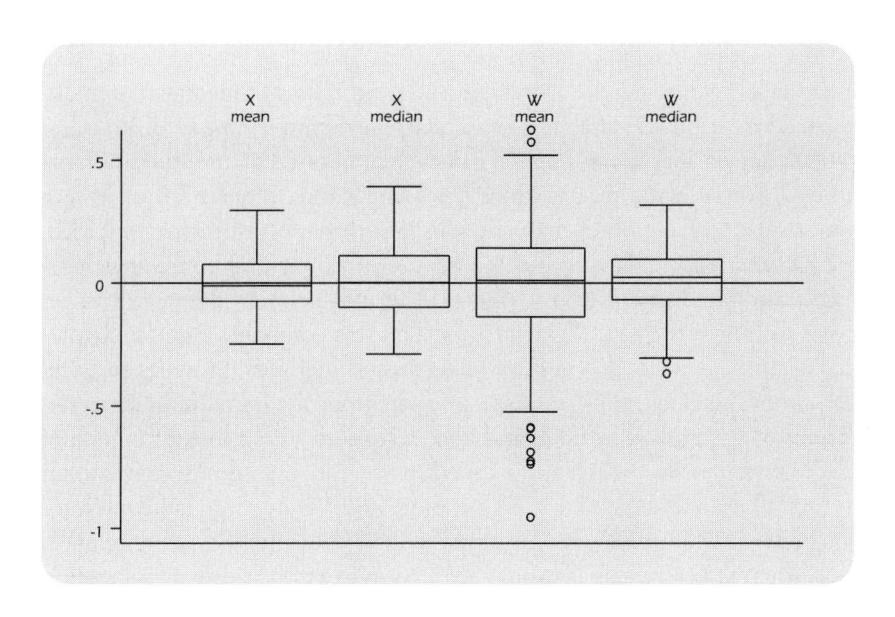


```
. quietly do midsim

. simul midsim, reps(200) dots
..........................................................................
..........................................................................
..............................................

. describe
Contains data
  obs:           200
 vars:             4                              1 Jan 1997 14:18
 size:         4,000 (96.1% of memory free)
---------------------------------------------------------------------------
   1. xmean        float  %9.0g
   2. xmedian      float  %9.0g
   3. wmean        float  %9.0g
   4. wmedian      float  %9.0g
---------------------------------------------------------------------------
Sorted by:

. summarize
Variable |     Obs        Mean    Std. Dev.       Min        Max
---------+-----------------------------------------------------
   xmean |     200    .0110439    .1055986   -.2496296   .2953244
 xmedian |     200    .0191151    .1312775   -.2899905   .3903356
   wmean |     200    .0043401    .2487608   -.7368132   .6213694
 wmedian |     200    .0164754     .132043   -.3206577    .316809
```

The means of these means and medians, across 200 samples, are all close to 0—consistent with our expectation that sample mean and median should both provide unbiased estimates of the true population means (0) of *x* and *w*. Also as theory predicts, the mean exhibits less sample-to-sample variation than the median, when applied to a normally distributed variable: The standard deviation of *xmean* (.106) is less than the standard deviation of *xmedian* (.131). When applied to the more outlier-prone variable *w*, however, the reverse holds true: The standard deviation of *wmean* (.249) becomes substantially larger than the standard deviation of *wmedian* (.132). Thus the median remains a relatively stable measure of center despite frequent outliers in the contaminated distribution, whereas the mean breaks down and varies more from one sample to the next. Figure 13.3 makes this comparison graphically.

Our final example extends this inquiry to regression methods, bringing together several themes from this book. Program *regsim.do* performs one iteration. It generates 100 observations of *x* (standard normal) and two *y* variables. *y1* is a linear function of *x* plus standard normal errors. *y2* is a linear function of *x* too, but adding contaminated normal errors. These variables permit us to explore how various regression methods behave in the presence of normal and nonnormal errors. Four methods are employed: ordinary least squares (**regress**), robust regression (**rreg**), quantile regression (**qreg**), and quantile regression with bootstrapped standard errors (**bsqreg**, with 200 repetitions). Differences among these methods were discussed in Chapter 9. Program *regsim* applies each method to the regression of *y1* on *x* and then to the regression of *y2* on *x*.

```
program define regsim
* Performs one iteration of a Monte Carlo simulation comparing
* OLS regression (regress) with robust (rreg) and quantile
* (qreg and bsqreg) regression.  Generates one n = 100 sample
* with x ~ N(0,1) and y variables defined by the models:
*
*   MODEL 1:       y1 = 2x + e1            e1 ~ N(0,1)
*
*   MODEL 2:       y2 = 2x + e2            e2 ~ N(0,1) with p = .95
*                                          e2 ~ N(0,10) with p = .05
*
* Bootstrap standard errors for bsqreg involve 200 repetitions.
*
     version 5.0
     if "`1'" == "?" {
          #delimit ;
          global S_1 "b1 b1r s1r b1q s1q s1qb
                      b2 b2r s2r b2q s2q s2qb";
          #delimit cr
          exit
     }
     drop _all
     set obs 100
```

```
     generate x = invnorm(uniform())
     generate e = invnorm(uniform())
     generate y1 = 2*x + e
     reg y1 x
     local B1 = _b[x]
     rreg y1 x, iterate(25)
     local B1R = _b[x]
     local S1R = _se[x]
     qreg y1 x
     local B1Q = _b[x]
     local S1Q = _se[x]
     bsqreg y1 x, reps(200)
     local S1QB = _se[x]
     replace e = 10 * e if uniform() < .05
     generate y2 = 2*x + e
     reg y2 x
     local B2 = _b[x]
     rreg y2 x, iterate(25)
     local B2R = _b[x]
     local S2R = _se[x]
     qreg y2 x
     local B2Q = _b[x]
     local S2Q = _se[x]
     bsqreg y2 x, reps(200)
     local S2QB = _se[x]
     #delimit ;
     post `1' (`B1') (`B1R') (`S1R') (`B1Q') (`S1Q') (`S1QB')
              (`B2') (`B2R') (`S2R') (`B2Q') (`S2Q') (`S2QB');
     #delimit cr
end
```

The program posts coefficient or standard error estimates from eight regression analyses, storing them as variables *b1* (coefficient from OLS regression of *y1* on *x*), *b1r* (coefficient from robust regression of *y1* on *x*), and so forth. All the robust and quantile regressions involve multiple iterations, however: typically 5 to 10 iterations for **rreg**, about five for **qreg**, and hundreds more for **bsqreg** with its 200 bootstrap resamplings. Thus a single execution of *regsim* could easily demand more than two thousand regressions—a slow job on most desktop computers. The following commands call for 10 repetitions, which take about 15 minutes to accomplish on a 100 MHz machine:

```
. quietly do regsim.do

. simul regsim, reps(10) dots
```

You might want to run a small simulation like this for yourself. For research purposes, however, we would need a much larger experiment such as 1000 repetitions and hence millions of individual regressions. Dataset *regsim.dta* contains results from this computation-intensive experiment. Results from the 1000 artificial samples are summarized below and graphed in Figure 13.4.

```
. use regsim, clear

. describe

Contains data from c:\data\regsim.dta
  obs:         1,000                          Regression simulations, n=100
 vars:            12                          5 Dec 1996 22:44
 size:        52,000 (95.3% of memory free)
-------------------------------------------------------------------------------
  1. b1          float  %9.0g                 OLS model 1 coefficient
  2. b1r         float  %9.0g                 Robust model 1 coefficient
  3. s1r         float  %9.0g                 Robust model 1 SE
  4. b1q         float  %9.0g                 Quantile model 1 coefficient
  5. s1q         float  %9.0g                 Quantile model 1 SE
  6. s1qb        float  %9.0g                 Quantile model 1 boot(200) SE
  7. b2          float  %9.0g                 OLS model 2 coefficient
  8. b2r         float  %9.0g                 Robust model 2 coefficient
  9. s2r         float  %9.0g                 Robust model 2 SE
 10. b2q         float  %9.0g                 Quantile model 2 coefficient
 11. s2q         float  %9.0g                 Quantile model 2 SE
 12. s2qb        float  %9.0g                 Quantile model 2 boot(200) SE
-------------------------------------------------------------------------------
Sorted by:


. summarize

Variable |     Obs        Mean   Std. Dev.        Min         Max
---------+-----------------------------------------------------
      b1 |    1000    2.001164   .0973893    1.716602    2.405942
     b1r |    1000     2.00112   .1022004     1.69302    2.490171
     s1r |    1000    .1038881   .0109089    .0694602    .1428644
     b1q |    1000    2.003288   .1278095    1.566468    2.594716
     s1q |    1000    .1250651     .02784     .047587    .2207107
    s1qb |    1000    .1356818   .0324703    .0611398    .2535811
      b2 |    1000    2.003155   .2480278    1.074015    3.100178
     b2r |    1000    2.000438   .1050054    1.698479    2.553535
     s2r |    1000    .1077567   .0117488    .0696244    .1516345
     b2q |    1000    2.002559    .134303    1.559719    2.622934
     s2q |    1000    .1310382   .0295157     .047587    .2704799
    s2qb |    1000    .1432131    .034812    .0542227    .2760452
```

All three regression methods produce mean coefficient estimates that are not significantly different from $\beta = 2$, as established by *t* tests such as:

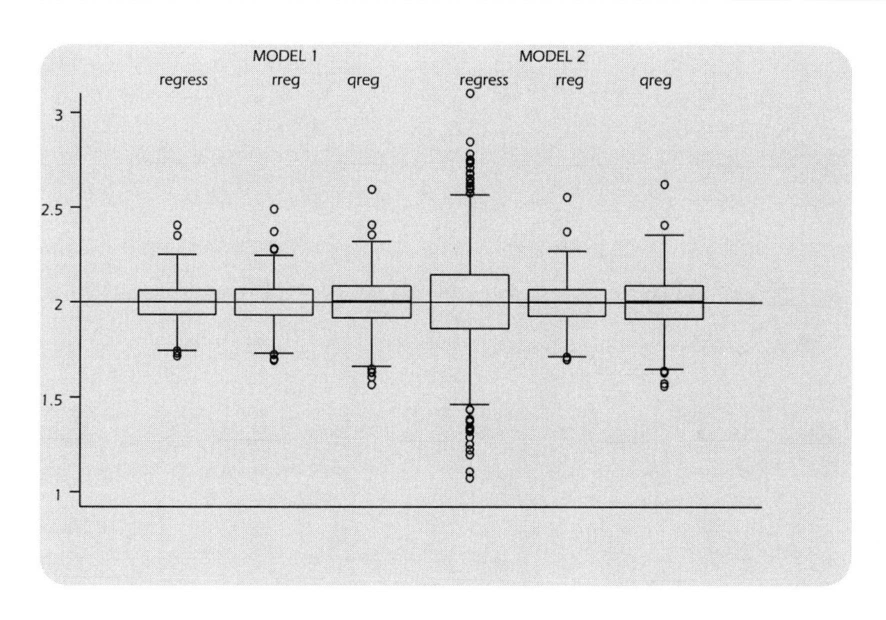

Figure 13.4

```
. ttest b1r = 2

One-sample t test                                     Number of obs =      1000

------------------------------------------------------------------------------
Variable |       Mean    Std. Err.        t      P>|t|     [95% Conf. Interval]
---------+--------------------------------------------------------------------
     b1r |    2.00112    .0032319    619.185    0.0000     1.994778    2.007462
------------------------------------------------------------------------------
Degrees of freedom: 999

                          Ho: mean(b1r) = 2

     Ha: mean < 2             Ha: mean ~= 2              Ha: mean > 2
       t =    0.3464             t =     0.3464             t =     0.3464
   P < t =    0.6354       P > |t| =     0.7291         P > t =     0.3646
```

Confidence intervals constructed around the mean estimated standard errors, however, suggest that these do differ significantly from the observed standard deviations of coefficients. The robust regression standard errors appear slightly higher (about 2%) compared with the observed standard deviations, suggesting that over these n = 100 samples, the asymptotic standard errors overestimate sampling variation. The quantile regression asymptotic standard errors, on the other hand, appear to slightly underestimate sampling variation. In this experiment, the bootstrap standard errors seem least accurate, because they exceed the observed standard deviations by about 6%.

Monte Carlo simulation has become a key method in modern statistical research, and it plays an increasing role in statistical teaching as well. These examples demonstrate how readily Stata supports Monte Carlo work.

References

Aldrich, John H. and Forrest D. Nelson (1984). *Linear Probability, Logit, and Probit Models.* Beverly Hills, CA: Sage.

Baker, Susan P., R.A. Whitfield, and Brian O'Neill (1987). "Geographic variations in mortality from motor vehicle crashes." *New England Journal of Medicine* 316(22):1384–1387.

Barron's Educational Series (1992). *Barron's Compact Guide to Colleges,* 8th Edition. New York: Barron's Educational Series.

Beatty, J. Kelly, Brian O'Leary, and Andrew Chaikin, Eds. (1981). *The New Solar System.* Cambridge, MA: Sky.

Belsley, David A., Edwin Kuh, and Roy E. Welsch (1980). *Regression Diagnostics: Identifying Influential Data and Sources of Collinearity.* New York: Wiley.

Brown, L.R., W.U. Chandler, C. Flavin, C. Pollock, S. Postel, L. Starke, and E.C. Wolf (1986). *State of the World 1986.* New York: Norton.

Brown, Lester R., Hal Kane, and David M. Roodman (1994). *Vital Signs 1994: The Trends That Are Shaping Our Future.* New York: Norton.

Chambers, John M., William S. Cleveland, Beat Kleiner, and Paul A. Tukey, Eds. (1983). *Graphical Methods for Data Analysis.* Belmont, CA: Wadsworth.

Chen, A.A., A.R. Daniel, S.T. Daniel, and C.R. Gray (1990). "Wind Power in Jamaica." *Solar Energy* 44(6):355–365.

Cleveland, William S. (1985). *The Elements of Graphing Data.* Monterey, CA: Wadsworth.

Cook, R. Dennis and Sanford Weisberg (1982). *Residuals and Influence in Regression.* New York: Chapman and Hall.

Cook, R. Dennis and Sanford Weisberg (1994). *An Introduction to Regression Graphics.* New York: Wiley.

Council on Environmental Quality (1988). *Environmental Quality 1987–1988.* Washington, DC: Council on Environmental Quality.

Cox, C. Barry and Peter D. Moore (1993). *Biogeography: An Ecological and Evolutionary Approach.* London: Blackwell.

Efron, Bradley and R. Tibshirani (1986). "Bootstrap methods for standard errors, confidence intervals, and other measures of statistical accuracy." *Statistical Science* 1(1):54–77.

Federal, Provincial, and Territorial Advisory Committee on Population Health (1996). *Report on the Health of Canadians.* Ottowa: Health Canada Communications.

Fox, John (1991). *Regression Diagnostics.* Newbury Park, CA: Sage.

Fox, John and J. Scott Long, Eds. (1990). *Modern Methods of Data Analysis.* Beverly Hills, CA: Sage.

Frigge, Michael, David C. Hoaglin, and Boris Iglewicz (1989). "Some implementations of the boxplot," *The American Statistician,* 43(1):50–54.

Greene, W.H. (1990). *Econometric Analysis.* New York: Macmillan.

Hall, Peter (1988). "Theoretical comparison of bootstrap confidence intervals." *The Annals of Statistics* 16(3):927–953.

Hamilton, Lawrence C. (1985). "Concern about toxic wastes: Three demographic predictors." *Sociological Perspectives* 28(4):463–486.

_____ (1990). *Modern Data Analysis: A First Course in Applied Statistics.* Pacific Grove, CA: Brooks/Cole.

_____ (1992a). *Regression with Graphics: A Second Course in Applied Statistics.* Pacific Grove, CA: Brooks/Cole.

_____ (1992b). "Quartiles, outliers and normality: Some Monte Carlo results." Pp. 9295 in Joseph Hilbe (ed.) *Stata Technical Bulletin Reprints, Volume 1.* College Station, TX: Stata Corporation.

_____ (1996). *Data Analysis for Social Scientists.* Belmont, CA: Duxbury.

_____ (1996). *Data Analysis for Social Scientists with StataQuest.* Belmont, CA: Duxbury.

Hamilton, Lawrence C. and Joseph M. Hilbe (1997). "Data analysis with Stata." Pp. 128–151 in Robert Stine and John Fox (eds.) *Statistical Computing Environments for Social Research.* Thousand Oaks, CA: Sage.

Hamilton, Lawrence C. and Carole L. Seyfrit (1994). "Interpreting multinomial logistic regression." Pp. 176–181 in Sean Becketti (ed.), *Stata Technical Bulletin Reprints, Volume 3.* College Station, TX: Stata Corporation.

Hanushek, Eric A. and John E. Jackson (1977). *Statistical Methods for Social Scientists.* New York: Academic Press.

Hilbe, Joseph (1991). "Data format conversion using DBMS/Copy and Stat/Transfer." *Stata Technical Bulletin* 3(September):3–6.

_____ (1996). "Windows file conversion software." *The American Statistician* 50(3):268–270.

Hoaglin, David C., Boris Iglewicz, and John W. Tukey (1986). "Performance of some resistant rules for outlier labeling." *Journal of the American Statistical Association* 81(396): 991–999.

Hoaglin, David C., Frederick Mosteller and John W. Tukey, Eds. (1983). *Understanding Robust and Exploratory Data Analysis.* New York: Wiley.

Hoaglin, David C., Frederick Mosteller, and John W. Tukey, Eds. (1985). *Exploring Data Tables, Trends and Shape.* New York: Wiley.

Hosmer, David W. and Stanley Lemeshow (1989). *Applied Logistic Regression.* New York: Wiley.

Iman, Ronald L. (1994). *A Data-Based Approach to Statistics.* Belmont, CA: Duxbury.

Jentoft, Svein and Trond Kristoffersen (1989). "Fishermen's co-management: The case of the Lofoten fishery." *Human Organization* 48(4):355–365.

Johnson, Anne M., Jane Wadsworth, Kaye Wellings, Sally Bradshaw, and Julia Field (1992). "Sexual lifestyles and HIV risk." *Nature* 360(3 December):410–412.

Johnston, J. (1984). *Econometric Methods,* 3rd edition. New York: McGraw-Hill.

Judson, D.H. (1992) "Performing loglinear analysis of cross-classifications." *Stata Technical Bulletin* 6(March):7–17.

League of Conservation Voters (1991). *The 1991 National Environmental Scorecard.* Washington DC.

Lee, Elisa T. (1992). *Statistical Methods for Survival Data Analysis,* 2nd edition. New York: Wiley.

Li, Gu–oying (1985). "Robust regression." Pp. 281–343 in D.C. Hoaglin, F. Mosteller, and J. W. Tukey (eds.) *Exploring Data Tables, Trends and Shape.* New York: Wiley.

Long, J. Scott (1997). *Regression Models for Categorical and Limited Dependent Variables.* Thousand Oaks, CA: Sage.

McCullagh, D.W. Jr., and J.A. Nelder (1989). *Generalized Linear Models,* 2nd edition. London: Chapman and Hall.

MacKenzie, Donald (1990). *Inventing Accuracy: A Historical Sociology of Nuclear Missile Guidance.* Cambridge, MA: MIT Press.

Mallows, C.L. (1986). "Augmented partial residuals." *Technometrics* 28:313–319.

Mayewski, P.A., G. Holdsworth, M.J. Spencer, S. Whitlow, M. Twickler, M.C. Morrison, K.K. Ferland, and L.D. Meeker (1993). "Ice-core sulfate from three northern hemisphere sites: Source and temperature forcing implications." *Atmospheric Environment* 27A(17/18):2915–2919.

Mayewski, P.A., L.D. Meeker, S. Whitlow, M.S. Twickler, M.C. Morrison, P. Bloomfield, G.C. Bond, R.B. Alley, A.J. Gow, P.M. Grootes, D.A. Meese, M. Ram, K.C. Taylor, and W. Wumkes (1994). "Changes in atmospheric circulation and ocean ice cover over the North Atlantic during the last 41,000 years." *Science* 263:1747–1751.

Nash, James and Lawrence Schwartz (1987). "Computers and the writing process." *Collegiate Microcomputer* 5(1):45–48.

National Center for Education Statistics (1992). *Digest of Education Statistics 1992.* Washington, DC: U.S. Government Printing Office.

National Center for Education Statistics (1993). *Digest of Education Statistics 1993.* Washington, DC: U.S. Government Printing Office.

Rawlings, John O. 1988. *Applied Regression Analysis: A Research Tool.* Pacific Grove: Wadsworth.

Report of the Presidential Commission on the Space Shuttle Challenger Accident (1986). Washington, DC.

Rosner, Bernard (1995). *Fundamentals of Biostatistics,* 4th edition. Belmont, CA: Duxbury.

Selvin, Steve (1995). *Practical Biostatistical Methods.* Belmont, CA: Duxbury.

Selvin, Steve (1996). *Statistical Analysis of Epidemiologic Data,* 2nd edition. New York: Oxford University Press.

Seyfrit, Carole L. (1993). *Hibernia's Generation: Social Impacts of Oil Development on Adolescents in Newfoundland.* St. John's: Institute of Social and Economic Research, Memorial University of Newfoundland.

Simon, Julian L. and Peter Bruce (1991a). "Resampling: A tool for everyday statistical work." *Chance* 4(1):22–32.

_____ (1991b). "Reply to Boomsma and Molenaar." *Chance* 4(4):30–31.

Statistics Greenland (1995). *Greenland 1995/96 Statistical Yearbook.* Nuuk: Home Rule Government.

Stine, Robert and John Fox, Eds. (1997). *Statistical Computing Environments for Social Research.* Thousand Oaks, CA: Sage.

Street, James O., Raymond J. Carroll, and David Ruppert (1988). "A note on computing robust regression estimates via iteratively reweighted least squares." *The American Statistician,* May, 43(2):152–154.

Tukey, John W. (1977). *Exploratory Data Analysis.* Reading, MA: Addison-Wesley.

Velleman, Paul F. (1982). "Applied Nonlinear Smoothing." Pp.141–177 in Samuel Leinhardt (ed.) *Sociological Methodology 1982.* San Francisco: Jossey-Bass.

Velleman, Paul F. and David C. Hoaglin (1981). *Applications, Basics and Computing of Exploratory Data Analysis.* Boston: Wadsworth.

Ward, Sally and Susan Ault (1990). "AIDS Knowledge, fear and safe sex practices on campus." *Sociology and Social Research* 74(3): 158–161.

Werner, Al (1990). "Lichen growth rates for the northwest coast of Spitsbergen, Svalbard." *Arctic and Alpine Research* 22(2):129–140.

World Bank (1987). *World Development Report 1987.* New York: Oxford University Press.

World Resources Institute (1993). *The 1993 Information Please Environmental Almanac.* Boston: Houghton Mifflin.

Zupan, Jeffrey M. (1973). *The Distribution of Air Quality in the New York Region.* Baltimore: Johns Hopkins University Press.

Index

I

K

L

M

N

O

P

Q

R

S